THE FATHERS OF THE CHURCH

MEDIAEVAL CONTINUATION

VOLUME 9

THE FATHERS OF THE CHURCH

MEDIAEVAL CONTINUATION

ALBERT THE GREAT

QUESTIONS CONCERNING ARISTOTLE'S ON ANIMALS

Translated by

IRVEN M. RESNICK
University of Tennessee, Chattanooga, Tennessee

and

KENNETH F. KITCHELL, JR.
University of Massachusetts Amherst, Massachusetts

THE CATHOLIC UNIVERSITY OF AMERICA PRESS
Washington, D.C.

The paper used in this publication meets the minimum requirements of the American National Standards for Information Science—Permanence of Paper for Printed Library Materials, ANSI Z39.48-1984.

LIBRARY OF CONGRESS CATALOGING-IN-PUBLICATION DATA
Albertus, Magnus, Saint, 1193?–1280.
[Quaestiones super De animalibus. English]
Questions concerning Aristotle's On animals / Albert the Great ; translated by Irven M. Resnick and Kenneth F. Kitchell, Jr.
p. cm. — (The fathers of the church. Mediaeval continuation ; v. 9)
Includes bibliographical references (p.) and index.
ISBN 978-0-8132-1519-8 (cloth : alk. paper) 1. Zoology—Pre-Linnean works. 2. Physiology—Early works to 1800. 3. Veterinary medicine—Early works to 1800. 4. Science, Medieval—Sources. 5. Aristotle. De partibus animalium. I. Resnick, Irven Michael II. Kitchell, Kenneth. III. Title. IV. Series.
QL41.A34513 2008
590—dc22

2007032021

CONTENTS

ACKNOWLEDGMENTS

The translators benefited from the assistance of numerous scholars and librarians who responded to various inquiries. It would be difficult to list them all here, and we would run the risk of offending by omission, but we would be remiss if we failed to express our appreciation to Dr. Henryk Anzulewicz of the Albertus-Magnus-Institut. In addition, we would like to thank the Oxford Centre for Hebrew and Jewish Studies for its hospitality and the assistance of staff at Oxford University's Bodleian Library.

Likewise, the support of the University of Tennessee at Chattanooga and University of Massachusetts Amherst, both financial and institutional, is greatly appreciated. The careful eye and sound knowledge of our editor, Dr. Carole Burnett, contributed to the volume's improvement in more ways than we can adequately express. Finally, our thanks to our families, who, once again, shared us with Albert.

GUIDES TO EDITORIAL CONVENTION AND ABBREVIATIONS

Standard Abbreviations Used in Citations

c.	chapter
CL	Classical Latin
col.	column
ed(s)	editor(s); edition; edited by
Ep.	*Epistula(e)*
fol.	folio
Fr.	French
Gr.	Greek
It.	Italian
lit.	literally
ll.	lines
MS(S)	manuscript(s)
n.	note
q.	question
r.	*recto*
s.v.	*sub voce,* under the heading or entry of
trans.	translated by
v.	*verso*
vol., vols.	volume or volumes

Abbreviations Used for Authors and Texts

A.	Albertus Magnus
DA	*De animalibus* (Stadler, 1916–20)
DCM	*De caelo et mundo* (1971)
De causis propr. elem.	*De causis proprietatum elementorum* (Hossfeld, 1980)

De nat. loc.	*De natura locorum* (Borgnet, 1890–99), vol. 9
De sensu et sens.	*De sensu et sensato* (Borgnet, 1890–99), vol. 9
De veg.	*De vegetabilibus* (Jessen, 1867)
DG	*De generatione et corruptione* (1971)
Metaph.	*Metaphysica* (1960–64)
Phys.	*Physica* (1987–93)
QDA	*Quaestiones de animalibus* (reference by book and question; Filthaut, 1955)
Ael. *DA*	Aelian *De natura animalium* (Scholfield, 1958–59)
Ar.	Aristotle (line references are to the appropriate Loeb editions whenever possible; note, however, that these are frequently inaccurate and that correlation to the actual Greek text will often be approximate)
Anal. Post.	*Posterior Analytics* (Tredennick and Forster, 1960)
Cat.	*Categories* (Tredennick, 1961–62)
DC	*De caelo* (Guthrie, 1945)
De long. et brev. vitae	*De longitudine et brevitate vitae* (Hett, 1935)
DG	*De generatione et corruptione* (Forster and Furley, 1965)
Eth. Nic.	*Nicomachean Ethics* (Rackman, 1947)
GA	*Generation of Animals* (Balme, 1972; Platt, 1910)
HA	*Historia animalium* (Peck, 1965–70; Balme, 1991; Thompson, 1966)
Metaph.	*Metaphysics* (Tredennick, 1961–62)
Part. An.	*Parts of Animals* (Peck, 1961)
Phys.	*Physics* (Wicksteed and Cornford, 1934)
De sensu et sens.	*On Sense and the Sensed* (Hett, 1935)
Averroes	
DCM	*De caelo et mundo*
DG	*De generatione et corruptione*
GA	*Generation of Animals*
Part. An.	*Parts of Animals*
Metaph.	*Metaphysics*

Avic.	Avicenna
Can. med.	*Canon of Medicine*
DA	*De animalibus* (reference by book, folio, and recto/verso; Stadler, 1916–20, 1: xiii)
Colon. Ed.	*Sancti Doctoris Ecclesiae Alberti Magni Ordinis Fratrum Praedicatorum episcopi Opera Omnia, ad fidem codicum manuscriptorum edenda apparatu critico notis prolegomenis indicibus instruenda curavit Institutum Alberti Magni Coloniense.* Ed. Bernhard Geyer (after vol. 37.2 [1978]: ed. Wilhelm Kübel). Monasterii Westfalorum: Aschendorff, 1951—
CSEL	*Corpus Scriptorum Ecclesiasticorum Latinorum*
Foerster	*Scriptores Physiognomonici Graeci et Latini* (1893)
Fonahn	*Arabic and Latin Anatomical Terminology* (reference by entry numbers)
Galen	(see Kühn)
De sympt. causis	*De morborum et symptomatum differentiis et causis libri sex*
De usu part.	*De usu partium*
In Hippocr. Epidem.	*In Hippocratis Epidemiarum*
In Hippocr. de humoribus	*In Hippocratis librum de humoribus*
Tegni	*Liber Tegni*
GB	*A Glossary of Greek Birds* (Thompson, 1966)
GF	*A Glossary of Greek Fishes* (Thompson, 1947)
GHW	*Grosser historischer Weltatlas* (Engel, 1970)
Hugh of St. Victor	*De bestiis et aliis rebus* (PL 177:13–164)
Isid. Orig.	Isidorus Hispalensis, *Origines* (Lindsay, 1985)
Kühn	*Claudii Galeni Opera Omnia* (1964–65)
Latham	*Revised Medieval Latin Word-List from British and Irish Sources* (1965)
Metam.	Ovid, *Metamorphoses*
PL	*Patrologia Cursus Completus.* Series Latina (reference by volume and column number; Migne, 1844–64)
Pliny HN	*Historia naturalis* (Rackham, 1940)

SZ	Kitchell, Kenneth F., Jr., and Irven M. Resnick. 1999. *Albertus Magnus "On Animals": A Medieval "Summa Zoologica."* 2 vols. Baltimore and London: Johns Hopkins University Press
ThC	Thomas of Cantimpré, *Liber de natura rerum* (Boese, 1973)
Vinc.	Vincent of Beauvais, *Speculum naturale*

Sigla

( )	indication of parenthetical expressions present in Latin text, whether or not appearing in parentheses there
< >	indication of material inserted into Latin text by editors of that text
[ ]	indication of material inserted by translators for purpose of clarification

BIBLIOGRAPHY

Below the reader will find bibliographical information concerning texts and sources cited in this work. For further research, the reader may also wish to consult *Albert the Great: A Selectively Annotated Bibliography (1900–2000),* Irven M. Resnick and Kenneth F. Kitchell, Jr., Medieval and Renaissance Texts and Studies 269 (Tempe: Arizona Center for Medieval and Renaissance Studies, 2004).

Al Hacen. *Optica.* See Witelo, *Opticae thesaurus.*

Albertus Magnus. *Quaestiones super Evangelium [CCXXX quaestiones super Evangelium 'Missus est'].* Edited by Auguste Borgnet (1890–99). *Opera Omnia* 37: 1–362.

———. *De anima.* Edited by Bernhard Geyer (1951–). *Opera Omnia* 7/1 (1968): xxii, 1–284.

———. *De natura et origine animae.* Edited by Bernhard Geyer (1951–). *Opera Omnia* 12 (1955): xx, 1–46.

———. *De bono.* Edited by Heinrich Kühle, Carl Feckes, Bernhard Geyer, and Wilhelm Kübel. Series edited by Bernhard Geyer (1951–). *Opera Omnia* 28 (1951): xxxii, 1–329.

———. *De caelo et mundo.* Edited by Paul Hossfeld. Series edited by Bernhard Geyer (1951–). *Opera Omnia* 5/1 (1971): xxiv, 1–341.

———. *De causis proprietatum elementorum.* Edited by Paul Hossfeld. Series edited by Bernhard Geyer (1951–). *Opera Omnia* 5/2 (1980): 49–106.

———. *De homine.* Edited by Auguste Borgnet (1890–99). *Opera Omnia* 35.

———. *De morte et vita.* Edited by Auguste Borgnet (1890–99). *Opera Omnia* 9: 345–71.

———. *De motibus animalium.* Edited by Auguste Borgnet (1890–99). *Opera Omnia* 9: 257–303.

———. *De natura loci, De causis proprietatum elementorum, De generatione et corruptione.* Edited by Paul Hossfeld. Series edited by Bernhard Geyer (1951–). *Opera Omnia* 5/2 (1980): xxx, 1–282.

———. *De principiis motus processivi.* Edited by Bernhard Geyer (1951–). *Opera Omnia* 12 (1955): xxiii–xxxii, 47–76.

———. *De sex principiis.* Edited by Auguste Borgnet (1890–99). *Opera Omnia* 1: 305–72.

———. *De somno et vigilia.* Edited by Auguste Borgnet (1890–99). *Opera Omnia* 9: 121–212.

———. *De spiritu et respiratione.* Edited by Auguste Borgnet (1890–99). *Opera Omnia* 9: 213–55.

———. *De vegetabilibus*. Edited by E. Meyer and K. Jessen (1867).
———. *Metaphysica. Libri I–V.* Edited by Bernhard Geyer (1951–). *Opera Omnia* 16/1 (1960): vii–xx, 1–300.
———. *Meteora [In IV libros Meteorum].* Edited by Auguste Borgnet (1890–99). *Opera Omnia* 4: 477–832.
———. *Opera Omnia ad fidem codicum manuscriptorum edenda, apparatu critico, notis, prolegomenis, indicibus instruenda curavit Institutum Alberti Magni Coloniense.* Edited by Bernhard Geyer (1951–).
———. *Physica.* Libri I–IV. Edited by Paul Hossfeld. Series edited by Bernhard Geyer (1951–). *Opera Omnia* 4/1 (1987): xxvi, 1–401.
———. *Physica. Libri V–VIII.* Edited by Paul Hossfeld. Series edited by Bernhard Geyer (1951–). *Opera Omnia* 4/2 (1993): xxvii–xxxiv, 401–772.
———. *Quaestiones super de animalibus.* Edited by Ephrem Filthaut. Series edited by Bernhard Geyer (1951–). *Opera Omnia* 12 (1955): xxxv–xlvii, 77–361.
———. *Summa de mirabili scientia Dei.* Edited by Auguste Borgnet (1890–99). *Opera Omnia* 31.
———. *Super Isaiam.* Edited by Ferdinand Siepmann. Series edited by Bernhard Geyer (1951–). *Opera Omnia* 19 (1952): vii–xxiv, 1–632.
———. *Topica.* Edited by Auguste Borgnet (1890–99). *Opera Omnia* 2: 233–524.
Alfred of Sareshel. *De motu cordis.* Edited by Baeumker, 1923.
Aristotle. *Categories.* Translated by Tredennick, 1961–62.
———. *De generatione et corruptione.* Edited by Forster and Furley, 1965.
———. *De incessu animalium.* Edited by Peck, 1961.
———. *De insomniis.* Edited by Drossaart Lulofs, 1947.
———. *De longitudine et brevitate vitae.* Edited by Hett, 1935. Pp. 388–409.
———. *De motu animalium.* Edited by Peck, 1961.
———. *De sensu et sensato.* Edited by Hett, 1935. Pp. 207–83.
———. *De somno et vigilia.* Edited by Hett, 1935. Pp. 316–45.
———. *De sophisticis elenchis.* Edited by Forster, 1965.
———. *Generation of Animals.* Edited by Balme, 1972; Platt, 1910.
———. *Historia animalium.* Edited by Peck, 1965–70; Balme, 1991; Thompson, 1966.
———. *Metaphysics.* Translated by Tredennick, 1961–62.
———. *Meteora.* Edited by Lee, 1952.
———. *Nicomachean Ethics.* Translated by Rackman, 1947.
———. *On Marvellous Things Heard.* Translated by Hett, 1936.
———. *Parts of Animals.* Translated by Ogle, 1912; Peck, 1961.
———. *Physics.* Translated by Wicksteed and Cornford, 1934.
———. *Posterior Analytics.* Translated by Tredennick, 1960.
———. *Topica.* Translated by Forster, 1960.
Arrizabalaga, Jon. "Facing the Black Death: Perceptions and Reactions of University Medical Practitioners." In *Practical Medicine from Salerno to the Black Death.* Edited by Luis Garcia-Ballester, Roger French, Jon

Arrizabalaga, and Andrew Cunningham. Cambridge: Cambridge University Press, 1994. Pp. 237–88.

Averroes. *Colliget. Libri VII.* Venice: Apud Junctas, 1562. Reprint, Frankfurt: Minerva, 1962.

Avicenna. *Liber canonis medicinae.* Translated into Latin by A. Bellunensis. Venice, 1595.

———. *De animalibus per magistrum Michaelem Scotum de arabico in latinum translatus.* Venice: Joannes and Gregorius de Gregoriis, de Forlivio, ca. 1500.

Baeumker, C., ed. *Alfred of Sareshel, De motu cordis.* Beiträge zur Geschichte der Philosophie des Mittelalters 23, 1–2. Münster: Aschendorff, 1923.

Balme, D. M., ed. and trans. *Aristotle. History of Animals. Books VII–X.* Loeb Classical Library. Cambridge, MA: Harvard University Press, 1991.

———, ed. and trans. *Aristotle. De Partibus Animalium and De Generatione Animalium I with passages from II.1–3.* Loeb Classical Library. Oxford: Clarendon Press, 1972.

Barach, Carl Sigmund, ed. *Excerpta e libro Alfredi Anglici De motu cordis, item Costa-Ben-Lucae De differentia animae et spiritus liber translatus a Johanne Hispalensi.* Innsbruck: Verlag der Wagnerischen Universitätsbuchhandlung, 1878.

Bardenhewer, Otto. *Pseudo-Aristotle. Liber de causis.* Freiburg: Herder, 1882. Reprint, Frankfurt: Minerva, 1970.

Binder, Wilhelm. *Novus thesaurus adagiorum Latinorum.* Stuttgart: E. Fischhaber, 1861.

Boese, H., ed. *Thomas Cantimpratensis: Liber de Natura Rerum.* Part 1. Berlin: De Gruyter, 1973.

Boethius. *Commentarii in librum Aristotelis Peri Hermeneias.* Edited by Charles Meiser. Leipzig: B.G. Teubner, 1877–80.

———. *De divisione.* PL 64: 875–91.

———. *Elenchorum Sophisticorum Aristotelis Libri duo An. Manl. Sev. Boetio interprete.* PL 64: 1007–40.

———. *Interpretatio Topicorum Aristotelis.* PL 64: 909–1007.

———. *In Isagoge Porphyrii commenta.* Edited by Brandt, 1906.

Borgnet, Auguste, ed. *B. Alberti Magni Ratisbonensis Episcopi Ordinis Praedicatorum Opera Omnia.* 38 vols. Paris: L. Vivès, 1890–99.

Bostock, John, and H. T. Riley, trans. and eds. *The Natural History of Pliny.* 6 vols. London: George Bell & Sons, 1893.

Boucé, Paul-Gabriel. "Imagination, Pregnant Women, and Monsters in Eighteenth-Century England and France." In *Sexual Underworlds of the Enlightenment.* Edited by G. S. Rousseau and Roy Porter. Chapel Hill, NC: University of North Carolina Press, 1988. Pp. 86–100.

Brandt, Samuel. *Anicii Manlii Severini Boethii In Isagogen Porphyrii commenta. Copiis a Georgia Schepss comparatis suisque usus recensuit Samuel Brandt.* CSEL 48. Vienna: F. Tempsky, 1906.

Büchmann, G. *Geflügelte Worte, der klassische Zitatenschatz.* Berlin: Verlag Praktisches Wissen, 1952.

Burnett, Charles. "Michael Scot and the Transmission of Scientific Culture from Toledo to Bologna via the Court of Frederick II Hohenstaufen." *Micrologus* 2: *Le scienze alla corte di Federico II.* Turnhout: Brepols, 1994. Pp. 101–26.

Corner, David. "The Earliest Surviving Manuscripts of Roger of Howden's 'Chronica.'" *The English Historical Review* 98 (1983): 297–310.

Courtois, V., ed. *Avicenna Commemoration Volume.* Calcutta: Iran Society, 1956.

Cooke, Harold P., and Hugh Tredennick, trans. *The Categories. On Interpretation. Prior Analytics.* Loeb Classical Library. Cambridge, MA: Harvard University Press, 1967.

Costa-ben-Luca. Edited by Barach, 1878.

De Asúa, Miguel. "Peter of Spain, Albert the Great, and the *Quaestiones de animalibus.*" *Physis* 34 (1997): 1–30.

Drossaart Lulofs, H. J., trans. *De Insomniis et De divinatione per somnum.* 2 vols. Leiden: Brill, 1947.

Eckert, Willehad Paul. "Albert der Grosse als Naturwissenschaftler." *Angelicum* 57 (1980): 477–95.

Engel, Josef. *Grosser historischer Weltatlas.* IIte Teil. Mittelalter. Munich: Bayerscher Schulbuch-Verlag, 1970.

Filthaut, Ephrem, ed. *De natura et origine animae, De principiis motus processivi, Quaestiones super de animalibus.* Series edited by Bernhard Geyer (1951–). *Opera Omnia* 12 (1955).

———. "Um die *Quaestiones de animalibus* Alberts des Grossen." In *Ostlender* (1952): 112–27.

Foerster, Richard, ed. *Scriptores Physiognomonici Graeci et Latini.* 2 vols. Leipzig: Teubner, 1893.

Fonahn, A. *Arabic and Latin Anatomical Terminology: Chiefly from the Middle Ages.* Kristiana: Dybwad, 1922.

Forster, E. S., and D. J. Furley, trans. *On Sophistical Refutations. On Coming-to-be and Passing Away; On the Cosmos.* Loeb Classical Library. Cambridge, MA: Harvard University Press, 1965.

Galen. *Opera Omnia.* Edited by Kühn, 1964–65.

George, Wilma, and Brunsdon Yapp. *The Naming of the Beasts: Natural History in the Medieval Bestiary.* London: Duckworth, 1991.

Geyer, Bernhard, ed. 1960–64. *Metaphysica.* 2 vols. Series edited by Bernhard Geyer (1951–). *Opera Omnia* 16/1–2 (1960–64).

———, ed. *Sancti Doctoris Ecclesiae Alberti Magni Ordinis Fratrum Praedicatorum episcopi Opera Omnia, ad fidem codicum manuscriptorum edenda apparatu critico notis prolegomenis indicibus instruenda curavit Institutum Alberti Magni Coloniense.* 1951– . Volumes subsequent to 37/2 (1978) edited by Wilhelm Kübel. Monasterii Westfalorum: Aschendorff.

Gilbert Porreta. *De sex principiis.* PL 188: 1257–70.

Goldstein-Préaud, Tamara. "Albert le Grand et les Questions du XIIIe siècle sur le *De animalibus* d'Aristote." *History and Philosophy of the Life Sciences* 3/1 (1981): 61–67.

Grant, Edward, ed. *Sourcebook in Medieval Science.* Cambridge, MA: Harvard University Press, 1974.

Guthrie, W. K. C., trans. *Aristotle on the Heavens*. Loeb Classical Library. Cambridge, MA: Harvard University Press, 1945.

Hall, Thomas S. "Life, Death, and the Radical Moisture: A Study of Thematic Pattern in Medieval Medical Theory." *Clio Medica* 6 (1971): 2–23.

Haskins, Charles H. "The 'Alchemy' Ascribed to Michael Scot." *Isis* 10 (1928): 350–59.

Hett, W. W., trans. *Aristotle, Minor Works*. Loeb Classical Library. Cambridge, MA: Harvard University Press, 1936.

———, trans. *Aristotle. On the Soul; Parva Naturalia; On Breath*. Loeb Classical Library. Cambridge, MA: Harvard University Press, 1935.

Hossfeld, Paul, ed. *Physica*. 2 vols. Series edited by Bernhard Geyer (1951–). *Opera Omnia* 4/1–2 (1987–93).

———, ed. *De causis proprietatum elementorum*. Series edited by Bernhard Geyer (1951–). *Opera Omnia* 5/2 (1980a): 48–106.

———, ed. *De natura loci, de causis proprietatum elementorum, De generatione et corruptione*. Series edited by Bernhard Geyer (1951–). *Opera Omnia* 5/2 (1980b).

———, ed. *De caelo et mundo*. Series edited by Bernhard Geyer (1951–). *Opera Omnia* 5/1 (1971).

Hyrtl, Joseph. *Das arabische und hebräische in der Anatomie*. Vienna: Wilhelm Braunmüller, 1879.

Isaac Israeli. *Omnia opera*. Louvain, 1515.

Isidore of Seville. *Origines*. Edited by Lindsay, 1911.

Jessen, Karl, ed. *Alberti Magni ex Ordine Praedicatorum De Vegetabilibus Libri VII*. Berlin: G. Reimer, 1867. Reprint, Frankfurt: Minerva, 1982.

Kitchell, Kenneth F., Jr., and Irven M. Resnick. *Albertus Magnus "On Animals": A Medieval "Summa Zoologica."* 2 vols. Baltimore and London: Johns Hopkins University Press, 1999.

Kühn, C. G., ed. *Claudii Galeni Opera Omnia*. 20 vols. Leipzig, 1821–33. Reprint, Hildesheim: Georg Olms, 1964–65.

Latham, R.. *Dictionary of Medieval Latin from British Sources*. London: Oxford University Press, 1975.

Lee, H. D. P., trans. *Aristotle. Meteorologica*. Loeb Classical Library. Cambridge, MA: Harvard University Press, 1952.

Lindsay, W. M., ed. *Isidori Hispalensis Episcopi Etymologiarum sive Originum Libri XX*. 2 vols. Oxford: Clarendon Press, 1911.

McVaugh, Michael. "The Translation of Greek and Arabic Science into Latin." In Grant, ed. *Sourcebook in Medieval Science*. Cambridge, MA: Harvard University Press, 1974. Pp. 35–41.

Meyer, E. H. F., ed. *De plantis libri duo Aristoteli vulgo adscripti. Ex Isaaci ben Honain versione arabica latine vertit Alfredus. Ad codd. mss. fidem addito apparatu critico*. Leipzig: L. Voss, 1841.

Meyer, Ernest, and Karl Jessen. *De vegetabilibus*. Berlin: G. Reimer, 1867. Reprint, Frankfurt: Minerva, 1982.

Migne, J.-P. *Patrologiae Cursus Completus. Series Latina*. 221 vols. Paris: Apud Garnière Fratres, 1844–55, 1862–64.

Nicholas of Damascus. Edited by Meyer, 1841.

Ogle, William, trans. *De partibus animalium.* In *The Works of Aristotle.* Edited by J. A.. Smith and W. D. Ross. Oxford: Clarendon Press, 1912.

Oliphant, Samuel Grant. "The Story of the Strix: Isidorus and the Glossographers." *Transactions of the American Philological Association* 45 (1914): 49–63.

Ostlender, Heinrich, ed. *Studia Albertina. Festschrift für Bernhard Geyer zum 70. Geburtstage.* Beiträge zur Geschichte der Philosophie und Theologie des Mittelalters, suppl. 4. Münster: Aschendorff, 1952.

Otto, A. *Die Sprichtwörter und sprichtwörtlichen Redensarten der Römer.* Leipzig: B. G. Teubner, 1890.

Peck, A. L., trans. *Aristotle, Historia animalium, Vols. 1,* 2. Loeb Classical Library. Cambridge: Harvard University Press, 1965–70.

———, trans. *Aristotle. Parts of Animals.* With *Movement of Animals and Progression of Animals,* trans. E. S. Forster. Rev. ed. Loeb Classical Library. Cambridge, MA: Harvard University Press, 1961.

Pelster, Franz. "Alberts des Grossen neu aufgefundene Quaestionen zu der aristotelischen Schrift *De animalibus.*" *Zeitschrift für katholische Theologie* 46 (1922): 322–34.

Peter Damian. *Epistulae.* Edited by K. Reindel. In MGH, *Die Briefe der deutschen Kaiserzeit.* 1983–93.

Petrus Alphonsus. *Disciplina clericalis.* PL 157: 671–706.

Pliny. *Natural History.* Translated by Botsford and Riley, 1893; Rackham, 1940.

Porphyry. *Isagoge.* Translated by Boethius (Brandt, 1906).

Pseudo-Aristotle. *Liber de causis.* Edited by Bardenhewer, 1882.

Rabanus Maurus. *De universo libri XXII.* PL 111: 9A–614B.

Rackham, H., trans. *Aristotle. The Nicomachean Ethics.* Cambridge, MA: Harvard University Press, 1947.

———, trans. *Pliny, Natural History.* 10 vols. Loeb Classical Library. Cambridge, MA: Harvard University Press, 1940.

Rath, Gernot. *Die arabischen Nomina Anatomica in der lateinischen Canonübersetzung.* In Courtois, 1956. Pp. 229–44.

Reindel, Kurt, ed. *Die Briefe des Petrus Damiani.* In *Monumenta Germaniae Historica, Die Briefe der Deutschen Kaiserzeit.* Teil 4: Nr. 151–80. Munich: Monumenta Germaniae Historica, 1993.

Resnick, Irven, and Kenneth F. Kitchell, Jr. "Albert the Great on the 'Language' of Animals." *American Catholic Philosophical Quarterly* 70 (1996): 41–61.

Resnick, Irven M. "Medieval Roots of the Myth of Jewish Male Menses." *Harvard Theological Review* 93.3 (2000): 241–63.

———. "Ps. Albert the Great on the Physiognomy of Jesus and Mary." *Mediaeval Studies* 64 (2002): 217–40.

Stadler, Hermann, ed. *Albertus Magnus De Animalibus Libri XXVI.* Beiträge zur Geschichte der Philosophie des Mittelalters 15 and 16. Münster: Aschendorfsche Verlagsbuchhandlung, 1916–20.

Thomas of Cantimpré. *Liber de Natura Rerum.* Edited by Boese, 1973.

Thompson, D'Arcy W. *A Glossary of Greek Birds*. London: Oxford University Press, 1936. Reprint, Hildesheim: Georg Olms, 1966.

———. *A Glossary of Greek Fishes*. London: Oxford University Press, 1947.

———, trans. and ed. *Historia Animalium*. Oxford: Clarendon Press, 1910.

Thorndike, Lynn. *Michael Scot*. London: Thomas Nelson and Sons, 1965.

Tredennick, Hugh, trans. *The Metaphysics*. 2 vols. Loeb Classical Library. Cambridge, MA: Harvard University Press, 1961–62.

———, trans. *The Posterior Analytics*. Loeb Classical Library. Cambridge, MA: Harvard University Press, 1960.

Vincent of Beauvais. *Bibliotheca mundi. Vincentii Burgundi . . . Speculum quadruplex, naturale, doctrinale, morale, historiale. In quo totius naturae historia, omnium scientiarum encyclopaedia, moralis philosophiae thesaurus, temporum & actionum humanarum theatrum amplissimum exhibetur.* Douai: Baltazaris Belleri, 1624.

Weisheipl, James A.. "Albert the Great and Medieval Culture." *Thomist* 4 (1980): 481–501.

White, Terence Hanbury. *The Book of Beasts, Being a Translation in full from a Latin Bestiary of the 12th Century*. New York: G. P. Putnam's Sons, 1954.

Wicksteed, Philip H., and Francis M. Cornford, trans. *Aristotle, the Physics*. 2 vols. Cambridge, MA: Harvard University Press, 1934.

Witelo. *Opticae thesaurus. Alhazeni Arabis libri septem, nunc primum editi. Eiusdem liber De crepusculis et nubium ascensionibus. Item Vitellonis Thuringopoloni libri X. Omnes instaurati, figuris illustrati et aucti, adiectis etiam in Alhazenum commentariis, a Federico Risnero*. Basil: Per Episcopios, 1572.

Yapp, Brunsdon. *Birds in Medieval Manuscripts*. London: British Library, 1981.

Zirkle, Conway. "Animals Impregnated by the Wind." *Isis* 25 (1936): 95–130.

INTRODUCTION

INTRODUCTION

Albert the Great's Questions concerning "On Animals"

During the twelfth and thirteenth centuries, a number of previously unknown works by (or attributed to) Aristotle became available in Latin in medieval Europe. In the twelfth century, many of these were translated from Arabic sources, especially by translators in Spain like Gerard of Cremona. In the thirteenth century, scholars increasingly sought Greek exemplars on which to base their translations.[1] Although some of the texts introduced new elements of Aristotelian logic, ethics, or metaphysics, a large body of material introduced Aristotle's books on natural science, including his biological works that circulated in Latin under the title *De animalibus,* that is, *On Animals.* This is the title that translator Michael Scot gave to the Arabic version in nineteen books of Aristotle's three treatises: *Historia animalium* (*History of Animals*),[2] *De partibus animalium* (*On the Parts of Animals*), and *De generatione animalium* (*On the Generation of Animals*). Michael Scot made his translation from Arabic ca. 1210.[3]

For a time, *De animalibus* was an immensely popular work and came to be read in two faculties of the medieval university, Arts and Medicine. It became a part of the medical curriculum be-

1. For an extensive list of texts translated in the twelfth and thirteenth centuries, see especially Michael McVaugh, "The Translation of Greek and Arabic Science into Latin," in Edward Grant, ed., *Sourcebook in Medieval Science* (Cambridge, MA: Harvard University Press, 1974).

2. Michael Scot's version includes the apocryphal book ten of *Historia animalium.*

3. For a good discussion of Michael Scot's role as translator, see Charles Burnett, "Michael Scot and the Transmission of Scientific Culture from Toledo to Bologna via the Court of Frederick II Hohenstaufen," *Micrologus* 2: *Le scienze alla corte di Federico II* (Turnhout: Brepols, 1994). For a more general discussion of his life and work, see Lynn Thorndike, *Michael Scot* (London: Thomas Nelson and Sons, 1965).

cause *De animalibus* concerns much more than animal anatomy, reproductive biology, and behavior; it also examines as part of the world of animals the human and therefore includes extensive materials useful to physicians on comparative anatomy and physiology, as well as diet and pharmacology. It appealed to the Arts faculty because in these books Aristotle examines the foundational or causal principles of Nature and discusses the character of scientific investigation, which proved interesting to both philosophers and theologians. The introduction of these works at the University of Paris in the thirteenth century did not proceed without controversy, as evidenced by the condemnation in Paris of Aristotle's books on natural science or philosophy in 1210. Nevertheless, it is clear that this condemnation was weakened over the next half century, and by 1254 masters in the Arts faculty at the university had made *De animalibus* a regular part of the curriculum.[4]

The earliest extant medieval commentary on *De animalibus* is by Peter of Spain (later Pope John XXI, d. 1277).[5] The most influential commentary on *De animalibus,* however, was produced by the Dominican theologian, philosopher, and scientist Albert the Great (d. 1280). Preserved in more than fifty-five manuscript copies, Albert's commentary was likely begun in the period 1256–60. He certainly worked on this project while he was Bishop of Regensburg, completing it by 1263, Weisheipl asserts, at Viterbo (where he had gone to resign his bishopric).[6] Regardless of the date one assigns to the completion of the work, it is clear that Albert occupied himself with *De animalibus* for many years, producing several versions, although perhaps never a final redaction.[7] This very influential commentary on Aristotle's nineteen books, which also includes extensive original investigations and speculations by Albert himself, joined to a bestiary

4. See especially Grant, 42–44.

5. See especially Miguel de Asúa, "Peter of Spain, Albert the Great, and the *Quaestiones de animalibus,*" *Physis* 34 (1997): 1–30.

6. James A. Weisheipl, "Albert the Great and Medieval Culture," *Thomist* 4 (1980): 499.

7. This suggestion has been taken up by Eckert, who suggests that "Albert hat sich jahrzehntelang mit den Tierbüchern beschäftigt. Eine Endredaktion war ihm jedoch nicht mehr möglich." Willehad Paul Eckert, *Angelicum* 57 (1980): 480–81.

portion at the end of the work, has recently been introduced and translated in its entirely for the first time by Professors Kenneth F. Kitchell, Jr., and Irven M. Resnick.[8]

In addition to his commentary on *De animalibus,* however, we have another work attributed to Albert the Great under the title *Quaestiones super de animalibus* (*Questions concerning* On Animals). This text represents a series of disputed questions on Aristotle's *De animalibus,* conducted in Cologne in 1258 and preserved in the report of Conrad of Austria from perhaps about the year 1260.[9] Historians had long known of this work from medieval catalogues. Until 1922, however, it was thought to be lost to us. In 1922 a manuscript containing the *Quaestiones* was discovered in the Milanese Bibliotheca Ambrosiana. Additional manuscripts containing this work, or fragments of it, were uncovered between 1932 and 1952 in other libraries. A critical edition of the Latin text was prepared by Ephrem Filthaut and appeared in 1955 in volume 12 of the *Opera omnia . . . Alberti Magni* in preparation in Cologne.[10]

In one sense, the *Quaestiones super de animalibus* (*QDA*) can be viewed as a useful companion piece to Albert's commentary on *De animalibus.* It represents Albert's attempt to introduce Aristotle's material to students in Cologne in his lectures during the year 1258. As such, *QDA* antedates his slightly later (and vastly larger) commentary, *De animalibus.* Although Albert did write down his commentary on *De animalibus*—indeed, Stadler's Latin edition is based on an autograph copy[11]—Albert did not himself write down these lectures that form the *QDA,* and this

8. Kenneth F. Kitchell, Jr., and Irven M. Resnick, *Albertus Magnus "On Animals": A Medieval "Summa Zoologica,"* 2 vols. (Baltimore and London: Johns Hopkins University Press, 1999). Hereafter this work will be abbreviated as *SZ.*

9. For a discussion of the discovery of this work, see especially Franz Pelster, "Alberts des Grossen neu aufgefundene Quaestionen zu der aristotelischen Schrift *De animalibus,*" *Zeitschrift für katholische Theologie* 46 (1922): 322–34.

10. *Alberti Magni . . . Opera omnia,* vol. 12: *Quaestiones super de animalibus,* ed. Ephrem Filthaut (Münster: Aschendorff, 1955). A discussion of the manuscript tradition can be found in the prolegomena, xxxv–xlvii. In addition, see Ephrem Filthaut, "Um die Quaestiones de animalibus Alberts des Grossen," *Ostlender* (1952), 112–27.

11. Hermann Stadler, *Albertus Magnus De Animalibus Libri XXVI.* Beiträge zur

presents the historian with an interesting problem. Although attributed to Albert the Great, the work is in fact a *reportatio;* that is, it contains what Albert taught about Aristotle's books on animals in Cologne in 1258, but the *QDA* itself represents the notes of Albert's student, Conrad of Austria, who heard Albert teaching. A few years later these questions were collected and redacted. Albert the Great can be considered the author of the work, but his imprint on the work is indirect rather than direct, and is mediated by Conrad of Austria. This means that one cannot properly distinguish the words of Albert from the words of the one reporting them. In addition, insofar as the text presents a quasi-Scholastic series of disputed questions based on Aristotle's *De animalibus*—with a question posed, followed in most instances by evidence *pro et contra* and some sort of response or solution to the question—neither is it always a simple matter to identify Albert's own position on a given question or to distinguish Albert's doctrine from other, contemporary sources for his discussion of this material.[12]

The identity of Conrad of Austria remains a thorny problem. One possibility is that Conrad is a Dominican of that name identified in a necrology from a provincial chapter of 1284–88. This Conrad was, before his entrance to the Order, a scribe to the duke of Austria who later joined the Order in Vienna (and thus is known as Conrad of Austria). Other candidates have been proposed as well, however, and Conrad's identity remains uncertain.

Regardless of the difficulties involved, *QDA* gives the reader access to a wealth of material from the middle of the thirteenth century that constitutes a sort of handbook of medieval science and medicine. A comparison to his commentary on *De animalibus* reveals on almost every page Albert's personality, and discloses his pressing interest in the animal world. This interest was not only theoretical, but was based in part on experience and experimentation. It is when recounting his own experienc-

---

Geschichte der Philosophie des Mittelalters 15 and 16 (Münster: Aschendorfsche Verlagsbuchhandlung, 1916–20).

12. For example, as De Asúa (1997) tries to show that Albert was dependent in part on the work of Peter of Spain.

es of animal behavior, anatomy, or biology that Albert's originality becomes evident and can enchant the reader.[13]

Until now, only a few pages of *QDA* have been available in English translation.[14] The translators propose to make *QDA* available here in a complete English translation for the first time. Although lacking the extensive annotation provided earlier in our translation of Albert's *De animalibus,* we will direct the reader to that work whenever possible for corresponding material and for a more detailed explanation of themes and issues treated in *QDA*.

A Note on References to Ancient and Medieval Works

Anyone consulting Ephrem Filthaut's 1955 edition of the *Quaestiones super de animalibus* is immediately impressed by the depth of scholarship it exhibits. In addition to a meticulous critical study of the manuscripts, Filthaut provides many notes, notable for the breadth of authors consulted. The translators, therefore, have generally relied on his citations. We have checked all possible references, but in the case of obscure works, often very difficult to obtain (for example, "Witelo *Optica,* Alhacen *Optica*" at *QDA* 1.29), we have deferred to Fr. Filthaut.

Citing Aristotle for general readers poses more difficulties than one would expect. It has long been customary to cite the works of Aristotle in reference to the nineteenth-century edition of Immanuel Bekker, who printed the Greek text of Aristotle's collected works in parallel columns. Thus a reference to his *Parts of Animals* might appear as *PA* 669b35 and refers to page 669 of the Bekker edition, line 35 of the right-hand column.

This is the most precise way to cite Aristotle, but is virtually

13. On Albert's experiential approach (with some comparison to the more theoretical approach of his contemporary Gérard du Breuil), see Tamara Goldstein-Préaud, "Albert le Grand et les Questions du XIIIe siècle sur le *De animalibus* d'Aristote," *History and Philosophy of the Life Sciences* 3/1 (1981): 61–67.

14. *QDA* 1.3 is available in English in Edward Grant, *Sourcebook in Medieval Science* (Cambridge, MA: Harvard University Press, 1974), 681–89.

useless for those lacking Greek or access to these particular volumes. Modern editions of the Greek of Aristotle continue the practice even when the number of lines does not correspond to the number of Greek lines in a modern printing. Thus the gap in *HA* 548a22–32 might in fact be more than ten lines. The problem is infinitely worse when the text is translated into English.

Moreover, Filthaut, in his edition of *QDA*, tends to cite sections of Aristotle's works that are generally linked to Albert's thought in a given passage. Thus *PA* 669b35–40 may refer to a portion of Aristotle concerned with the same subject matter Albert is treating rather than being an exact quotation.

We have therefore adopted the following practice as best suited to the purposes and readership of this volume. Instead of citing *HA* 548a22–32, which may or may not correspond exactly with the Greek edition or translation consulted by a reader, we will rather cite both by book number and chapter number, followed by approximate Bekker numbers, for example, *HA* 5.15–16 (548a22–32). Interested readers will do well to consult Loeb editions, published by Harvard University Press, which have facing pages of Greek and English text. This will more readily enable those with English texts to compare, in general terms, parallel passages of Aristotle and Albert. When Albert is actually quoting Aristotle, or another author, we have attempted to cite exact lines, but the reader should not expect to find line-for-line correlation with the Greek.

Citing Avicenna poses similar difficulties. Many of the Latin editions of his works have not been translated, and few exist in modern editions. Here we have generally followed Filthaut, after consulting the editions of the *Canon* and *De animalibus* cited in the bibliography.

In addition to citing the Stadler edition of Albert's *De animalibus,* we frequently refer the reader to our own heavily annotated edition of the work,[15] using the abbreviation *SZ*. Interested readers will find in-depth references for further research in these volumes.

15. Kitchell and Resnick, 1999.

As both Stadler and Filthaut point out, citing Galen presents many difficulties, not the least of which is determining what version of Galen Albert had before him at a given time.[16] The references given in this translation are those of Filthaut. Kühn's edition remains the standard for consulting the ancient text.[17]

16. Filthaut (1955), xlvi; Stadler (1916–20), 2.viii.

17. C. G. Kühn, ed., *Claudii Galeni Opera Omnia,* 20 vols. (Leipzig, 1821–33; reprint, Hildesheim: Georg Olms, 1964–65).

QUESTIONS CONCERNING ARISTOTLE'S ON ANIMALS

Here begin the questions on the books *On Animals.*

BOOK ONE

Question 1: Whether this book has animals for its subject.

FIRST ONE ASKS whether this book has animals for its subject.

1. It seems not, because every science has to do with things that are universal and incorruptible, but every animal is individual and corruptible, and therefore, etc.

2. Moreover, if it were about animals, then it would concern either some or all animals. But it is not more about some than others, and it would therefore be about them all. But this is impossible, since then it would seem to be infinite and the infinite cannot be known by humans.

3. Besides, an animal is composed of body and soul. Therefore any treatment of animals can be threefold: either in terms of the soul, in terms of the body, or in terms of the soul as it is related to the body. But the soul is treated in the book *On the Soul.* Moreover, the body is sufficiently treated in the book, the *Physics,* and in other subordinate sciences like medicine. But the soul as related to the body is adequately treated in our small books on natural matters [*Parva naturalia*]. If this book, then, were about animals it would be fruitless and superfluous, because the whole subject of what is in the animal is treated elsewhere, as has been proven. Therefore, etc.

But the opposite is made clear by how the Philosopher decided his subject matter[1] and the title of the book.[2]

1. Latin: *determinatio.* The root of this word is *terminus,* and thus it suggests a boundary.

2. Ar., *HA* 1.1 (486a5).

To this one must say that this book can be said to be about animals because it primarily concerns the animated, movable body as applied to individual species of animals. This is so because such a study [*scientia*] is related to science [*scientia*] in the same way a specific subject is related to a [broader] subject. But this study is part of natural science and its subject will therefore be part of the broader subject. But the broader subject is that of the movable body. Therefore, the subject of this study is a specific, movable body.

Thus one must understand that "everything is intelligible to the extent that it is separable from matter."[3] But since sciences are a characteristic [*habitus*] of the intellect, they have to be distinguished with respect to their objects, and this is why, to the extent that something is separated from its matter in diverse ways, it is naturally destined to pertain to a different science. For this reason, because some things are separated from matter with respect to their being [*esse*] and definition [*ratio*], like metaphysical objects, and some are separated with respect to their definition but not their being, like mathematical objects, and some, like natural objects, are separated from matter neither with respect to their being nor their definition, and are not separated in general but only in particular, then these sciences are essentially different—metaphysics, which concerns the first sort of being, and mathematics, which is about the second, and physics, which is about the third.[4]

And likewise, within the same science the parts are distinguished with respect to a greater or lesser degree of separation from matter. And since that which is more universal is further abstracted from matter, it therefore belongs to a different part of science to make a determination about the subject in general and in particular. This is why the book on *Physics*,[5] in which a determination is made about movable body in general, and the book *On Heaven* [*and Earth*],[6] in which a determination is made about movable body restricted in terms of place [*contracto ad ubi*], concern one part of natural science, and consideration of

3. Ar., *De anima* 3.4 (429b21–22); cf. A., *De anima* 3.2.16, 3.3.1.
4. See A., *Metaph.*, 1.1.1.
5. See A., *Phys.*, 1.1.4.
6. See A., *DCM* 1.1.1; *DG* 1.1.1.

a movable body defined by being animated, consequently, belongs to another part. And because consideration of the soul or of the ensouled is threefold: one way, considered absolutely, and another way as it is formed in the body or parts of the body, and this can still be considered in general, and a third way can be about that soul when applying it and its natures [*rationes*] to individual species of plants and animals, and because the first is treated in the book *On the Soul*,[7] and the second in the short books that follow it, this is why it is necessary to treat the third in another part of natural science, and this is treated in the books *On Plants* and *On Animals*, which are in our hands.

Thus, properly speaking, the subject of this book is body, animated with a sensitive soul, as applied to individual species of animals. This is why it has to occupy the last place in the order of the books of natural [science].

1. On to the arguments. To the first, one must reply that animals can be treated in two ways, either with respect to the being [*esse*] they have apart from the soul or with respect to the being they have in the soul, that is, in the intellect. If they are considered in the first way, then each one is individual and corporeal, and then there can be no science of animals. But if in the second way, then there can be a science of animals since there is nothing in the intellect but what is universal and separable from matter.

2. To the second, one must reply that this is a science of all animals, although not as they are considered in and of themselves, because this way they have no fixed number but are rather infinite as far as we are concerned and do not fall under our understanding, but rather as they are reduced to determinate species and are united and come together in a common nature.

3. To the third, one must reply that although there could be a threefold treatment of an animal and each one could be treated in a different fashion in its own book, nevertheless the treatment that is considered here is not treated elsewhere. For here the soul is treated in relation not to each and every body, but only to the body that is bound to a species of animal. The soul

7. See A., *De anima* 1.1.1.

is considered absolutely, however, in the book *On the Soul,* and body is treated without reference to the soul in the *Physics,* and in the other books of natural science the soul is considered nevertheless as it is formed for a body in general. And therefore this book is not unnecessary.

Question 2: Whether a variety of organic parts is necessary to the animal.

"Certain parts of bodies, etc."[8] One asks whether an animal needs a variety of organic parts.

1. It seems not, because the more perfect the form, the more operations it can perform and with fewer intermediaries. For the sun, owing to its perfection, can act by means of its own light in more operations than some inferior thing. But the form of man is more perfect than the form of an element or a stone; therefore, it can act in more operations and with fewer intermediaries and, as a consequence, requires less diversity in its matter.

2. Likewise, if an animal needed a variety of parts, then this would be the case either for the sake of its essence or for the sake of its operation. But it is not for the sake of its essence, because it is simpler than the essence of other mixed bodies and freer from matter. Nor is it for the sake of its operations, because a plant has more parts but nevertheless has fewer operations.

3. Likewise, the Philosopher asks in *On Heaven and Earth,*[9] why some spheres have one star and some have several, and some have several motions and some have fewer. And he says that the sphere that is more perfect has fewer motions, and this is why the starry heaven, which is the cause of being and permanence, moves with a single motion or by fewer motions than some other, inferior sphere. Therefore, since these inferior things imitate the superior as far as they are able, then, to the extent that something is more perfect, to that extent it will have fewer motions. And as a consequence the human, since he

8. Ar., *HA* 1.1. (486a5–8). See Avic. *DA,* 1 (fol. 29raA); A., *DA,* 1.1.2.12 (*SZ* 1: 49).

9. Ar., *DC* 2.12 (291b28–292a1, 292b19–293a4); A., *DCM* 2.3.15.

is the most perfect animal, will have less need for diverse parts since the diverse parts exist only on account of diverse motions.

On the contrary. As the Philosopher says in the third book of *On the Soul,*[10] "Nature does nothing in vain, nor is it lacking things necessary to it." But nature provides each animal diverse organic parts. Therefore, these are necessary to it.

One must say that an animal needs a variety of organs. The reason for this is that the more perfect the form, the more operations it can perform. But matter is, to a certain extent, an impediment to operations; for a form bound to matter is perforce constrained or limited by it. Thus a form separated from matter, like intelligence, can be almost infinite with respect to a material form.[11] Therefore, a form bound to a uniform matter has a uniform activity. Thus, each part of fire is fire and each part warms like the whole fire. Since, then, the soul is a more perfect form than the form of an element or a stone, it is the source of several operations. But it cannot perform many operations when it is united to matter, unless its matter is diversified, because it can only perform a uniform operation through a uniform matter. And this is why, if the whole animal body were just like the eye, it would not hear nor would it smell, and if its power were proportionate it would see by means of the whole body.[12] And this is why it is necessary that the body, which is the animal's matter, be diversified into parts, so that diverse operations may be performed by means of the diverse parts, because if the body were undiversified [*unigeneum*] in its parts then the animal could perform only undiversified actions.

Nevertheless one must understand through the arguments that there is a difference between these inferior things and the superior ones.[13] For among superior ones it is the case that the more perfect something is, the better it attains its end with fewer motions. Thus the supreme sphere attains its end by a single motion, and the inferior one by several. And similarly, when the

10. Ar., *De anima* 3.11 (434a30); A., *De anima* 3.4.2, 3.5.1.

11. This reading relies on MSS P & E.

12. See Ar., *De anima* 2.1 (412b18f.).

13. Throughout this section A. uses *superior* and *inferior,* implying quality (better, worse) as well as, perhaps, distance (higher, lower).

first cause understands itself it understands all things, whereas inferior substances understand in a less comely manner. But it is different among these inferior things, because the more perfect something is the better it attains its end through several motions. And the reason for this is that these inferior things are joined to matter, and the form is the principle of acting or moving and operating. Therefore, just as the more distant something is from matter the more it shares in the perfection of the form, so too the more distant it is from matter the more it has of [form's] operation. And this is why, when it comes to material things, the human, since he is the most perfect animal, attains his end through several operations. Thus, the order is reversed between material and immaterial things. And the whole reason is that in material things perfection is attained in accordance with the remoteness from matter, but in immaterial ones it is attained in accordance with their nearness [*approximatio*] to the simplest principle.

1. On to the arguments. To the first argument one must reply that the more perfect the form the more it extends its operation to several things. Nevertheless it is not necessary that it have several operations. Rather, that it have several operations stems from the perfection of a material thing. Now, an immaterial thing or the first cause can affect more in a single act than a material thing can with several. Thus the proper or characteristic act of the sun is to give light, but through this act it extends itself to a multitude of others.

Thus, one must reply to the form of the argument that a plurality of operations can be observed in two ways, either from the standpoint of the agent's power or from the standpoint of the things affected. In the first way, an animal has several operations; in the second way, the sun does. And this is why an animal requires several parts, with which it operates as if with instruments, just as the sun does, because just as the hand and the foot and the eye are the animal's instruments, so too these inferior things are the sun's instruments, because a human can generate a human only through a human and a cow can generate a cow only through a cow.

2. To the second argument one must say that the animal es-

sence is the root cause of diversity in the parts and its operation is just like a final cause. Now because the animal soul is more perfect with respect to its powers than is the form of inanimate things, and each power requires a different organ, for this reason from the beginning [*originaliter*] a diversity of organs is required owing to the essence of the soul itself, just as diversity in the matter is required owing to a diversity of form, and in the end [*finaliter*] diversity is required owing to the diverse operations which are performed only through diverse organs.

3. To the third argument one must say, just as we said, that the motion of inferior and superior things is different. Nevertheless, if the argument is made about a star, just as already has been done concerning its motion, then one will conclude the opposite, that it is the case among the spheres that a superior sphere has a greater number of stars than does a lower one. Therefore it will be the case among inferior things that a more perfect body will have a greater plurality of operations and powers, and a plurality of powers demands a plurality of organs. This is why it follows that an animal has more need of diverse parts than does something inanimate. And this is true.

Question 3: Whether an organic member that has been cut off can be restored.

One asks whether an organic member that has been cut off can be restored.[14]

1. It seems so, because organs exist in animals just as branches do in plants. But a branch that has been cut off can be regenerated. Therefore, for the same reason, organic members can be regenerated in animals.

2. Likewise, organic members are composed of homogenous parts, but homogenous parts can be regenerated, as is clear in flesh; therefore, etc.

3. Again, nourishment is converted into the substance of the one nourished; therefore any member that is lost can be restored by nourishment.

14. A., *DA* 1.1.2.13–20 (*SZ* 1: 49–52).

The opposite is clear to the senses, since a hand that is cut off or an eye that is plucked out is not regenerated.

One must say that organic (that is, official) members in animals cannot be restored, because the more noble a thing is the more care nature bestows on its production. Thus it disposes a more noble matter upon something capable of being produced more nobly. So the animal's semen is nobler and more subtle than the plant's seed [*semen*], and this is why nature ordained a determinate place for the animal's semen, namely, the testicles, but not for the plant's seed.[15] Now, however, the organic members are generated from the spermatic semen by means of the power delegated to such parts for the purpose of being a formative power. Therefore, if such a member is cut off, there is no matter remaining in the body from which such a member can be produced nor is there such a productive power in the body, since the productive power of the hand is in the hand and that of the foot is in the foot, and once the hand has been generated this productive power of the hand disappears. Thus such members cannot then be regenerated owing both to a defect in the matter and a defect in the agent.[16]

1. On to the arguments. To the first, one must reply that branches that have been cut off can be regenerated because among animated things a plant is nearer to matter and to inanimate things. Thus it is particularly undiversified both in the whole and the part. So no matter is required to produce a branch beyond that which is the principle of nourishment. Thus, in plants the principle of nourishment and of the generation of branches is one and the same. But this is not the case for organic members in animals.

2. To the second, one must reply that some parts, for example, some nerves, bones, and some others like these, cannot be restored, just as the organic parts cannot. For those that partake more of form and less of matter, like those produced mostly from the spermatic semen, cannot be restored. But those that have a more material condition or are nearer to matter, yet do

15. A., *DA* 1.1.6.82–84 (*SZ* 1: 75–77).

16. Reading *tum . . . tum* for the unintelligible *tum . . . cum.*

not arise only from the spermatic moisture but rather from the nutrimental moisture, can be restored. Examples include flesh, hair, and nails. And yet there is some flesh, like the flesh on the face, that cannot be restored, and this is because flesh such as this is produced from the spermatic moisture.

Or, in another way. Flesh is of two types: one type with respect to species, and another with respect to matter. The first type is not regenerated, but the second can be, and even comes back the same. Of the first type are the flesh of the foreskin, the lips, the jaws, and the like.

3. To the third, one must reply that although nourishment can be converted into the substance of the one nourished, nevertheless, once a hand has been cut off, the power is absent that can convert nourishment into a likeness of the hand. And this is why such members cannot be restored through nourishment, as has been clearly declared, etc.

Question 4: Whether any animal breathes air.

"And the types of animals, etc."[17] Here one asks whether any animal breathes air.

1. It seems not, since respiration occurs for the sake of cooling the heart and lungs. But air is warm and moist, and water is cold and moist. Therefore a breathing animal needs to draw in water more than air.

2. In the same way, an animal drawing in water derives its life from the water; an indication of this is that if it should be out of the water, it dies immediately. But an animal cannot live on air; therefore it cannot take in air.

The Philosopher says the opposite.[18]

One must say that some animals have a very hot heart. Nature gave these lungs, which are a sort of fan for the heart,[19]

17. Ar., *HA* 1.1 (487a11–12). See Avic., *DA* 1 (fol. 29raA); A., *DA* 1.1.3.25 (*SZ* 1: 54).

18. Ar., *HA* 1.1 (487a28f.); see Avic., *DA* 1 (fol. 29raA); A., *DA* 1.1.3.26–29 (*SZ* 1: 54–56).

19. A., *DA* 1.3.2.543–44, 13.1.1.9, 13.1.6.46 (*SZ* 1: 254–55; 2: 988, 1005).

and by their expansion air is drawn in and by their contraction it is expelled, as one sees in a fan or bellows. But in one way it is true that the heart is a particularly impassible member, for according to the Philosopher in the third book of *On the Parts* [*of Animals*], "it is not susceptible to weakness."[20] Water and earth change things materially a great deal, and this is why neither water nor earth is drawn to cool the heart, because these things are particularly material. Neither is fire drawn there, for it would increase the heart's heat. But, on account of its thinness and conformity, air is drawn in. Those animals, however, that do not have a proper heart or lungs but rather have some material object comparable to the heart, as certain fish do, draw in water to cool it, just as those that do have lungs draw in air. And the Philosopher says this in the text that "some draw in air and expel air" and "some, like fish, draw in water and expel it through gills."[21]

1. On to the arguments. To the first, one must reply that although air may be warm and moist, it nevertheless has less heat than the heart. And it is apparent that very hot water can be cooled by the infusion or admixture of tepid water. Additionally, the air in which we live is especially cold owing to the commingling of vapors and its proximity to earth and water.

2. To the second, one must reply that an animal that draws in water does not derive its life from this water, because "we are nourished by the same things upon which we exist," according to Aristotle in the book *On Generation*.[22] So, since an animal is mixed, it is therefore necessary that its nourishment be mixed. Thus, those proposing that the chameleon and the mole, the herring and the salamander live on pure elements do not speak the truth, because nourishment has to be in a proximate disposition to the one to be nourished, and a simple [element] is not proximate in power to the mixed, but is remote from it. So fish do not live on sea water or on water alone, but on some mixture of water and earth.

20. Ar., *Part. An.* 3.4 (667a33); see A., *DA* 13.1.4.37 (*SZ* 2: 1000).
21. Ar., *HA* 1.1 (487a17–18, 28–29).
22. Ar., *DG* 2.8 (335a10–11).

Question 5: Whether an aquatic animal, once its form is changed, can become terrestrial.

Next one asks whether an aquatic animal, once its form is changed, can become terrestrial.

1. It seems not. For it is said in the fourth book of the *On Meteorology,* in the old translation: "Let alchemists know that species cannot be transmuted, but altered."[23] But there is no less diversity among animals than among metals, and therefore, etc.

2. Again, no transition occurs from one genus to another genus. But according to the Philosopher at the beginning of this book,[24] fish and birds "differ in genus," and as a result one cannot pass into the other.

The Philosopher says the opposite.[25]

One must reply that "animal" can be considered in two ways: either with respect to matter or with respect to form. If with respect to form, then one cannot be transmuted into another. For just as white does not become black, so one species of animal cannot be changed into another. If, however, we consider "animal" with respect to matter, then one can be changed into another. For it happens that the matter existing under one species has a great affinity to the matter of another species, and this is why, when the form of the one withdraws, the form of another can be introduced to the matter. This is evident in dead bodies, when vermin and hornets [*strabones*] are generated from them.[26] It is also evident in the generation of certain animals

23. The reference to the "old translation" suggests that William of Moerbeke's new translation from the Greek was already available.

24. Ar., *HA* 1.1 (486a23).

25. Ar., *HA* 1.1 (487b3f.). See Avic. *DA* 1 (fol. 29raA); A., *DA* 1.1.3.30 (*SZ* 1: 56).

26. "Hornets": The text is surely corrupt here, reading *strabones,* or "squinters." The mention of vermin and squinters coming from corpses is evocative, but clearly corrupt. The manuscripts offer no help, but the solution undoubtedly lies in *scrabones,* a variant and late form of *crabones,* hornets. On the hornet, *Vespa crabro,* see ThC 9.16, Vinc. 20,157, Ald. *Aen.* 75, and Isid. *Orig.* 12.8.4 (*scabrones*). Hornets and wasps were said to arise spontaneously from the bodies of various dead animals. See *DA* 26.13(14) (*SZ* 2: 1747) for *crabro,* and see *DA* 8.4.6.190 (*SZ* 1: 750) for a description of wasps generated from dead bodies.

that occurs through an intermediary, as is apparent for frogs. For certain black animals with tails are generated from the semen of frogs that later, after they have cast off the tails, adopt feet, and whereas earlier they live in water, they later live on land. The reason, however, that an aquatic animal becomes terrestrial and not vice versa can be that the matter of the aquatic animal can be thickened by the coldness of the water so much that it becomes suited to the form of a terrestrial animal. But the longer it remains a terrestrial animal, the more it tends to dryness and as a consequence it becomes more removed from the nature of an aquatic animal.

1. On to the arguments. To the first, one must say that one species of metal never becomes another, even though by means of alchemy it can happen that one species appears as another. It is the same among animals: one species does not become another, although after one species another species can be introduced to the same matter.

2. To the second, one must reply that an aquatic animal, a walking animal, and a flying animal do not differ in genus, because there is not another genus contained under "animal." Nevertheless the Philosopher does say that "fish" and "bird" do differ in genus, because they differ in their general properties; for example, their movement occurs by means of various differences. It has been posited, then, that fish and bird differ in proximate genus [*genus propinquum*], yet they do not differ in the logical genus [*genus naturale*] or in the predicamental genus [*genus praedicabile*], because they agree in their matter and whatever agrees in matter cannot pass from one genus into another.

Question 6: Whether motion is present in every animal.

Now one asks about the motion of animals. And first, whether motion is found in every animal.[27]

1. It seems so. Motion occurs on account of some need. But every animal needs something, for example nourishment, which is often at some distance from it. Therefore, etc.

27. See A., *DA* 1.1.3.31–36 (*SZ* 1: 56–58).

2. Again, motion follows sense. This is why Avicenna maintains that wherever sense is absent, there is no motion.[28] Therefore, since every animal has sense, it will have motion.

3. Again, the cause of an animal's immobility is proximity to its food source [*coniunctio alimenti*]. Conversely, therefore, the cause of mobility is separation from its food source. But every animal has a food source separated from it; therefore, etc.

The Philosopher says the opposite, and this is also obvious to the senses.[29]

One must reply that locomotion is of two types: one is the motion of expansion and contraction, and the other is progressive motion. The first follows upon the sense of taste or touch and is present in every animal because there is no animal so immobile that it does not retreat as if it were pricked if something harmful approaches, whereas, if something agreeable approaches, it expands and spreads itself over it. The second type of motion is found only in those possessing sensations that take place by means of an external medium. For unless an animal has sight or smell or hearing, progressive motion may injure it more than assist it, since it might put itself in the way of things that corrupt as well as things that benefit. Therefore, since not every animal has sight, hearing, or smell, not every animal has progressive motion.

1. On to the arguments. To the first, one must reply that need is of two types: namely, for completion or for preservation. Now a human needs many things for his preservation, whereas immobile animals have a need for completion but do not need as many things for their preservation. And when one says that every animal needs nourishment that is at some distance from it, it must be said that although the nourishment of every animal is distinct substantially before its consumption, nevertheless in many cases it is in close proximity to it, and this is why motion is not needed for them to obtain it.

2. To the second, one must reply that motion—whether expansion or contraction or progressive motion—follows sensa-

28. Avic., *De anima* pars 2, c. 3 (fol. 7rbA); Ar., *De anima* 2.5 (416b33–34).
29. Ar., *HA* 1.1 (487b6–8). See A., *DA* 1.1.3.32 (*SZ* 1: 56).

tion. Those animals are called immobile that do not move progressively.

3. To the third, one must reply (and it is already clear) that the cause of motion is the separation of the food source with respect to substance and location. But the food source is not separated from every animal, as will be made clear.

Question 7: Whether an immobile animal is always aquatic and not terrestrial.

Second, one asks whether an immobile animal is always aquatic and not terrestrial.

1. It seems not. The more form something has, the more activity [*operatio*] it has. But water has more form than earth. Therefore, an aquatic animal has more form than a terrestrial animal and, as consequence, more activity. Therefore, if there is some aquatic animal that is immobile, a terrestrial one will be much more so.

2. Again, although plants are terrestrial they are immobile. Therefore, for the same reason, even terrestrial animals can be immobile.

The Philosopher says the opposite.[30]

One must reply that an aquatic animal can be immobile more than a terrestrial one for two reasons. One is that every animal requires nourishment. Now it is certain that water is in continual movement and flux. So what now is absent from the water can be made present through the water's motion. And this is why some animals, which live on the dregs of the sea, do not need to move to seek a food source, because it is adequately carried to them by the water's motion. But this is not the case for the ground, since it is immobile, as it were, and this is why, if there were some animal fixed on the ground, it could not be preserved very long.

1. With regard to the arguments: To the first, one must say that although water may have more of form than earth, never-

30. Ar., *HA* 1.1 (487b6f.).

theless certain aquatic animals have more crude earthiness, and so they are born to have less in the way of motion and activity. And the cause of this has already been stated.

2. To the second, one must reply that animals, unlike plants, are not born to be nourished by such an earthy food source, and so plants can continually receive their nourishment from the same place or piece of earth without motion, but animals can not. This, then, is the second cause of immobility for aquatic animals, that life depends on heat and moisture and not the dryness that is in the earth. And because earth has more dryness and water more moisture, an aquatic animal can live longer in the same place than a terrestrial animal.

Question 8: Whether some animals should live in a social group.

Further one asks whether some animals should live in a social group.

1. It seems not. For "man is a city-dwelling [*civile*] animal" and a social one according to the Philosopher in book one of his *Politics*,[31] and this is why human life is governed by politics and economics. If, then, all animals were sociable, they would have to be governed by politics and economics, which is not true because politics is a human virtue.

2. Again, if some affect [*passio*] befits one on account of some medium, then once the medium is posited, the affect is also posited. Now, however, animals would only live in a social group if they did so to work better on things suitable to them and flee harmful ones. But this is useful for every animal. Therefore, every animal will be sociable.

The Philosopher says the opposite.[32]

One must reply that some animals are gregarious or socia-

31. Ar., *Politica* 1.2 (1253a7–9); see *Eth. Nic.* 1.7 (1097b11). A.'s *civilis* with its root in *civis,* citizen, faithfully reproduces Ar.'s *politikos* with its root in *polis* or "city."

32. Ar., *HA* 1.1 (487b33f.). See Avic., *DA,* 1 (fol. 29rbB); A., *DA* 1.1.3.37 (*SZ* 1: 58–59).

ble and some are solitary and some conform to both modes. In order to understand the evidence for this, one must know that although there are four interior sensitive powers—namely, the common sense, the imaginative, the estimative, and the recollective [*memorativa*]—the estimative power is capable of receiving intentions which sense does not receive, so that those animals that have a better estimative power can better guard themselves and care for themselves. Thus there are some flying animals that, owing to the dryness of their brain—in which the estimative power flourishes—are always in social groups, like the crane and the bee. For cranes travel from one region to another, and for this reason, for the sake of avoiding dangers on the way, they gather themselves into a single group. And bees act in a similar fashion in order to make honey better, and ants to gather seeds, and to resist those that would take their honey or seeds. For sweet and fat foods have a large number of predators. And it is the same for terrestrial and aquatic animals.

But some are solitary animals, like birds of prey. These choose to exist alone, lest a gathering of others cause them to fall short in their prey. But the human conforms to each mode. For he lives with others when he has in mind politics, which is concerned with the governance of cities, or economics, which is concerned with the disposition of the household. When, however, he is intent on the ethical or monastic life, which consists principally in the contemplation of truth, he lives or chooses to exist in a solitary fashion.

One must understand that some gregarious animals establish for themselves a leader or king, by whom they are ruled. For example, the bees establish a certain bee as leader or prince because of the size of his body. This one always lives in the hive to guard the honey and to repel bees from another swarm, and the other bees serve him, and when he goes out all the others follow him just as faithful soldiers follow the prince, as is evident in the summer when they flee from the hive.[33] But the cranes likewise raise up a prince over themselves, not because of his size or power. Rather, when they fly by day they have one who leads

33. See A., *DA* 8.6.1.226 (*SZ* 1: 767).

the others in a procession, and this one, just like a prince, goes ahead of the others to spy out singular dangers. At night they elect another one to guard them from dangers and keep track of the night watches.[34] Thus it is said that a crane holds one stone in his [raised] foot so that he will not fall asleep, since if he falls asleep it falls out, waking him up, and he does this instinctively [*habet a natura*].

But many gregarious animals do not have a prince, like ants and doves, who seek food while gathered into a group as they fly.[35] Thus they do this by means of their own estimative power.

1. On to the arguments. To the first, one must say that a human is a social animal by nature, but his socializing is the result of discretion's mediation. But for the other animals it is the result of a natural instinct. And this is why other animals, properly speaking, do not have politics or economics.

2. To the second argument one must reply that although it may be useful for every animal to live in a social group, in order better to pursue things that are suited to it and to flee from things harmful, nevertheless different animals move differently by means of different estimative powers. Now when doves seek nourishment, it seems better for them to be in a social group, and the same is so for ducks and geese [*auca*].[36] So, when they see a hawk or falcon, they gather into one group on account of their fear of birds of prey and of enemies. But among birds of prey it seems better to exist in solitude, because they only fear birds of their own kind, by whom they are prevented from seizing their prey.

34. See A., *DA* 8.2.4.66, 8.6.1.226 , 23.1.24.113(49) (*SZ* 1: 694–95, 767; 2: 1626).

35. See Ar., *HA* 1.1 (488a10f.).

36. The editor provides a corrected reading of *aucis* for *ancis* (see *Corrigenda*, p. 360). For the *auca*, see A., *DA* 23.1.22(6) (*SZ* 2: 1556). On the social character of ducks, for purposes of self-defense, see also *DA* 1.1.5.56 (*SZ* 1: 66).

Question 9: Whether every animal has a fixed type of nourishment.

Further one asks whether every animal has a fixed type of nourishment.

1. And it seems so. For every animal desires nourishment. Therefore, appetite is fixed for a fixed type of nourishment. Since, then, every animal has a fixed appetite, it also has a fixed type of nourishment.

2. Again, this is apparent by induction [*per inductionem*]. For the bee enjoys honey, but would not enjoy iron, and the hawk enjoys flesh but would not enjoy wood or stones or things of this sort.

The Philosopher says the opposite.[37] This is apparent in the human, who does not fix any nourishment for himself.

One must say that some animals have a fixed nourishment and some do not. For some, for example the hawk and falcons and others of this sort, are very warm and dry and this is why they are nourished only by things very moist and difficult to digest, like raw meats. And some, like bees, have a delicate and frail complexion, and this is why they are nourished only by sweet things. But the human has not only a natural but also a rational appetite, and this is why he does not have a defined nourishment. Rather, by use of his reason he can convert the unsuitable into something suitable and the suitable into something unsuitable. And in this sense some animals, like the dog, imitate the human owing to the indiscriminate nature of their stomach.

1. On to the arguments. To the first, one must reply that not every animal has a defined appetite, since in the human appetite exists with reason, which has the power to incline to contraries.

2. To the second argument one must say that although some animals desire things defined by a natural instinct, nevertheless

37. Ar., *HA* 1.1 (488a14f.). See Avic., *DA* 1 (fol. 29rbB); A., *DA* 1.1.3.41 (*SZ* 1: 60).

nothing prevents those which are prudent or capable of reasoning or obedient to reason from having a nourishment that is not fixed.

Question 10: Whether an animal is born to seek out nourishment more at night than by day.

Further one asks whether an animal is born to seek out nourishment more at night rather than by day.

1. It seems not. Because nothing is seen without light. Therefore one can discover nourishment better during the day than at night.

2. Again, night is better suited for rest than is the day. Therefore each animal has to seek nourishment in the day rather than at night.

The Philosopher says the opposite.[38]

One must say that, for many reasons, some animals seek nourishment more at night than during the day. One can be that a given animal does not tolerate the light of day, like the bat and the owl, but at night its sight is strengthened, etc. Another reason is that they can apprehend their nourishment more stealthily, just as a sheep cannot see a wolf as well at night as it does during the day. Thus, the lion can more quickly catch[39] the stag, and the wolf the sheep, at night rather than during the day. And another cause for this is that the food source of some animals is better revealed at night than during the day, for example mice, which are the cat's food source. And another cause can be that many animals fear humans, who guard the animals' food source. This is why some, like mice, wolves, cats, and ferrets, hunt more at night than during the day, since humans do not guard them at night.[40]

Nevertheless, one must attend to the fact that some animals

38. Ar., *HA* 1.1 (488a25f.). See Avic., *DA* 1 (fol. 29rbB); A., *DA* 1.1.3.43 (*SZ* 1: 60–61).

39. The Latin here allows both a sense of "apprehend" in the sense of "see" and, more literally, "capture."

40. On the ferret see *DA* 22.2.1.101(46) (*SZ* 2: 1509).

seek out a food source only during one time of the year, and during the other part of the year they rest and keep themselves in caves, as serpents and lizards do, because they do not tolerate the cold of winter owing to the fragility of their body and the weakness of their heat. Neither does the sparrow or the cuckoo, but rather in winter, asleep and almost dead, they lie concealed. But then their nature returns to the muscles and is fostered in the muscles.[41] This is just as is said of the bear, that for forty days it does not leave its bed nor eat nor drink, but only licks its paws, and it is nourished on that which it sucks up from the paws. Likewise, around the time of the winter solstice bees also lick their feet for forty days and are nourished from that, as is said.

1. On to the arguments. To the first, one must reply that light for some animals is more harmful than helpful, as is the case for the bat owing to the weakness of its vision, which daylight disperses and nighttime unifies, and for this reason etc. And, similarly, some animals see without external light, like the cat, the dog, and others, like the horse.

2. To the second, one must say that night is more a time of rest for gentle animals, but night is more appropriate for animals that lie in wait, because light is hateful to those doing bad things, as is clear in the wolf's case.

Question 11: Whether every flyer also walks.[42]

Further, one asks whether every flyer also walks.

1. It seems not. When something is divided in terms of essential differences, what is contained under the one is not contained under the other. The ass is not contained under "rational

41. The received Latin does not make sense (*sed tunc natura eorum recurrit ad musculos et de posito tunc in musculis natura fovetur*). If one changes *de posito* to *deposita* then the phrase reads, "nature having been deposited then in the muscles," and what we have is a likely doubling of the first phrase ("nature returns to the muscles"). What is most likely is that *deposita* became corrupted, and another, later hand added *natura eorum recurrit ad musculos* as an explanation. In any case, the sense is clear enough.

42. See A., *DA* 1.1.3.33–34 (*SZ* 1: 57–58).

[animal]" and neither then is man contained under "irrational [animal]," but "animal" is divided into "flyer" and "walker." Therefore, whatever is a flyer is not a walker.

2. Again, not every walker is a flyer. Therefore, not every flyer is a walker.

The Philosopher says the opposite.[43]

One must reply that every flyer is a walker. And the reason for this can be that a flyer lives on things that exist on the ground. And this is why if it flew continuously or if it were stuck to the ground [by lacking feet] and could not make any forward progress, then it would not be able to pursue the food suitable to it. And this is why it has feet, so that it can get from place to place by flying, but by walking hither and thither it can seek its food. And, besides this, a flyer cannot raise its body from the ground suddenly. That is why birds first support themselves on their feet before they take flight, and some cannot fly without running beforehand, much as cranes first run a few steps and then fly.

Again, another cause may stem from their composition, because all animals are composed of heavy and light elements. But in walkers the light elements are material, and the heavy ones in their composition are rather like formal elements, and this is why they have their place on the ground. But in flyers the light ones are the formal elements, and the heavy ones are material, and so by virtue of the light ones they possess flight, and by virtue of the heavy they possess the ability to walk.

Moreover, one must understand that animals that have swift wings also have weak feet. And the reason for this can be that wings are made from subtle and light matter. And in the same way feet have a certain lightness in comparison to other parts of the body. For otherwise, if they had great weight, then the animal would neither be able to walk nor lift its feet without difficulty since the entire body would be heavy. And this is why the subtle material, which ought to be pushed out to the feet, is transmitted to the wings.

Or another reason can be predicated of the end, since such birds live off small flies and small animals and do not live much

43. Ar., *HA* 1.1 (487b20f.).

off earth-born things. And for this reason nature provides them with speed in their wings but weakness in their feet.

1. On to the arguments. To the first, one must reply that flyer and walker are not opposite *differentiae* unless they are understood as exclusive, and this is why they are compatible with one another.

2. To the second, one must reply that among walkers exclusively only the light elements exist in the way of matter, and this is why they only walk, but in flyers the light elements exist in the way of form, and the heavy elements exist in the way of matter, and this is why they can do both: both walk and fly.

Question 12: Whether every animal emits a call [vox].[44]

One asks whether every animal emits a call.

1. It seems so. Because a call is an animal sound and occurs through the instrument of taste. For the tongue "is suited to two works of nature," as has been said in the second book of *On the Soul,* namely, "for taste and speech."[45] Therefore, just as every animal has [a sense of] taste, so, it seems, it will have a call.

2. Again, calls are given to animals in order to express mutual affections. But this is useful for every animal, so that one may express its notion [*conceptum*] to another, namely, grief or joy, etc. But nature is not deficient in necessaries, and therefore, etc.

The opposite is obvious to the senses.

One must reply that not every animal emits a call.[46] And the reason for this is that a call is caused by air that has been breathed in striking against the vocal artery. But not every animal breathes, and not every animal has a vocal artery. And this is why not every animal produces a call. And so bloodless animals, of which there are four genuses [*genera*] according to the

44. A., *DA* 1.1.3.45–47 (*SZ* 1: 62). On *vox* and animal language, see Irven M. Resnick and Kenneth F. Kitchell, Jr. (1996): 1–21.

45. Ar., *De anima* 2.8 (420b16–17).

46. Ar., *HA* 1.1 (488a31–32).

Philosopher in the first book, do not produce a call.[47] One genus is the cephalopods [*malakiae*], which have a soft external flesh and hard internal flesh, like the octopi and the cuttlefish.[48] And another genus, to the contrary, has a soft outer shell and soft flesh on the inside, like the crabs. A third genus has a hard shell, like the shellfish [*ostrea*]. A fourth genus contains animals with sectioned bodies, like the bees, wasps, and flies. And similarly some animals with blood do not emit a call, like fish, because they do not breathe air and, as has been said, a call is caused by a movement of air.

One must further understand what is necessary to produce a good or delightful call: that the organs would be well aerated and, consequently, that they be of a warm and moist nature and not a dry and earthy nature. And because the human has a warm and moist nature, he has a charming call,[49] and those whose organs are more moist although, nevertheless, temperately so, sing more sweetly, and therefore humans living in moist regions speak more sweetly than humans living in dry regions. And therefore, because the nightingale has an organ that is warm and moist and itself is the most temperate among the birds it sings better and more sweetly. But an ass does not sound well but rather brays rudely, on account of the earthiness of its organ and because it has a melancholy complexion. Neither do rapacious animals like the sparrow hawk [*nisus*] sound well on account of the dryness of the tongue, which is proved by the fact that their tongue is black.

1. On to the arguments. To the first, one must reply that not every animal sound is a call [*vox*] but only the sound caused in

47. Ar., *De anima* 2.8 (420b5f.).

48. The editor of *QDA* chose the reading *lupi*, "wolves," over *pulpi*, but this was ill advised, as *pulpus* is a variant for "octopus" (see Fr. *poulpe* and It. *Polpo*). In *DA* A. does discuss several "water wolves" but all are ferocious creatures like sharks and pike and even, incongruously, a seal. See *DA* 1.1.3.33; 1.1.8.99; 2.1.7.79; 5.1.2.15, 34; 7.2.1.98 (*SZ* 1: 57; 82; 323; 494–95; 630–31) with explanatory notes. For *malakiae* see book 4, n. 5, below.

49. The timbre of the mere *vox* is pleasing in and of itself, without any consideration of *sermo*, ordered speech. We, of course, are more prone to call our "call" a voice, but A. uses the same word for both.

the vocal artery. Further, not all animals have such an organ, not every animal has a tongue, and it is not necessary that every one that has the sense of taste have speech [*loquelam*] because, as is said in the second book of *On the Soul,* "taste is for the sake of being" and "speech is for the sake of well-being."[50] Thus it is absent from imperfect animals, for example, the shellfish [*conchilia*], or at least it is not apparent to us, and yet nevertheless they have the sense of taste.

2. To the second, one must reply that many animals are imperfect in comparison to the human and the cow, and this is why they do not require many things that are necessary to the perfect animals. Thus, a call would not be very useful for the crab because it would not seek out the things necessary to itself any better by having one, etc.

Question 13: Why all animals except the human are very noisy during intercourse.

One asks further why all animals except the human are very noisy during intercourse.[51] Thus the cock, when it needs to mate, crows and leaps around and clucks about,[52] and the horse and others act in a similar way also. A human, however, as if preparing for war, arms the penis [*priapum*] silently and stealthily and has intercourse.

And one must reply that voice [*vox*] is caused in many ways, by either an external or an internal cause. In humans, voice originates from an external cause. In fact, an internal cause is twofold: namely, either natural or unnatural. It originates from an unnatural cause in the swan, because when it dies is when it sings sweetly. But it originates from a natural cause when internally warmth and moisture have been multiplied. But sperm is warm and moist, and before coitus it descends as if drop by drop through the animal's members and comes to the testicles. And this is why brute animals sing out a great deal and cry as if enraged on account of the delight they perceive at that time

50. Ar., *De anima* 2.8 (420b16–20).

51. See A., *DA* 1.1.3.47 (*SZ* 1: 62).

52. "Clucks": A. uses *bombizat,* more properly a word for bees.

in the individual members, since they are following their natural complexion and sense or nature.[53] But the human animal is discreet and prudent, thanks to reason, and has discretion, and therefore can abstain on account of shame from those acts toward which coition is ordered, and this is why he does not make a lot of noise. To the contrary, if he can, he lies down hiddenly, and the more hiddenly the more sweet it is. And this is true.

Avicenna, however, says that at that time man enjoys such pleasure that he cannot speak, [54] for the heart, as it happens, contracts from such surpassing pleasure that some cannot speak since they are afraid.[55] And it is evident from this cause why the human and other animals are quite bold before coition but afterward become more timid, because boldness originates from the heat and motion of the blood. And this is the reason why animals are quite bold then and move about a great deal, owing to the motion of the heat and the spirits.[56] But fear stems from coldness and the congealing of the spirits, something that occurs after coition on account of the evacuation of these spirits, and this is why they are fearful then. Thus, an adulterer does not fear before coition to approach another's wife even with him present, owing to the strong movement of the spirits, but after coition, because he is weakened, he is fearful and flees from the presence of a rat or cat.

Moreover, based on these things a cause can be presented why a castrated man is less compliant, but the contrary occurs in other animals.[57] The cause for this, that is, the explanation [*ratio*], is that a man is a particularly temperate animal, but as a result of castration the seminal or spermatic material, which is warm and moist and tempered, and on which a man's life and complexion depend, is diminished in a man. And this is the reason that, as a result of castration, a man's complexion withdraws, that is, falls away from, a balanced state [*temperamentum*] and, as a result, his power is lessened and he becomes ef-

53. A., *DA* 1.1.3.47 (*SZ* 1: 62).
54. See Avic., *Can. med.* 3.21.1.10.
55. See A., *DA* 1.3.4.588 (*SZ* 1: 270).
56. A., *DA* 12.2.1.103 (*SZ* 2: 932).
57. A., *DA* 8.5.3.216–19 (*SZ* 1: 761–63).

feminate. But in other animals there are many moistures and superfluities and the heat is weaker owing to these superfluities, and this is why their moistures and superfluities are diminished as a result of castration and their heat is, as it were, forced or reflected back upon individual members and the animals thus become more compliant.

Nevertheless, there is a penetrating heat and a consuming moisture in the cock, and therefore, when it is castrated, its heat is diminished and its moisture becomes more abundant, and this is why it then becomes fat. But in other animals the heat is weak, and as a result of castration their superfluous moistures are consumed by that weak heat, which can then be repelled by not much at all, and this is why all castrated animals become thinner. And if they do become fatter, they preserve their fat much longer than uncastrated animals, because the heat is effeminate and not as consuming, and when they grow fatigued, they do not recuperate as quickly as uncastrated animals.

Question 14: Whether brute animals are differentiated according to their habits.[58]

One asks whether brute animals are differentiated according to their habits.

1. It seems not. Habits exist in a rational being, just as in a subject, by means of participation. But brute animals do not participate in reason. Therefore, good habits are not found in them.

2. Again, happiness is the goal of good habits. If then a brute animal should have good habits, the brute animal would be happy, but the opposite is said in the *Ethics*.[59]

The Philosopher says the opposite.[60]

One must say that good habits, properly speaking, do not ex-

58. For various uses of *mores* to suggest an animal's lifestyle, habits, or behavior, see also *DA* 8.1.1.1 n.1 (*SZ* 1: 667).

59. See A., *DA* 1.1.4.53–55 (*SZ* 1: 64–65); Ar., *Eth. Nic.* 1.10 (1099b32f.).

60. Ar., *HA* 1.1 (488b12f.). See Avic., *DA* 1 (fol. 29rbC); A., *DA* 1.1.3.50–51 (*SZ* 1: 63–64).

ist in brute animals, but only a certain imitation of good habits. Just as the turtle-dove imitates chastity and the sheep imitates gentleness as a result of each one's nature, so also the lion imitates fortitude and generosity. But they do not practice these for the sake of a good or bad end, but do this by a natural instinct. Thus, they do not observe the conditions required for good habits. For generosity means to give from one's own goods and not from someone else's, but a lion practices generosity by giving of his prey and thus from something that is someone else's. Similarly he has fortitude who fights for the sake of the public good, but a lion only practices fortitude from an internal necessity [*ex impetu animi*] and therefore for his own good and not for the good of other animals. And the same holds true for other traits. Thus, self-governance [*regimen propriae vitae*] is not attributed to brute animals, because this consists in three things with respect to one's own life: namely, in the contemplation of truth, to which ethics is ordered; in the running of the household, which economics teaches; and in the separation and association of citizens and cities, which politics teaches. And this is why if some animals or birds are gregarious and sociable, in this respect they imitate political life. Yet they are often deceived. Thus, cranes, having spotted another crane when hunting descend to it, and in this way they are seized by a fowler, and the same is true for pigeons [*columbae*].[61] And similarly, bees do not share with other bees the provisions they make in hives or hollows, and this is why they lack an ordered life style [*regimine vitae*] in the proper sense of the word.

The arguments prove that brute animals do not, properly speaking, have good habits in the same way they are found in humans, and this is true, but they have them only in an imitative way, and this is how the Philosopher understood it.

61. In this work we regularly translate *columba* as pigeon and *palumba* as ring-dove. Note, though, that the two were often confused, as at DA 5.1.6.41 (*SZ* 1: 408). On the difficulty in keeping the two straight, *GB* 238f., 300f.

Question 15: Whether every animal that has a vessel to receive the first superfluity, which is excrement, also has a vessel for the second superfluity, which is urine.

"And all animals have two members," etc.[62] In response to this, one asks whether every animal that has a vessel to receive the first superfluity, which is excrement, has a vessel for the second superfluity, which is urine.

1. It seems so. For an animal is nourished by a pure nutriment. But the nutriment is purified in the second digestion, which occurs in the liver, as well as in the first digestion, which occurs in the stomach. Therefore, if it has a vessel to receive the superfluity of the first digestion, it will also have something to receive it in the second digestion.

2. Again, in well-ordered activities there is no transit from one extreme to another except through an intermediary. But there are three superfluities in digestion: namely, excrement, urine, and sweat. But every animal that has excrement has sweat. Therefore, it will also have urine.

The Philosopher says the opposite.[63]

Question 16: Whether every animal that has urine has excrement.

Second, one asks about the converse, whether every animal that has urine has excrement.

1. It seems not. For Avicenna says that a fetus [*fetus*] in the uterus has urine and sweat, but not excrement.[64] But a child [*puer*] in the uterus is an animal, and therefore, etc.

2. Again, excrement is left over from the digestion that occurs in the stomach. But according to both Avicenna and the

62. Ar., *HA* 1.2 (488b29–30). See Avic., *DA* 1 (fol. 29rbC); A., *DA* 1.1.5.59–62 (*SZ* 1: 67–68).

63. Ar., *HA* 1.2 (489a2–7). See Avic., *DA* 1 (fol. 29rbC).

64. Avic., *DA* 9.5 (fol. 43rbA); *Can. med.* 3.21.1.2.

Philosopher[65] a child in the womb is not nourished from its stomach but through the umbilical cord, and it therefore will not have excrement.

The Philosopher says the opposite.[66] For he says that "every animal that has a bladder has a stomach," but not contrariwise.

To the first question one must reply that not every animal that has a vessel reserved for the first superfluity has a vessel reserved for the second superfluity. And the reason for this is twofold. One reason is that in every animal there is a second superfluity just as there is a first. But in some animals the second superfluity descends into something necessary for them, just as in birds the second superfluity descends to the areas for wings and feathers. And this is why they do not have a designated vessel capable of receiving it; instead, in such animals, as quickly as the urine is isolated in the liver, by nature's wisdom it is resolved into fumes and vapors and travels to the body's surface to generate feathers.

Another cause is that some animals are warm and moist, like the human and the sheep, and some are warm and dry, like birds. In warm and moist animals moisture abounds in the liver, and this is why the superfluity that is left over after the digestion that occurs in the liver is sent to a designated vessel such as a bladder. But in warm and dry animals the entire moisture barely suffices to temper their excessive heat. So the superfluity of the second digestion is consumed by their own complexional heat, and this is why birds do not urinate but only defecate. Thus, briefly: food is taken in to restore what has been lost, But drink is taken in for two reasons: so that together with it the food may better penetrate to the members, and in like manner, so that the members will be moistened or the heat tempered. And because it is more necessary for that which has been lost to be restored than for the members to be moistened or for the heat to be tempered, not every animal therefore has the same necessity for drinking that it has for eating.

65. Ar., *HA* 7.8 (586a31f.); Avic., *DA* 9.5 (fol. 43rbA); *Can. med.* 3.21.1.2; A., *DA* 9.1.7.67–68 (*SZ* 1: 798–99).

66. Ar., *HA* 1.2 (489a5f.).

And besides this, the birds of prey live on raw and moist flesh, and this is why they do not need to drink in the way walkers do, and because they live and they fly in the clean, pure, and subtle air and this air, according to Galen in the *Tegni,* satisfies their thirst more than water.[67] This is why the air moistens them in their chests, an air that is cold in act but which internally is some measure of moisture mixed with food. For this reason, etc. Thus, universally across their genus, flyers are warmer than walkers; nevertheless, in a species a given walker is warmer than a given flyer.

And further, since the entire moisture which some animals take in barely suffices to moisten their members or temper their heat, they do not, for this reason, emit urine. But because a second digestion presupposes a first, and that which is digested first is not converted entirely into the substance of a food source, the Philosopher therefore says that "every animal that has a bladder has a stomach."[68] From this a solution to each question is thus evident.

1. On to the arguments. To the first, one must say that nourishment is purified in the second digestion. But what is superfluous there is either immediately consumed by the strength of the heat in such animals, or it is sent to moisten certain designated parts, like the feathers and wings on birds. Thus every animal has a second superfluity, but not every one has a vessel reserved for it. And the reason for this has already been seen.

2. One must reply to the second in the same way, that even if there occurs a passage from the first digestion to the second, nevertheless it is not necessary that there be a designated vessel for the superfluity of the second digestion.

To the arguments for the second question one must say that the Philosopher is speaking about an animal that is nourished by means of a stomach existing outside the womb, and living on things produced from the earth. But the fetus in the uterus is nourished by the umbilical cord and on menstrual blood, which needs no purification and does not have the first superfluity,

67. Galen, *Tegni* 2.
68. Ar., *HA* 1.2 (489a5–6).

but only the second and the third, and the second and third digestions suffice for the purification of its menstrual blood. And this is why the fetus in the uterus does not have the first superfluity, etc., and this is why he does not assert that it does.

Question 17: Whether touch is in the flesh or in something analogous to it.

Now one asks about the members serving the senses. And first, whether touch is in the flesh or in something analogous to it.[69]

1. It seems so. Because touch is in any given part of the body. But nothing is more abundant in any given part than is flesh or something analogous to it. Therefore, etc.

2. Again, touch is either in the flesh or in nerve or bone. But there are many animals that do not have nerves and bones, in whom, nevertheless, touch is present. Therefore, etc.

3. Again, touch consists in the mean proportion and balance of tangibles; but this is found particularly in flesh. Therefore, etc.

The Philosopher demonstrates the opposite in book two of *On the Soul,* because "a sensible placed on a sense is not perceived,"[70] but a tangible, like heat, placed on the flesh is perceived. Therefore, flesh is not the organ of touch.

To the question one must say that touch can exist in something in four ways. It may do so just as it does in an organ and in a principle, and thus "touch is in something that is like the heart," as is said in *On Sense and the Sensed.*[71] In another way, touch is in something which makes the operation of touch manifest, and thus touch is in the brain. In the third way it is in something that transmits the tactile power, and in this way it is in the nerves. In the fourth way it is in something as if in a medium, and in this way it is in the flesh. Thus, when the Philosopher says that "touch is in the flesh,"[72] he does not mean that it

69. See A., *DA* 1.1.5.66 (*SZ* 1: 69–70); *De anima* 2.3.31.
70. Ar., *De anima* 2.11 (423b24–25).
71. Ar., *De sensu et sens.* 2 (439a1–2).
72. Ar., *HA* 1.4 (489a23–25).

is in the flesh as in an organ, unless the term flesh is extended to blood and nerve and cartilage and such things.

1. On to the arguments. To the first, one must reply that touch can be considered in two ways: either with respect to substance or with respect to operation. With respect to substance, it is in every part of the body, because sensation is in every part of the body simply as touch and touch alone. If it is considered with respect to operation, then it is in a determinate part, for example, in the nerves or in something similar to nerves. Thus the Philosopher says concerning sight in *On Sense and the Sensed* that "sight is not in the outer surface of the eye, but in something interior."[73]

2. To the second, one must reply that although nerve and bone properly understood are not in every animal, nevertheless each one has something analogous to them, just as, in the same way, not each one has flesh but each one has something analogous to flesh.

3. To the third, one must reply that some nerves are more earthy and for that reason insensitive, but others are better proportioned, like the nerves of the muscles. Now muscle is just like nerve-filled flesh, and such nerves are especially sensitive. So an injury to them is especially painful. Therefore, some nerves are reduced more to balance and to the mean than is flesh as properly understood, unless the term "flesh" is extended to other things, as the nerves of the muscles are more tempered, etc.

Question 18: Whether sensation exists in uniform members [membris similibus] *and whether operation exists in non-uniform members.*

One asks whether sensation exists in uniform members and whether operation exists in non-uniform members.

1. It seems not. For in a living body two things are present: distance from its opposite and organization. By its distance from its opposite it is disposed to life; and by organization it is dis-

73. Ar., *De sensu et sens.* 2 (438b8–10).

posed to operation. But sensation is an operative power. Therefore, sensation exists in an organic part.

2. Again, the animal power is divided into a motive and a sensitive power. But the motive power exists in the organic parts. Therefore, so too does the sensitive power.

The same seems to be true with respect to vision, taste, and hearing. For they exist in organic parts and are thus present in the non-uniform parts.

The Philosopher says the opposite.[74]

One must say that sensation exists in uniform parts, and the operative power in non-uniform parts. The reason for this is that sense is a passive power and so it is related to matter. But the operative (that is, active) powers are related to form. Therefore, since whatever is received is related to the one able to receive it, the material parts will correspond to the sensitive power and the formal parts to the operative power. But uniform parts are material when compared with the non-uniform. And for this reason, etc.

Nevertheless, one must understand that, of all the senses, touch especially operates through a uniform part. The reason for this is twofold: one stemming from the sense of touch itself, and the other from the object. The reason stemming from the sense itself is that it is especially diffused throughout the body. For it is just like a guardian for the machine that is the animal body, because it guards it by guarding it from heat, cold, and dryness, and so too from other things. And thus it is reasonable that it should be in a part that can be more extended throughout the body. And a homogenous part—like nerve, or flesh, or something analogous to them—is one such as this. The reason stemming from the object appears to be much the same case, because touch discerns the extremes [*excellentias*] of tangible objects.[75] Therefore, it is necessary that it exist in particular balance. But greater balance is observed in uniform than in non-uniform parts.

74. Ar., *HA* 4.4 (489a23f.).

75. As Ar. notes at *De anima* 424a5f., each of the senses has its peculiar object, and the object has its own qualities. The object of touch (the "tangible") is, for example, hot or cold. But touch itself is neither hot nor cold, although it

1. On to the arguments. To the first, one must reply that act is double: there is a first and second act. The first act is like life, sense, and other such things. And a part corresponds to an act to the extent that it is distant from its opposite. The second act is something like to live, to sense, and the like. And this too is double: one is proximate, and the other is remote. The proximate is something like to live and to touch, without which there is no life. The more remote is to hear, to see, to smell, and the like. And this is why to live and to touch, owing to their proximity to a first act, can exist in uniform parts, better than to see or to hear, which are in organic parts.

2. To the second, one must reply that although the animal power may be divided into motive and sensitive, nevertheless the motive is active and the sensitive is passive, and this is why the explanation [*ratio*] is not the same for each.

3. To the third, the cause has already been given why touch can exist better in uniform parts than can hearing. This is owing to its proximity to the first act. Thus, the first sense is the foundation for the others, and the organs for the others are founded on the organ for touch, which would not be so were touch not based in something uniform.

Question 19: Why some animals have feet and some do not.

One asks why some animals have feet and some do not.[76]

It seems that many that have feet, like flies and bees, were born more to lack them than were fish, which are more perfect animals.

---

must be, potentially, either one. Insofar as the sense of touch itself is neutral, it occupies a kind of "mean" or midway point, and has the power to sense what lies on either side of it. The "excesses" of these qualities, however, destroy or corrupt the sense itself. If a sound is too sharp or too flat, it destroys the hearing; if a "tangible" is too hot or cold, it is painful and will injure the sense of touch. Thus, for the sense to perform properly, a proper mixture and *ratio* of the sensible qualities is essential. On this point, and for a definition of *excellentiae sensibilium,* see A.'s *De anima* 2.4.9. It is in this way, namely, that a sense must occupy the mean between extremes or the excess of sensible qualities, that touch is midway between the *excellentiae sensibilium.*

76. A., *DA* 1.1.7.86–89 (*SZ* 1: 77–78); Avic., *DA* 1 (fol. 29raC).

One must say that animals with ringed bodies have a weak nature, and this is why what they lack in composition nature provides in organization, providing them with many feet by which a ringed body can be supported. But fish stay in the water, which is not a solid body, and this is why water cannot support any sort of a weighty body walking on it. And this is why no fish has feet, except for one that lives equally in the water and on land or which has naturally to seek its nutriment on the bottom of the water. Thus, because nature operates for the sake of an end, and "does nothing in vain and is not lacking in things necessary,"[77] many questions that address these sorts of issues can be solved by reference to nature's intention and its desire.

Question 20: Why the rational animal has only one species.

Next one asks why the rational animal has only one species, although there are many species of irrational animals.[78] Because as many ways as an individual of a set of opposites is divided, so too for the rest. Therefore, etc.

One must reply that the intellect does not use a bodily organ and this is why it cannot be diversified substantially according to the diversity of the one receiving it. But sense uses a bodily organ, and diversity in matter requires diversity in form. And this is why to the extent that the body, which is the organ for the sensitive soul, is variously proportioned, so too is the sensitive soul diversified. But this diversification of the body is due to the diversity of the soul in a formal sense. And this is why there can be many species of irrational animal, on account of the diversity of the sensitive soul. And again, although there can be many ordered to one end, the end is nevertheless indivisible. But all inferior animals are ordered to "rational animal," and this is why although there are many brute, irrational beasts there will be nevertheless one species of rational animal.

77. Ar., *De anima* 3.9 (423b21–22), 3.12 (434a31).
78. Ar., *De anima* 3.4 (429a18f.).

Question 21: Whether physiognomy can be established on the basis of parts of the body.

"And it is necessary that we recall, etc." In this part the Philosopher makes a determination about anatomy and physiognomy.[79]

One asks whether it is possible to obtain a physiognomy concerning moral traits from the parts of the body.[80]

1. It seems not. For according to the Philosopher in the third book of *On the Soul,* the intellect does not use a corporeal organ. Therefore, its operation does not occur through a mediating organ. But *mores* do not occur without the intellect's operation; otherwise, these would be found in brute beasts; therefore, *mores* cannot be known from body parts.

2. Likewise, this does not follow. According to the Philosopher in the *Topics,* this or that person is naturally chaste and is therefore chaste.[81] But this conclusion would hold if *mores* could be known through the disposition of the body; therefore, etc.

3. Likewise, according to the Philosopher in the second book of the *Ethics,* nature does not accustom itself to contraries; but *mores* can be made accustomed to contraries. Therefore, *mores* do not follow the dispositions of nature or matter.

In this chapter, the Philosopher implies the opposite.

One must reply that two things are necessary for *mores,* that is, virtues, namely, the sensitive appetite and right reason; but the sensitive appetite uses a corporeal organ, and for this rea-

79. See A., *DA* 1.2.1.110 (*SZ* 1: 87).

80. See A., *DA* 1.2.2.126–30 (*SZ* 1: 93–95). Since physiognomers sought to deduce moral traits such as bravery, rectitude, cowardice, or lecherousness from physical traits, *mores* here can probably be safely translated as "moral traits." For the remainder of this passage, however, the original Latin will be left intact to assist the reader in forming his or her own opinion.

81. Ar., *Topica* 2.11 (115b14–17); trans. Boethius (PL 64: 934A); A., *Topica* 2.2.8. "Naturally chaste": it was something of a commonplace that a cold complexion naturally supports chastity. For discussion with reference to the Virgin Mary's chastity, see Ps. Albertus Magnus, *Quaestiones super Evangelium,* q. 18, 6–8; for the arguments to the contrary, see *Quaestiones super Evangelium,* q. 18, 9–11. For a discussion of this question, see also Resnick (2002).

son a corporeal organ is necessarily required for *mores*. This is why the Philosopher determines in the *Ethics* that moral virtues exist in the rational appetite through participation, that is, in the sensitive appetite. But right reason is required for the completion of virtue. For if the appetite were not regulated by reason, an error would occur in *mores*. But now it is the case that some are naturally disposed to fortitude and some to liberality and some to chastity, who nevertheless can be changed or inclined to the opposite by habitual action. And similarly certain ones are naturally disposed to vice, as melancholiacs are to envy and cholerics to wrath, who yet can be habituated to the contraries by the discernment of the intellect. As a result, natural aptitudes to *mores* or their opposites can be recognized from the parts of the body, with respect to humans. But a habit [*habitus*] existing in the soul cannot be recognized from them. For the Philosopher relates that the disciples of Hippocrates showed an excellently portrayed likeness of the man to the best physiognomer, Philotinus, and asked him the natural *mores* of the one whose image it was, and after inspecting the image he said that it was an image of one who is unstable and incontinent. And the disciples were amazed and rejected what he had said of such a great man, and reported to Hippocrates what he had said, and he responded that the physiognomer had spoken the truth. But he said himself that by means of the discernment of the intellect and the love of study and the natural *mores* of virtue, he had changed into the contrary. Thus he was by nature one thing, although he was good by operation.[82]

Thus, briefly: One can know by the parts of the body to which *mores* a man is naturally disposed, although one cannot know what *mores* he will exercise and use.

1. On to the arguments. To the first, one must reply that although the intellect does not use an organ, nevertheless it requires sensitive organs for its operation, and this is why the intellect's operation is often changed by the various dispositions of the body. For it is impeded in madmen owing to an injury to

82. Various versions of the tale are collected in Foerster, 2:187–91. Compare with *DA* 1.2.2.127 (*SZ* 1: 93) where the name given is Phylemon.

the organ of phantasy and so the body's dispositions contribute a great deal to *mores.*

2. To the second, one must reply that someone who is chaste is properly one who has the habit of chastity and not one who has the potential or aptitude for it. And it therefore does not follow: "He is naturally chaste, and therefore he is chaste."

3. To the third, one must reply that *mores* do not follow upon the dispositions of the body by necessity; but that there rather is a certain inclination toward *mores* owing to the diverse disposition of the body.

Question 22: In respect to which part is physiognomy especially carried out?

Is it in respect to the heart or the brain?

Second, in respect to which part is physiognomy especially carried out? Is it in respect to the heart or the brain?

1. It seems that it is in respect to the heart, because the operative virtues flow from the heart as their root; but *mores* proceed from the operative virtues. Therefore, they are especially known in respect to the heart.

2. Likewise, the motive virtues proceed from the brain, from the posterior part, and the sensitive virtues from the anterior part. If then a physiognomy of *mores* occurs in respect to the brain, it occurs especially in respect to the posterior part, from which the operative virtues flow. But this is not true, as the Philosopher's determination makes clear;[83] therefore, etc.

The Philosopher's determination makes the opposite clear, for physiognomy is especially accomplished by means of the face and the anterior parts that are near the brain.

One must say that *mores* can be linked to a given part in two ways, either as if to a root and virtue or as if to a power and the origin of the *mores.* And thus physiognomy can be accomplished especially in respect to the heart, because the heart is the first part and the root of an animal, and that is why it is situated in

83. Ar., *HA* 1.8–9 (491b9f.).

the middle of the body just like a king in his kingdom.[84] And thus a person will vary in *mores* with respect to the heart's possessing more or less heat or dryness and whether it is large or small.

They can also be linked to that in which they especially appear or become apparent. And thus physiognomy can especially occur with respect to the brain. For, although the motive virtues come forth from the heart as their root, this nevertheless happens with the brain playing a mediating role. And this is why a physiognomy of the natural *mores* of a person can be obtained especially through the brain or the parts adjacent to it.

1. On to the arguments. To the first, one must reply that although the motive powers [*vires*] proceed from the heart as their root, in a more immediate way they come from the brain.

2. To the second, one must reply that the head is not completely a sphere, but is oblong in part. The brain is under the sinciput and under the occiput there is more emptiness, though there is no absolute emptiness there, but rather this part is bonier than the anterior part. And as a result, the sensitive powers are situated nearer the brain, and this is why physiognomy is obtained in the senses and the anterior part of the brain more than in the posterior part.

Questions 23–24: Why the suture in the bone of a human skull is more noticeable than in the bone of other animals.[85]

Why several sutures are visible on the head of a man and only one on a woman's head.

Next one asks why the suture in the bone of a human skull is more noticeable than in the bone of other animals.

It seems that the human bone has to be more solid because a human's brain requires more protection; therefore, etc.

And the same applies for why several sutures appear on the head of a man and only one on a woman's head.

84. See A., *DA* 1.1.5.73 (*SZ* 1: 71–72).

85. Ar., *HA* 1.7 (491b1–8); Avic., *DA* 1 (fol. 29va–bE); A., *DA* 1.2.1.112 (*SZ* 1: 88).

1. (a)[86] To the first, one must reply that the human brain is large with respect to the size of his body when compared to other animals.[87] Thus, the skull is often filled with vapors rising to the brain, on account of which his brain especially requires ventilating, and for this reason there is a suture on the human head.

(b) Again, the veins and arteries proceed from the brain and it is through these that the sensitive and motive powers are diffused to the individual members. And this is why it is fitting that there should be a suture in the bone, so that it would allow transit and stability for the veins and arteries.

(c) Third, the brain is wrapped in two pellicles,[88] the outer one of which is hard.[89] And this is why there are sutures on the bone, so that the brain is not injured by this pellicle but instead attaches more firmly to bone and flesh.

2. To the second, one must reply that there are more vapors in a woman than in a man, and it is easier to divide something in length than in breadth, and this is why a circular suture appears on a woman's head, whereas on a man's head there are several sutures, and they are divided in a more longitudinal fashion.

3. And in addition to this, owing to a defect in her heat the vapors in a woman's head cannot penetrate the bone to the extent that vapors in a man's head can, and this is why women's heads are more subject to pain than the heads of men. And so too to the other.

86. The numbering that follows is inserted by the translators to help correct and clarify the incorrect numbering of the original text.

87. A., *DA* 1.3.1.535 (*SZ* 1: 250).

88. A., *DA* 12.2.4.130 (*SZ* 2: 942).

89. "Hard": *dura,* hence the name of the membrane, used even today, the dura mater.

Questions 25–28: Why a large forehead indicates a slow person, prone to foolishness, and a small one indicates a good disposition.

Why eyebrows that are straight indicate femininity and pliability whereas ones arched toward the nose indicate discretion and intellect.

Why those arched toward the temples indicate weakness and a bad disposition.

Why a lot of hair hanging to the ground indicates envy and wantonness.

Next one asks why a large forehead indicates a slow person, inclined to foolishness, and a small one indicates a good disposition.

Second, one asks why eyebrows that are straight indicate femininity and pliability whereas ones arched toward the nose indicate discretion and intellect.

And why those arched toward the temples indicate weakness [*mollities*] and a bad disposition.

And why a lot of hair hanging to the ground indicates envy and wantonness.[90]

To the first, one must reply that the forehead is near to the brain, and this is why the size of the forehead is an indicator of the brain's size. Now the brain is moist and cold and phlegmatic,[91] and as a result there is a heaviness there, because [the cold] thickens the spirits and greatly blocks the heat, which is the instrument of agility. And this is why a large forehead indicates slowness.

Again, if it is broad it is an indication that the sensitive powers flowing from the brain through the forehead are greatly dispersed and as a result this is a sign of weakness, because a power is not strengthened by unification in anything very broad. And this is why a broad forehead is a sign of foolishness.

90. Ar., *HA* 1.8 (491b9–26); Avic., *DA* 1 (fol. 29vbE–F); A., *DA* 1.2.2.136–38 (*SZ* 1: 96–97).

91. A., *DA* 1.3.1.518–20; 12.2.2–3, 113–19 (*SZ* 1: 244–45; 2: 936–38).

Nevertheless, one can distinguish that size or breadth can arise from either a natural or an unnatural cause. If it arises from an unnatural cause, then it is not a sign of slowness or foolishness. Such a thing may occur in children because of the nurse's inept swaddling, because by being bound incorrectly, the forehead becomes incorrectly shaped. If it should arise from a natural cause, either it arises from the strength of an active power, and then it is a sign of a good disposition, and this occurs when the neck and other members near it are proportioned to the forehead. If, however, it arises from a large quantity of matter and from the weakness of the active power, then it is a sign of a bad disposition, for example, when the head is thick and the neck is thin. And so too for the others.

To the second question one must say that each thing is distinguished by its end. Thus if some animal is deficient with respect to its end, it is a sign of an error in its principles. But eyebrows exist to safeguard and to adorn the eyes, and so that a superfluity may be emitted hiddenly through the pores and eyebrows and converted into hairs. For otherwise the superfluity would fall into the eyes. This is why eyebrows have to be shaped naturally, according to the shape of the eyes, because eyebrows ought naturally to curve in at that very place where the visual rays are especially directed. And this is why if they are well curved in toward the cone of the nose, they indicate a good disposition with respect to discretion, intellect, and good vision. If, however, they curve toward the temples, they indicate a weakness and bad disposition, because then they already depart from their end. And if they are linear, lying in a straight line, they are a sign of moisture, because the moist easily receives every impression, and as a result they are an indication of femininity and pliability. But if they droop and are very hairy, since hair is especially generated from an earthy matter which is hard and dry, then they are signs of melancholy, and melancholy is a cause of envy, because melancholics are envious and disparagers.

This physiognomy ought to be applied to irrational animals, because brute beasts follow nature, and where these sorts of dispositions are naturally found in beasts, then the effect on their characters [*mores*] will be of this sort. For the bear, which has drooping eyebrows, is envious and subject to anger. And so too for the others, like the pig, which has hairy eyebrows and is wanton.

Questions 29–31: Whether the eye has a different nature according to its colors.

Why the black in the eye varies in different people and not in the other animals except in the horse and, rarely, in the dog.

Why the eye has a round shape.

Consequently one asks about the eyes, and first concerning the eyes' anatomy.[92] First, whether the eye has a different nature according to its colors.

1. It seems not. Something that is colorless ought to be receptive to color, and the eye is receptive to color. Therefore, it ought not have its own color.

2. Again, whiteness is caused from an abundance of light. Therefore, where there is more visual light, there will be more whiteness. But the visual power flourishes more in the mean than in the extreme. Therefore, whiteness and blackness, for the same reason, are poorly placed or ordered in the eye.

Second, one asks why the black in the eye varies in different people and not in the other animals except in the horse and, rarely, in the dog.[93] And the contrary to this seems true, because a human is more fixed and stable, as the Philosopher says in that chapter.[94] Therefore his accidents ought to be permanent, etc.

Third, one asks why the eye has a round shape. It seems that it ought to be triangular, "because everything that is seen is seen under an angle," according to the author of the book on *Perspective*.[95] Therefore, it ought to be triangular, with its base toward the thing seen and the acute angle in the visible eye.

92. A. *DA* 1.2.3.140–43 (*SZ* 1: 98–99); Ar., *HA* 1.8 (491b20f.).

93. Ar., *HA* 1.10 (492a5f.); A., *DA* 1.2.3.163; 1.2.7.205, 19.1.3.9–14 (*SZ* 1: 106–7; 123–24; 2: 1336–39).

94. Ar., *HA* 1.6 (491a20f.).

95. Alhacen, *Optica* 1.19, 2.36–38. Witelo, *Optica* 3.4–21.

To the first, one must say that the eye is composed of three humors and seven tunics.[96] The principal humor of the eye is called the glacial humor, in which the power of the eye especially flourishes. The second humor is internal and toward the brain. It is the nutriment for the earlier one and is called the vitreous humor, because it resembles liquid glass. The third humor is external, toward the eye's surface, and it is the matter purged from the glacial humor. It is called the albumen, because it resembles the albumen in the egg. Thus the glacial or crystalline or hail-like [*grandineus*] humor, which is the same thing, is located between the vitreous humor on the inside and the albumen on the exterior side.

There are three interior tunics and four exterior tunics. The first interior tunic is called the retina, and it wraps about the glacial humor and proceeds from or originates from the optic nerve. The second one is called the *secundina* and it proceeds from the interior web of the brain, which is called the pia mater. The third is harder than the others, and is called the *sclirotica* and it proceeds from the dura mater of the brain. This one is inmost toward the brain and is harder than the others in order to prevent the rheums and noxious things from entering from elsewhere. The first exterior tunic is called the "spider's web" [*tela aranea*] owing to its subtlety, and it is joined with the interior of the retina and divides the glacial from the albumen-like [*albugineus*] humor. The second is called the *uvea,* which does not cover the entire eye but has an aperture in the middle shaped rather like a grape [*uva*] missing the part by which it hangs from the vine. And this exists so that the visible species can approach the glacial humor more freely. The third is called the *cornea* and it is hard and difficult to cut; thus it is called the *cornea* [horn-like] owing to its hardness and color. And it is permeable just like horn. And this one is joined to the inner side of the *sclerotic,* and the *uvea* is joined with the *secundina.* The fourth tunic is called the *conjunctiva,* and it arises from the subcutaneous flesh on top of the cranium, and it is fatty and moist so that it may connect, lubricate, and fatten the other interi-

96. See A., *DA* 1.2.7.197–204 (*SZ* 1: 120–23) where slightly different terminology is used.

or tunics. And this is why its circular portion is nearer to white flesh, but because vision requires a consolidation of the visual rays and since white causes fragmentation whereas black consolidates, nature ordained that the part nearer the glacial humor, that is around the pupil, should be black, in order to consolidate the rays. Thus the moisture that is in the middle of the eye has no color apart from the color of water in order to receive all the species of colors as water does. Now the conjoined part is black so that the rays may be consolidated. But the more remote part is white owing to its fattiness, which it receives from flesh as a poultice or lubrication for the tunics. The *tela aranea* arises from the surface of the optic nerve, the *uvea* arises from the skin immediately touching the cranium, and the *cornea* arises from another fleshy tunic of the cranium, existing under the flesh and not very far toward the bottom.

1. On to the arguments. To the first, one must say that the pupil, through which we see, is not a color, although parts joined to it have color.

2. To the second, one must reply that although whiteness may be caused by light it is nevertheless necessary that it exist in a bounded body.[97] But the vision's light exists in the eye in a permeable, transparent, or unbounded body, and this is why there cannot be a middle part that is white, because, if there were, it would scatter the vision.

To the second question one must reply that the black in a human's eye varies in differing ways owing to the different levels of participation in the influence from the brain.[98] Thus the black will be greater or less to exactly the extent that the spirit coming to the black of the eye varies. And because a human's brain varies a great deal—sometimes tending to gray, sometimes to black, sometimes to an off-white mottled color—this is why the black in a human's eye varies. But of all the animals the horse's brain particularly resembles a human brain. Thus the horse, like a human, suffers illness in the head and in other parts from

97. The English is left as vague as the Latin. It is unclear whether "it" refers to the light or the whiteness.

98. See A., *DA* 1.2.7.205, 19.1.3.9–12 (*SZ* 1: 123–24; 2: 1336–38).

the brain's rheum. And I once had a horse that suffered from rheum, and mucous [*coryza*] flowed to the nostrils drop by drop. I fumigated the horse often after having spread dried cow dung over coals and after having covered its head, so that, though he had seemed on the verge of death, he escaped. And this is why the black in the horse's eye varies just as it does in the eye of a human. And I had a beautiful little dog at Cologne that had one white eye and one black. And perhaps this was owing to a large quantity of albumen-like humor and the good quality of the spirits and of the visible [*visivae*] light.

To the argument one must say that a human is more stable in his substantial [qualities] than are other animals. And although the eyes of different people will differ, there is no less stability in a human on this account, because the eye's disposition in the same person is only changed by some accident. So a human is more stable in substantial qualities than in accidental ones, whereas it is the opposite in others.

To the third question one must reply that a spherical shape is especially receptive of, and is better suited for, vision, just as we see in a mirror that those things that are viewed in a flat mirror seem to have their proper size.[99] This is not the case, however, for those things viewed in a convex mirror, because the things seen in a convex mirror do not appear on the surface of the mirror but appear in that spot where the rays that are reflected or refracted from the mirror come together perpendicular [*cathetis*] with the eye, just as we said in the third book of the *On Meteorology* in the chapter on the rainbow.[100] Therefore, a spherical shape is best suited for vision because it can hold many things and is less easily impaired and less easily injured, because it touches a plane surface only at a single point.

To the argument one must reply that, although any object is seen in a pyramidal angle, as it were (with the base lying at the thing seen and its apex at the eye), nevertheless the eye itself cannot be angular, because it is not pyramidal but rather of a spherical shape.[101] And the rest are evident.

99. A., *DA* 1.2.7.197 (*SZ* 1: 120). 100. A., *Meteora* 3.4.12–15.

101. This translation combines readings from variant manuscripts to bring sense to a troubled passage.

Questions 32–33: Why bulging eyes indicate a disturbed power of discretion and why sunken eyes indicate acute vision.

Why an overlarge aperture and immobility in the eye indicate immodesty, and mobility indicates inconstancy.

Second,[102] one asks about the physiognomy of the eye. And first, why bulging eyes indicate a disturbed power of discretion and why sunken eyes indicate acute vision.[103] Second, one asks why an overlarge aperture and immobility in the eye indicate immodesty, and mobility indicates inconstancy.

To the first, one must reply that the eyes originate in the brain, which is moist. Thus bulging eyes are a sign of a lot of moisture coming from the brain, and this is an indication of a disturbance in the brain's powers. In addition, the further a power is stretched out from its source, the weaker it is, and this is why bulging eyes are an indication of a weak power. But sunken eyes are nearer to the brain and to the first principle of sight, and the [visual] rays are better unified in these. Thus they attest to an acute visual power.

To the second question one must reply that the eye's immobility indicates that nature does not control the eye very well. But if a person should freely keep his eyes always unmoving, then he himself acts badly, and he does this freely, and this is an indication that he is not ashamed of acting badly. And this is why an overlarge aperture and immobility in the eye (as when one perceives that a person is not moving his eyes and keeps them fixed for a long time) is a sign of shamelessness. Movement of the eyes arises from excess heat, because heat is easily movable. According to Avicenna,[104] this is a sign of inconstancy and of a deceiver [*insidiator*][105] passing from image to image, and these people are deceivers and robbers, etc.

102. It is worth pointing out that all major MSS read "second" both here and in the next sentence.

103. Ar., *HA* 1.10 (491b34–492a12). Avic., *DA* 1 (fol. 29vbF); A., *DA* 1.2.3.143–67 (*SZ* 1: 99–108).

104. Avic., *DA* 1 (fol. 29vbF).

105. Lit., "one who lies in wait," and thus also a thief or one in ambush.

Questions 34–37: Why ears may be located on various parts of the head although the eyes, nevertheless, are located close together.[106]

Why ears are twisted on the inside.

Why all animals have movable ears, but the human alone has immovable ears.

Why melancholy animals have long ears.

Now one asks about the ear.[107]

First, why ears may be located on various parts of the head although the eyes, nevertheless, will be located close together.

Second, one asks why ears are twisted on the inside.

Third, one asks why all animals have movable ears, but the human alone has immovable ears.

And fourth, why melancholy animals, like the ass, the hare, the mule, the stag, and the cow, have long ears.

To the first, one must reply that sound, which is the proper sensible for the ears, multiplies in air moved in a circular fashion in the manner of wave-like water, as can be seen when circles emanate out from the place where a rock is thrown into the water. And this is why the ears are located on opposite sides of each other, so that sound may be heard from everywhere, because if they were on one side they would not hear from all sides, nor would the sound, should it travel a long distance, be heard everywhere, etc. This is why it diffuses itself spherically. But "anything that is seen is seen at an angle" whose base is in the thing seen and whose apex [*conus*] is at the eye, and a visual discrimination does not occur before it approaches the optic nerve. And this is why the eyes are situated together, so that what is seen by one eye and by the other at straight lines can establish an angle at the optic nerve, because the more acute the angle of the apex [*acutior conus*] the better is the object seen. And this is

106. Following the *corrigenda,* p. 360, reading *sunt* for *sint* following *oculi in propinquo . . .*

107. Ar., *HA* 1.11 (492a13–b4); Avic., *DA* 1 (fol. 29vbH); A., *DA* 1.2.4.168–75 (*SZ* 1: 109–12).

why a small eye is much better, because the visual rays are better united in it and the angle [*angulus*] is more acute, etc.

To the second question one must reply that sound is amplified with a tremor of the air. And this is why the interior of the ear is twisted and winding, so that a sort of bending back of the air may occur and so that the sound will redound better in the ear.

Another reason is that the air is warmed by the bending and is altered and purified in this twisting so that no cold approaches the temporal bone [*os petrosum*] of the brain and injures it, which is sensible, because cold from outside is the cause of rheum and especially damages the brain itself, which has a weak heat.

To the third question one must reply that the ears descend from the brain along intermediary nerves. But now it is the case that too much moisture in the nerves is a cause of immobility, just as occurs in paralytics, whose members are not born for motion owing to too much moisture of the nerves, which relaxes them. But among all the animals the nerves descending to the ears of a human are the moistest, and this is why they are more immovable. This is demonstrated by the fact that the human brain is colder and moister, and this is to temper the heat of the heart; therefore, etc.

In addition, there are no muscles in the ears of a human, which mediate voluntary motion in an animal. But there are muscles in the ears of other animals, and this is why they are movable at will.

To the fourth question one must reply that melancholy animals are naturally timid owing to their weak heat and shortage of spirits.[108] Indeed, a small amount of heat is unable to warm a large heart, just as a small fire cannot warm a large furnace. This is why it suits them to hear from a greater distance, so that, like the roebuck [*capreolus*], they may more quickly take flight, and this is why wise nature made their ears double and gave them long ears—for the sake of flight, which is their protection, so that they will perceive hunters at a distance.

108. See A., *DA* 12.2.1.103 (*SZ* 2: 932).

Another reason is that cartilage exists and is made from melancholy matter, and the ear is composed of cartilage. Then argue: The more the melancholy matter abounds, the more reasonable it is that the melancholy members will be stretched out. And this is why melancholy animals have long ears.

Question 38: Whether sneezing is a sign of good or evil.

One asks about the nose. And first, one asks whether sneezing is a sign of good or evil.[109]

And one must say that sneezing can occur at the end or the beginning of an illness. If at the end, it is an indication of health, because the digestive power is weak in the brain owing to its coldness and the sneeze occurs from the fact that the superfluous vapors collected in the area of the brain are expelled by nature. If therefore a digestive and expulsive nature, which is weak in the brain, has power over superfluous vapors, it is a sign that the digestive nature, which is stronger in other parts [of the body], is that much more powerful. And this is why sneezing at the end of an illness is a sign of convalescence. But if it occurs at the beginning of an illness, it is a sign of flow and movement of the humors. And this is why sneezing occurs frequently and it is a sign of an overabundant material and of a future illness. In health, it is a good sign if there is no rheum or fullness in the head, because it argues for the formation of power.

Question 39: Whether long and conspicuous lines on the hands may be signs of a long life, and whether short ones are signs of a shorter life.

Consequently one asks whether long and conspicuous lines on the hands may be signs of a long life, and, contrariwise, whether short ones are signs of a shorter life.[110]

And it seems not. Because if the lines are not conspicuous, it is a sign that the hand is well filled [with flesh] and this, as a

109. Ar., *HA* 1.11 (492b5f.); Avic., *DA* 1 (fol. 29vbG); A., *DA* 1.2.8.214–20 (*SZ* 1: 126–27).

110. Ar., *HA* 1.15 (493b30f.). See A., *DA* 1.2.26.493–94 (*SZ* 1: 235).

result, attests to the strength of its power. Short lines, then, are not signs of a short life.

One must reply that the hand follows the nutritive and generative power. Therefore a large-sized hand is a good sign in the male because it indicates a large-sized penis, but it is a bad sign in a woman because it is an indicator of a large-sized womb in which the sperm is not well concentrated and, as a result, the sperm does not become fertile. Thus very often women having a broad and ample vulva and a large opening and a large womb are naturally sterile. But if the opening of the vulva is large for the sake of a large penis during coition, then the rule does not hold but instead it fails. Ovid [says]: "The time passes by, and we grow old through silent years."[111] We desire old age; when it comes, it is a bad wind. Always desiring—that is the vulva—to be spread wide, sometimes lying open it is opened," namely, when it is opened by a penis just as if with a key.

Again, in the palm of the hand there is a good bit of movable flesh, but in the wrist [*rasceta*] of the hand there is a good bit of flesh that stays in one place.[112] Now, nature is more solicitous of the flesh with respect to the *species* than with respect to the matter. When therefore the palm of the hand is well-lined, it is a sign that there is a lot of flesh there with respect to the *species,* because the hand is heavily lined only as a result of nature's great solicitude. And when nature is solicitous of the extreme parts, it is a sign that it is probably more solicitous of the interior parts. And this is why a good and conspicuous line on the hand is a sign of abundance in the parts with respect to the *species,* and few lines are a sign of defect. And this is why the Philosopher says that conspicuous lines are a sign of a long life, because there is a lot there from the principles of life, and con-

111. Ovid, *Fasti* 6, 771, where the line actually begins slightly differently. See *Metam.* 3, 162.11.60. The editor of the Latin text correctly suggests that the Ovid citation and subsequent explanation are uncharacteristic of A. and believes them to be interpolations.

112. *Rasceta:* From the Arabic, the word indicates that part which intervenes between the long bones and the actual parts of the foot and hand. In the hand, it can be thought of as the wrist, while in the foot it is not quite the same as the ankle. See especially *DA* 1.2.12.281, 285–90, 298 (*SZ* 1: 150, 152–54, 156–57); Rath (1956), 236; Hyrtl (1879), 198; Fonahn, 2771f.

trariwise one that is not lined, etc., just as has been said. And the rest is clear.

Question 40: Whether parts in an animal have a location like the parts of the universe.

Next one asks whether parts in an animal have a location like the parts of the universe.

1. It seems not. For it happens that the front of a man is oriented toward the right side of heaven and the right is toward the left. Therefore, the location of parts in the animal does not correspond to their location in the universe.

2. Again, that seems to be the root from which other things arise. Now, however, the sensitive and motive powers flow from the head. Therefore, the head is the root of a human. But the root of a living thing is below, as is clear in plants. Therefore, below in an animal is not below for the universe, etc.

The Philosopher says the opposite.[113]

One must reply that because local motion is properly directed toward somewhere, differences of position are perceived in the universe according to diverse principles of local motion. So, that by which heavy things begin to be moved is said to be above in the universe and that by which light things begin to be moved is said to be below. But the right [side] in the universe is the one from which the movement of the heavens begins and this is from the east. But the "front" in the universe is that toward which the stars are moved after they rise, and this is our hemisphere. In an animal, "above" and "below" are understood with respect to the movement of food, and because an animal receives its food through the mouth, but plants through the root, this is why the head is "above" in the animal and the root is "above" in the plant. But right and left are understood with respect to local motion, since that part is to the right side in an animal by which the animal begins locomotion. But front and back are understood with respect to sensitive motion. The front

113. Ar., *HA* 1.15 (494a20f.). See A., *DA* 1.2.26.498–99 (*SZ* 1: 237).

part is the one in which the senses are located, because the animal moves toward that part. "Above" in a human, then, is directed toward the "above" of heaven, and the right side in a human is understood with respect to the principle of local motion, just as is the right side of heaven. And the front of the human is that part to which motion proceeds, just as it does in heaven. And this is why the Philosopher says that the parts of the human are situated according to the location of the parts in the universe. But this is not the case in other animals; in them "above" is not in the direction of the "above" of heaven, nor is "below" toward "below," but rather in between these.

1. To the first argument one must reply that this occurs even though the front of the human is toward the "above" of heaven; nevertheless, "above" is understood in the same sense in both human and in the universe.

2. To the second, one must reply that the root of an animated thing is properly said to be its "above" and not its "below," as the argument accepts. This is why, etc. For although the plant's root is "below," with respect to the universe, it is nevertheless "above" with respect to the plant, etc.

Questions 41–43: Whether the brain is naturally cold and moist.

Why the brain is divided into front and rear and not right and left.

Why a human has such a large brain in proportion to his body size.

Next one asks whether the brain is naturally cold and moist.

1. It seems not. Because, according to the Philosopher in the second book of *On the Soul,* sense and movement do not occur without heat.[114] But sense and motion proceed from the brain. Therefore, it is reasonable that the brain should be warm.

2. Again, when the nerves are very moist, the members are

114. Ar., *De anima* 2.4 (416b29f.); 2.5 (416b33f.).

paralyzed. In order that motion arise from the brain and without paralysis, it is necessary, it seems, that the brain be dry because, as Galen says in his *On Illness and Accident*,[115] the nerves' dryness assists motion, as is apparent in madmen. Therefore, the brain is warm and dry.

Second, one asks why the brain is divided into front and rear [and not right and left], since other members are divided into right and left, like the eye and ear, the hand, foot, etc. Therefore, the same scheme should apply to the brain.

Third, one asks why a human has such a large brain in proportion to his body size, while others have a smaller one, like the horse and others. Now, the human is "the most noble animal," as it says in the text.[116] Therefore, he ought naturally to have larger formal parts and smaller material ones. But the material parts are cold and moist in relation to the warm and dry. Therefore, as is apparent, he ought to have a smaller brain than the others, since it is cold and moist and these are the material ones, etc.

To the first argument one must reply that the vapors and fumes ascending from the nutritive parts ascend to the brain. If, then, the brain were warm in its own right [*de se*], it would be immediately inflamed owing to the large quantity of vapors ascending the furnace of the entire body, as is the case in a furnace, and if it were woody and dry, it would immediately burst into flame from the smoke.[117]

And this is the reason why nature bestows on it a cold and moist complexion, so that the warm organs, and especially the heart, may be tempered by its coldness, and so that its coldness might be tempered by the heat of the others.

Again, two pathways come from the heart to the brain. If the

115. Perhaps a reference to *De morborum differentiis*. See Kühn, *Galeni opera omnia*, 6: 836–80 (especially c. 12, pp. 872–73).

116. Ar., *HA* 1.6 (491a19–20); 2.12 (503b32).

117. "These members (namely, the heart and the liver), therefore, are in the bodies of animals of necessity, for the sake of the principal heat which exists in them. The inner members of an animal must be like a furnace in which the natural heat is produced, and this heat must be conserved in these two members." *DA* 13.1.6.52 (*SZ* 2: 1007).

brain were hot, its heat would be turned back upon the heart, and it would kindle the heart's heat, and, as a result, in a very short time the animal would be suffocated. And this is why it is necessary for the brain to be naturally cold.

Further, the brain is the receptor for all the sensible species, and that which is moist receives them best, and this is why it is necessary for the brain to be moist. And because the spirits cross through the brain, it is necessary for it to be fatty, in order to strengthen the spirits by means of its oiliness and fat. And because fat is viscous, the brain is viscous, in order to retain well the species it received, just as bird lime [*viscus*] holds a bird.

To the first argument one must reply that, although the brain is neither warm nor dry in itself, nevertheless it receives warmth and dryness from the heart (to which, in location, it is opposed) and from the fumes and spirits coming to the brain, and therefore it has heat sufficient for the sensitive powers and dryness sufficient for the motive powers.

To the second argument one must reply that sense and motion arise in the brain. But sensation is an affect [*passio*] and motion is an action, since to sense is to undergo and to move is to act.[118] This is why it is necessary that these powers flow out from different parts. But the senses, like sight, smell, and taste, flow from the anterior part, and this is why the posterior part of the brain is necessarily left for the motive powers. And this is the reason why the brain is divided into two parts, of which the first is moister owing to the influence of the sensitive powers [*vires sensitivae*] while the posterior part is dryer owing to the motive powers [*virtutes motivae*].

With this, a response to the argument is clear.

To the third question one must say that a human is especially temperate. This is the reason why nature saw to it that the human brain was in proportion to the heart, so that each may be returned to a balanced temperament by the other, and this is why they are located opposite one another. But other animals are not so temperate, and this is why in other animals the brain is not proportioned to the heart, and this is the reason why the

118. Ar., *De anima* 2.5 (416b33f.).

human brain is larger than is the brain in other animals. Another reason is that there are more internal senses, like reason, etc., in the human than in others, and because internally they have organs, or chambers, through which they act; for this reason it is necessary that the brain be larger than in other animals.

Questions 44–47: Whether the heart is pyramidal in shape.

Why its apex [conus] *does not rise upward in quadrupeds, as it does in fish.*

Why it is not located in the middle of the human [body].

Why it affords very poor nourishment although it is the best member.

Next one asks whether the heart is pyramidal in shape.[119]

It seems not, because a more perfect organ deserves a more perfect shape, and the heart, since it is a more perfect member, will have a more perfect shape. But this is a spherical or round shape, and therefore, etc.

Second, one asks why its apex [*conus*] does not rise upward in quadrupeds, as it does in fish, whereas in quadrupeds its base rises upward.[120]

Third, one asks why it is not located in the middle of the human [body] as it is in other animals, since the human should be most suitably proportioned.

Fourth, one asks why it affords very poor nourishment although it is the best and a noble member.

To the first, one must reply that among all the members the heart is hottest. But in a foreign matter heat tends to form a pyramidal shape, as is clear from fire in wood, which is in a foreign matter, and this is why the heart has a pyramidal shape, and also so that it might infuse life better to the rest of the body.

To the second, one must reply that the heart is opposite the

119. See Ar., *HA* 1.17 (496a4f.); Avic., *DA* 2.1 (fol. 30vbB), 3.1 (fol. 31raA–vaB); A., *DA* 1.3.4.575–93 (*SZ* 1: 266–72).

120. *Conus*: In the *DA*, A. uses *acumen* for this pointed feature of the heart; see, e.g., 1.3.4.583 (*SZ* 1: 268).

brain so that the one may be tempered by the other. But in quadrupeds the heart is warmer than in fish, and this is why, if the base of the heart were below and the apex were above, the heart would overheat the lower members and would not be adequately tempered by the brain. Therefore, the heart's aperture is in the part facing the brain in quadrupeds, in order to be tempered better by the brain's coldness and, conversely, in order better to temper the brain. But there is not as much heat in fish. Therefore, if the heart's aperture faced the brain, the heat would be very quickly extinguished by the brain's coldness and the coldness of the water that enters into it.

To the third, one must reply that the maximum balance is required in the human, and this is why the heart is located on the left, because it infuses power toward its right side, so that what is on the right will acquire by its influence what the left side acquires by its location. And in addition to this the spleen, which is receptive of melancholy, is on the left side,[121] and this is why the left side is not overly cooled by the spleen, nor the right side overly heated by the liver and the gall bladder, because the heart tends downward toward the left side from the liver on the right, etc.

To the fourth, one must reply that the heart is a member that endures a long time and has a nature that retains its disposition, and this is why it is naturally solid and very compact. Things of this sort afford poor nourishment because they resist digestion and they fatigue the power of digesting, and this is why, etc.

Question 48: Whether the upper intestines should be thinner and longer and more delicate than the lower ones.

One asks whether the upper intestines should be thinner and longer and more delicate than the lower ones.

It seems not. For undigested food occupies a larger volume than digested food. Therefore, the upper intestines ought to be broader and thicker than the lower ones.

One must reply that in the human the powers are regulat-

121. A., *DA* 1.3.5.608–9 (*SZ* 1: 278–79).

ed by reason. Thus his imaginative power not only extends to the nutritive power but concerns other things as well. And this is the reason that, in order for the food to remain for a longer period of time in the human, nature has ordained that the upper intestines be long and thin. They are long so that the pure food will not be absorbed by them all at once but so that to the contrary, what is not absorbed in one part will be absorbed in another. Otherwise, the human would have to eat continuously as brute beasts do. But feces from the food are thick and earthy, and this is why they demand thicker and wider vessels, and this is why the upper and lower intestines differ with respect to being thin and thick, broad and long and short, as it evident in the study of anatomy.

Question 49: Why the human lungs are more solid and are not divided into lobes.

Next one asks why the human lungs are more solid and are not divided into lobes as in the flyers.[122] And it seems contradictory, because the heat is greater in the human heart than in that of another animal. Therefore, as a consequence, the human needs better ventilation in the lungs than do the others.

One must say that the heat is weak and dispersed in the flyers because they are ventilated a good bit by means of their feathers and plumes and through their pores. And this is why it is necessary that thinner spirits come to their hearts, because thick spirits would immediately extinguish the heart's heat. And this is the reason that there are many lobes in their lungs, so that the spirit that is drawn there will be better purified and rendered thinner before it is dispatched to the heart. But there is more heat in the human heart, and for this reason it can sustain cruder spirits, and because of this its lungs are not divided because an air that is rendered thin and pure is not sufficient to temper the heart, and the same is true for many of the brute beasts, like the cow, etc.

122. Ar., *HA* 1.16 (495a30–b1); A., *DA* 1.3.2.540 (*SZ* 1: 252).

Questions 50–51: Why the kidneys and the eyes and ears are divided into pairs.

Why the right kidney is higher and larger than the left.

Further one asks why the kidneys and the eyes and ears are divided into pairs, but the other organs are not.[123]

Second, one asks why the right kidney is higher and larger than the left.

To the first, one must reply that the kidneys in the human and in animals exist for the sake of the complete digestion of the watery fluid emitted by the liver. For the liver is a watery and very moist member, and the wateriness in the liver is mixed in with blood. Thus this wateriness is dispatched along two veins to the two kidneys, so that pure wateriness may be separated out from the blood. And this is why, in those with weak kidneys, the urine looks like water used to wash raw meat, because the kidneys cannot entirely purify it from the blood. The kidneys are located near to the vertebrae of the back. Thus, if they were joined together as one member, that member would be burdened by its size and would bring harm upon the other members, and this is why nature doubled them with one on one side of the vertebrae and the other on the other. And in the same manner nature made two eyes and two ears and two hands and two feet, so that, if one should happen to lose one of them, the remaining one could perform its task. This is not so for the other members.

To the second, one must reply that fat is caused by poorly digested moisture.[124] Thus colder people, because they have a poorer digestion, are fatter than warmer people, as pigs are compared to other warm animals. In the same way, because the heat consumes and resolves moisture, as one can see in emaciated, choleric people and in many others owing to their lack of moisture. But the heart infuses its power on the right side. Thus there is naturally more heat on the right side than on the left, and, as a result, there is more heat in the right kidney. And for

123. Ar., *HA* 1.17 (496b34f.). See A., *DA* 1.3.6.612–14 (*SZ* 1: 280–81).
124. A., *DA* 1.1.6.85, 12.2.2.106–8 (*SZ* 1: 77; 2: 933–34).

this reason, because it is the trait of something warm to be located above and to consume moisture, this is why the right kidney is above and is larger, and the left is just the opposite, etc.

Questions 52–53: Why the exterior members are divided into pairs, while the interior members are not.

Why the eye and the ear are divided into pairs, but not the tongue, except in serpents.

Next one asks why the exterior members, like the eye and ear, are divided into pairs, while the interior members, like the heart and the liver, are not.

Second one asks why the eye and the ear are divided into pairs, but not the tongue, except in serpents.

To the first, one must reply that the interior members have the basis of a principle with respect to the exterior members, and a principle exists in one way, whereas things from a principle exist in several ways, and this is why the interior members are not paired as much as the exterior ones, because a principle ought to be simple and one.

To the second, one must reply that the tongue is fitted for two operations of nature, namely, for taste and for speech. But speech occurs through a mediating motion, and one motion impedes another motion which is not ordered to the same end. But among motions that are ordered to the same end, one assists the other, as Galen says in the first book of *On Critical Days* [*De Crisi*].[125] Thus if there were two tongues and both spoke at the same time, then the voice of one would impede the voice of the other, just as the voice of one person impedes the voice of another. And this is why animals that make sounds do not have two instruments for calling or giving voice. Serpents, however, live on very cold nourishment, for they live off earth, and this is why they need an instrument that is better disposed for perceiving flavors, because the earth is not very flavorful. And this is why nature gave two tongues to these, so that they might better distinguish the flavors of their food. Thus serpents appear

125. Not in the place indicated (see Kühn 9: 550ff.).

to have a divided tongue at least on the end, even though these are not two tongues absolutely.[126] Nevertheless, we have seen in *** a serpent having two tongues which could even be petted in a lap.[127] Nevertheless, some of the peripatetics say that they have one tongue in place of a stinger, where the venom is, and especially in the teeth, so that they taste with one and they puncture or fight with the other.

Question 54: Whether all animal members lead back to a single member.

Now one asks whether all animal members lead back to a single member.

1. It seems not. Because those things that have a different regimen do not lead back to some one member. But diverse members have a different regimen. Therefore, they do not lead back to some one member.

2. Again, those things which have a diverse end have no return to one member. But the members have diverse ends, for the end of the foot is to walk, and the end of the eye is to see; therefore, etc.

To the contrary. The lesser world imitates the greater world. But all the parts in the greater world lead back to a single thing. For individual things receive powers and influences from other things, and all receive these from a first, that is, from a first cause. Therefore, it will be the same in the lesser world, which is the human, and in every other animal that is a part of the world.

One must say that in every animal there is one first member, which is first by generation and by causality, and from which

126. A., *DA* 2.2.2.98, 12.3.5.205 (*SZ* 1: 332; 2: 973).

127. None of the MSS offers a reasonable reading for the lacuna in the text. It would appear to be a place name, and one reading, that of P, *in regno,* translates as "in the kingdom" and puts one in mind of Frederick II's love of animals. Likewise, *quem etiam in gremio fovi potest post,* itself of dubious meaning, is of decidedly odd Latinity and may also be corrupt. See A., *Super Is.* 40:11, ed. Ferdinand Siepmann (1952), 412, ll. 68, 82.

all others receive influence and power because in all things ordained to one end it belongs to the same agent to establish order among them. But all the members of an animal are ordered to one end— to the preservation of the species—and ordered by the same agent, which is the power in the seed. This is why it is necessarily required that the members be ordered among themselves. But it belongs to things that are ordered to reach a first principle, for otherwise the process would be infinite, a thing which neither nature nor art nor reason nor science endures; therefore, etc.

1. To the first argument one must reply that regimen is understood in two ways: as individual [*speciale*] or common. Different members have a different regimen in the individual, yet nevertheless they have one regimen in common. And this suffices for the basis of the principle, etc.

2. To the second, one must reply that an end is understood in two ways: proximate and mediated. Thus, although there may be one or another proximate end for any part whatsoever, nevertheless the mediated end is the same. So just as the eye exists immediately for the sake of sight, or the ear for the sake of hearing, and so too for the others, nevertheless each is further ordered to the conservation of the individual. And thus they all have one mediated and remote end, which is to conserve the animal in its existence, etc.

Question 55: What is that first thing to which all the animal members lead back?

Further one asks what is that first thing to which all the animal members lead back.

It seems to be the brain.[128] Because, although the senses are what make an animal an animal, the senses have their origin from the brain, and therefore, etc.

But it also seems to be the liver.[129] Because that member seems to be the principal one through which all members are

128. A., *DA* 1.1.5.73; 1.2.14.301; 12.1.1.14 (*SZ* 1: 71–72, 158; 2: 936).
129. A., *DA* 1.1.5–6.75–76 (*SZ* 1: 72–73).

preserved and conserved and in which things that have been lost are restored. But this is accomplished by the liver, in which the nutritive power flourishes; therefore, etc.

But it also seems to be the testicles. Because that member appears to be the principal one through which a thing acquires being. But this happens through the generative members, as is clear. Therefore, etc.

To the contrary. Life is an animal's first act. Therefore, that in which life first exists is the first animal member. But according to the Philosopher[130] life takes root in the heart, and he says that the heart is first to live and last to die, and Avicenna[131] says that the heart is the first root of the powers. Therefore, the heart is the principal member.

One must reply that the animal body is divided into four regions. One is of the animal parts in which the sensitive and motive powers exist, and this is the head region. Another is of the natural parts in which the powers serving nutrition exist, like the power of the stomach, liver, veins, kidneys, and things of this sort. A third region is that of the spiritual parts, to which the heart and the lungs and the trachea [*vocalis arteria*] belong. And a fourth region is that of the members suitable for generation. Thus an animal is considered in four different ways, and in this sense there are various principal members positioned in the human. For some consider him from the standpoint of generation, and these have to posit that the members in which the semen exists, that is the testicles, are principal. Others consider him from the standpoint of nutrition, and these have to establish the liver as the principal member. Others consider him from the standpoint of sense and motion, and these posit that the brain is the principal member. And some consider him from the standpoint of the spiritual members, and these posit that the heart is the principal member. Therefore, there is a no small controversy between physicians and the Philosopher. For the physicians claim that the brain is the principal member because they pay attention to the senses and because they are

130. Ar., *Part. An.* 3.3 (665a11f.).

131. Avic., *Can. med.* 1.1.5.1; *DA* 3.1 (fol. 31a–b).

workers on things that are sensible. But the Philosopher posits the heart alone.[132] And Avicenna says that the physicians must follow the Philosopher because he speaks more truthfully.[133] For the heart is located in the middle of the animal just like a prince in his kingdom, and, just as the prince sends his commands and rules the individual parts of his kingdom through his ministers, so does the heart send life and power to the individual members by means of their organs. Now it sends sensation to the eyes and the ears through the brain as if through a minister, and it sends motion to the hands and feet similarly through the brain. But on each member it bestows the power of digesting, expelling, and attracting by using the liver just as if it were a bailiff. But it gives the power of reproduction to another external member by means of the seminal vessels. And therefore, according to the truth of the matter, the heart is the principal member. But the physicians consider the matter according to the senses, as it appears to them, and not according to the reality, and this is why, etc. Thus in the second book of *On Heaven and Earth* the Philosopher likens the superior bodies to the heart.[134]

This reason also appears persuasive, because all the corporeal and material things in the universe lead back to spiritual ones, and therefore it seems similarly reasonable in animals that all members lead back to spiritual ones, among which the first one is the heart.

This is also clear to the senses because the dispositions and sufferings [*passiones*] of individual members are borne back to the heart. For no injury occurs in any part of the body without that affect redounding to the heart. Thus the heart suffers with members in pain and shares the enjoyment of those experiencing pleasure.

To the argument one must reply that sense and motion, nutrition and generation can be related to some member in two ways, either with respect to its first root and true origin or with respect to its immediate principle of operation. If in the first way, then all these lead back to the heart, which is the first and

132. Ar., *Part. An.* 3.3 (665a11f.); *De somno et vigilia* 2 (456a4f.).
133. Avic., *Can. med.* 1.1.5.1.
134. See Ar., *DC* 2.3 (286a9f.).

true principle of generation, and upon which the entire machinery of the body is constructed like a house on its foundation, and from which all the members first receive life. If in the second way, then sense and motion lead back to the brain because it is clear that senses, motions, and operations proceed from it. And thus nutrition leads back to the liver according to sense, and generation leads back to the seminal vessels, namely, the testicles.

BOOK TWO

Question 1: Whether the cause of similarity or dissimilarity among the members of various animals occurs as a result of the complexion of mixables or on account of the soul.

"CERTAIN MEMBERS of animals," etc.[1] With respect to this second book, one asks whether the cause of similarity or dissimilarity among the members of various animals occurs as a result of the complexion of mixables or on account of the soul.[2]

1. It seems that this is not on account of the soul. This is because a cause of diverse things in a species produces a diverse effect in the species. But the souls of different animals differ in their species.[3] Since, then, in the species their members are similar, that similarity cannot occur on account of the soul.

2. Likewise, the likeness between the heads of Sortis and Plato is greater than the likeness between Sortis's head and his hand.[4] If, then, the similarity occurred on account of the soul, the head and the hand would agree more in appearance than would the head with the head.

1. Ar., *HA* 2,1 (497b6–9). Cf. A., *DA* 2.1.1.1–2 (*SZ* 1: 286–87).

2. "Complexion of mixables": a seemingly redundant phrase referring to the mixing of the four primary elements that result in the body's complexion. The four elements—air, fire, earth, and water—combine in material bodies and are associated with the four primary qualities—the hot, cold, moist, and dry. Complexion, then, results from the interaction of these qualities, to produce composite bodies that are hot and dry (choleric), hot and moist (sanguineous), cold and moist (phlegmatic), cold and dry (melancholic), or, when the four qualities are equal, temperate or balanced.

3. Perhaps: "differ in kind."

4. The enigmatic "Sortis" is almost surely a corruption of "Socrates," whose name is commonly used in such statements. Cf. *DA* 1.1.2.15, 18.1.4.35, 38 (*SZ* 1: 50; 2: 1297–99).

3. Likewise, before the organization in the seed, the soul exists only virtually [*virtute*]. Therefore the similarity or dissimilarity of members after the organization cannot be on account of the soul.

On the contrary, the form of a substantial whole is the form of any given part. Therefore, just as diversity in the whole occurs from the form of the whole in the species, so too does diversity in the parts.

One should say that similarity and dissimilarity among the members is effectively on account of the soul of the one generating, and formally from the form of the mixed.[5] That a fetus has just this sort of head or just this sort of foot occurs passively on account of the seed and actively on account of the power of the soul of the one generating. For a certain power of the one generating is split off from it at the same time as the seed is, and this acts in the sperm by its power and acts to produce one like the one generating. Therefore, because the seed derives from the superfluity of the nourishment of the one generating and is receptive to the form of the one generating, and because the power in the seed is able to produce a soul like itself, the power therefore organizes the semen itself so that the fetus's parts become like those of the one generating. But the seed itself is heterogeneous in power, although it is homogeneous in act. I call something heterogeneous that has one of its parts converted into flesh specifically and another is converted into bone and another is converted into the principal members. This is because members are formed variously on account of the variation existing in the mixing together of the parts of the seed, just as will be shown below.[6] Thus, similarity among the members of different individuals in the same species occurs materially from the proportion of the mixables, and formally on account of the form of the species, and effectively from the agent, which is the power in the sperm.

5. A., *DA* 15.2.1.63, 16.1.14.72, 18.1.4.30, 18.1.5.42 (*SZ* 2: 1112–13, 1194–95, 1295, 1301).

6. A.'s internal reference is unclear, but may point to *QDA* 18.1–6.

1. One must reply to the first argument that if similarity and dissimilarity in the members occurred solely on account of the soul, then a different soul would produce dissimilar members in the species. Now, however, a proportion of the mixables is required for the members, and for that reason, if the proportion is the same in the one being acted upon, nothing prevents a likeness in species[7] from being produced from diverse generating entities.

2. One should respond in the same way to the second argument, because the power in the seed is the active agent of different things in different ones, because it acts in the power of the one generating. Therefore, from the same material, or from one part of the material, it forms a hand, and from another part it forms a foot or an eye, but does so from a different proportion of the mixables.

3. To the third argument one should say that while the soul exists virtually [*in virtute*], the members are not organized. Rather, that power which acts to introduce the soul organizes them itself. And organization occurs on account of the powers of the soul to be introduced, because different organs are necessary for different tasks. This argument, however, clearly proves that for similarity in species a like proportion of the mixables is necessary. And this we concede.

Question 2: Whether the lion's neck is continuous and not hollow.

Further one asks whether the neck of the lion is continuous and not hollow.[8]

1. It seems that it is not. For vertebrae, that is, joints, are in the neck of another animal, so that it may move or turn its head this way or that more freely. Since the lion is a very fast and quite ra-

7. Perhaps: "in appearance." As already suggested above in n. 3, *species* can mean "kind," "appearance," or the more technical term "species."

8. Belief in the lion's having a neck of solid bone is as old as Ar. *HA* 497b17; *Part. An.* 686a23. Cf. *DA* 2.1.1.3, 3.2.7.150, 12.1.7.89, 22.2.1.108.58 (*SZ* 1: 287, 417; 2: 927, 1513).

pacious animal, all the more so does it require joints in its neck so that it may move it this way and that as it leaps after its prey.

2. Likewise, marrow is the nourishment for bones, but marrow is in the hollow of a bone. If, then, the neck of the lion lacked this cavity, there would be no marrow in it and, as result, it would be deprived of nourishment.

But the Philosopher says the opposite.[9]

To this one must answer that the lion's neck is continuous and without a joint for two reasons. One arises on account of the end, because a lion is a rapacious animal living on prey and it is a very combative animal, pitting itself against animals of great power and size, and solid members are thus required, of necessity. On this account, its neck is continuous, because if it were jointed through vertebrae it would be easily broken. Another reason is because of the material. For the lion is a choleric animal, and for that reason is very audacious and confident. Now a choleric person dares to rise up against three or four others even if they are stronger than he is, and it is the same for the lion. This is the reason heat especially abounds in the lion. And this is why although the moderate heat in other animals converts flesh or marrowy food into nourishment for the bones, this is not so in the lion. Rather, its great heat consumes the whole marrow and that is why its bones are solid, and without marrow or a cavity. This is also the reason, on account of their solidity throughout, that rubbing the lion's bones together can create fire, the same as from striking together rock and iron.[10]

1. One should respond to the first argument that because of the lion's fortitude and exceeding boldness it is necessary that its neck be continuous.

2. To the second argument one must respond that the lion's bones are warmed or use nourishment just as its other parts do. But on account of its great heat the marrow is not preserved in the bones, but rather is dispersed through the parts and is consumed in part by the excessive heat.

9. Ar., *HA* 2.1 (497b13f.).

10. Cf. *DA* 3.2.1.74, 12.2.8.160 (*SZ* 1: 381; 2: 955); Ar., *Part. An.* 2.9 (655a14f.).

Question 3: Whether the right foot is more unfettered and better suited for motion than the left.

One asks whether the right foot is more unfettered and better suited for motion than the left.[11]

It seems that this is not so. Because warm and dry things are the better suited to motion. But in the human the left side is warmer and dryer, because the heart, which is the principle of heat, leans to the left side, and on the left side lies the spleen, which is the receptacle of melancholy, but on the right side is the liver in which moisture abounds.[12] Therefore, the left side is better suited for motion than the right.

The Philosopher says the opposite.[13]

One should say that in all but the human, the left foot is less suited for motion than the right, just as the left hand is less suited for motion than the right, because all animals use their front feet in the place of hands on the human. An indication of this is that they are better fitted for motion, just as a person's hands are more dexterous than his feet. Nevertheless, the elephant is an exception, because on account of its body's size it has larger and less flexible front feet so that it may more readily bear the weight of its body and the burden placed upon them.[14] Now in this sense it is true that a warm and dry part is better suited for motion than a cold and moist one. This is why members that are paralyzed are not suited for motion, on account of their moisture. But the heart, which is the principal member in the animal, infuses its power into the right side, and the heart is the warmest member. For this reason, right sides are better suited for motion. Nevertheless, in the human the heart leans to the left side, insofar as it can do so, and as a result the left side will be warmer than the right, and then the person has a left side more suited to motion than the right, as in many cases it happens that a person is ambidextrous.[15] And for this reason the members on the right side in males are larger, on account of

11. See A., *DA* 2.1.1.4 (*SZ* 1: 287).
12. See A., *DA* 1.3.4.581 (*SZ* 1: 268).
13. Ar., *HA* 2.1 (497b19f.).
14. Ar., *HA* 2.1 (498a1f.); cf. A., *DA* 2.1.1.5 (*SZ* 1: 287–88).
15. See A., *DA* 2.1.1.10 (*SZ* 1: 298–90).

the greater influx from the heart on that side, but in females the members on the left side are larger, on account of the superfluity of the moisture and the weakness of their digestion.

One must respond to the argument that although the heart may incline to the left side in terms of its position, nevertheless it infuses its heat and power more to the right side. For this reason, it inclines more to the right side in terms of its influence.

Question 4: Why a human's chest is broader, with respect to its size, than that of other animals.

Further one asks why a human's chest is broader, with respect to its size, than that of other animals.[16]

For it seems that the chest should come together more narrowly and subtly, since it is of a more subtle complexion.

One must respond that a broad human chest can occur for three reasons. One is the abundance of the heat in the heart, distributing and expanding the parts near it. Another reason is that the human walks upright, and other animals walk with their stomach facing the earth.[17] And for this reason in other animals a wide chest might hinder movement, but in the human it contributes to movement because the chest is located more equidistantly above the feet. This is valid as long as the chest is wide and not stretched into an arc, because then it would be extended in front of or behind the feet. As it is, it is placed directly over them. The third reason is to guard the heart and the respiratory members. A broad chest performs this better than a narrow one.

Question 5: Whether the motive power proceeds from the posterior or anterior part of the brain.

One inquires further about the motive powers. And it is asked whether the motive power proceeds from the posterior or anterior part of the brain.

16. See A., *DA* 2.1.1.11–12 (*SZ* 1: 290–91).

17. Lit.: "over their belly" [*super ventrem*]. Just below A. will state that humans walk *super pedes*, "over their feet."

1. It seems that it proceeds from the anterior part because motion follows sense, according to Avicenna.[18] But sense proceeds from the anterior part, and therefore motion will as well.

2. Likewise, motion flows from the brain, and therefore from that part to which the brain inclines. But the brain is located on the anterior part, as was said above, and therefore, etc.[19]

3. Likewise, powers are known through their objects. But the same thing is the object of sense and motion. Therefore, the sensitive and motive powers are in the same part.

To the contrary: one and the same thing is not the principle of contraries. But the anterior part is the principle of the sensitive power, which is passive, and therefore it is not the principle of the motive power, which is active.

One must say that the motive power proceeds immediately from the posterior part of the brain. For just as moisture is necessary for the reception of sensation, so too is dryness necessary for motion. But the first part of the brain is moister, and the posterior is drier. Therefore the motive nerves proceed more properly from the posterior part than from the anterior part, and the opposite is so for the sensitive nerves.

Likewise, sense operates through apprehending. But we apprehend things better and more easily which are in front of us. But movement is from behind, because motion is from a starting point to an end point. This is why a movable immediately leaves behind the starting point at which it formerly was. And for this reason, rationally, the motive power is in the posterior part, whereas the sensitive power is in the anterior part. An indication of this is that if there is some injury to the spinal chord [*nucha*] descending from the brain through the vertebrae, the animal immediately ceases to move.[20]

1. To the first argument one must respond that although motion follows sense it is not necessarily the case that it proceeds

18. See Avic., *De anima* p. 2 c.3 (fol. 7vb).

19. See *QDA* 1.41–43.

20. *Nucha:* Derived from the Arabic for the "marrow" of the backbone, that is, the actual spinal cord itself. First defined at *DA* 1.2.11.257 (*SZ* 1: 141) but used regularly throughout.

from the same part. Now motion follows sense because what we sense first, later we acquire by motion.

2. To the second argument one should respond that although the motive power proceeds from that part to which the brain inclines, nevertheless the position of the brain is not indivisible.[21] And for this reason it can proceed from one part or from the other, although it does so more suitably from the posterior, because this is drier.

3. To the third argument one should respond that although the object of the motive power and the sensitive power may be the same with respect to the thing, nevertheless it differs in *ratio*, just as the sensible and the appetible can be the same under a different *ratio*.[22] And sense and appetite are therefore different powers.

Question 6: Whether the human foot naturally bends to the front, and the hand to the rear.

Further one inquires into the bending of the motive members in animals. And first, whether the human foot naturally bends to the front, and the hand to the rear.

1. It seems not. Four-footed beasts have anterior feet in place of the hands in the human. But beasts bend their anterior feet to the front; therefore, similarly, the human bends the hand to the front.

2. Likewise, bipedal birds are just like the human, but birds bend their feet to the front. Therefore, so too will the human, by the same reasoning.

But the Philosopher says the opposite.[23]

One must say that human feet naturally bend to the front

21. The text actually identifies this as the response to the third argument, and the one following as the response to the second, but the editor has attempted to correct this error.

22. *Ratio* is a term that conveys numerous and varied meanings. In this instance, it seems to indicate "definition," "meaning," or "with respect to the power of reason."

23. Ar., *HA* 2.1 (498a3f.); cf. Avic., *DA* 1 (fol. 29vbG); A., *DA* 2.1.2.13–17 (*SZ* 1: 291–93).

and the hands to the rear, because in every animal the feet extend to that part in which there is more of the body. Now brute beasts extend their rear feet to the front of the body because more of the body is to the front, and the front feet extend and bend to the rear for the same reason, so much so that in the rear feet the knee bends to the rear, but the feet extend toward the front. In the same way, in the front feet the bend is to the front, and the curvature and extension is to the rear. But in the human it is the case that the thigh [*coxa*] is joined to the trunk [*clibanus*] and bending occurs to the rear, and for that reason it necessarily extends to the front.[24] The lower leg [*crus*] is joined to the lower part of the thigh at the knee, and so with respect to the knee the larger part of the body is to the rear, and for that reason it bends to the front and the curvature is to the rear. Nevertheless, the joining of the feet in the human is the opposite. But the human hand is ordered for grasping, and for that reason it naturally curves to the chest. Wings are ordered in the same way in birds, so that the birds move forward using them. For this reason, the curvature in wings is to the front, so that, when extending the extremities of the wings to the front, they come in front of the body, just as sailors when extending the oars reach them out in front of the ship.

Thus, one must respond briefly to the question that the curvature is toward that part toward which or on which there is more of the body.

1. On to the arguments. To the first, one should say that although in quadrupeds the feet are in place of hands in humans, hands work differently from feet and for that reason they bend differently.

2. To the second argument one should say that birds bend their feet to the front because more of the body is on the front, but the wings, contrariwise, bend to the rear. But the human bends the foot to the rear because more of the body is in the rear.

24. This somewhat cryptic passage is made worse by the fact that for A., *coxa* can mean "hip" or "thigh." Neither, of course, properly bends. One suspects trouble in the text. *Clibanus:* lit., "oven." "Now all that lies between the neck and the base of the belly at the groin is called the trunk." *DA* 1.2.1.110 (*SZ* 1: 87).

Question 7: Whether an animal naturally has to place its right foot forward.

It is asked whether an animal naturally has to place its right foot forward.

1. It seems not, because something naturally uses that which suits it when it needs more of that thing. But when a man defends himself, or when he fights, he places his left foot forward. Therefore, he does not naturally place his right foot forward.

2. Likewise, it is not a matter of placing that foot forward on which he supports himself better; but rather he must support himself more on the right foot, since it is stronger, and it is therefore not placed forward.

The Philosopher indicates the opposite.[25]

One should say that every animal properly disposed with respect to its interior organs has naturally to place its right foot forward. And the reason for this is that warmer ones are better fitted for motion. But the heart, which is the principal member, infuses its heat and power more into the right side, and for that reason the right foot is better suited for motion, and the animal places it forward. The left foot, however, is more stable and firmer, just as the Philosopher has it in the second book of the *On Heaven and Earth*,[26] and for that reason the body supports itself on it more, as is evident among fighters. If it is a male, he places his right foot forward, but, if a female, she places her left foot forward because that part is warmer and rendered better suited for motion. Nevertheless, if there are animals in which the heart inclines greatly to the left, and the liver likewise, these will place the left foot forward.

1. Thus, one should reply to the first argument that when a man is defending himself, he needs to support himself on his foot, and to deflect the blow, and for that reason he sometimes places his left foot forward inasmuch as his right foot is better

25. Ar., *HA* 2.1 (498b5f.). Cf. Avic., *DA* 1 (fol. 29vbG); A. *DA* 2.1.2.19 (*SZ* 1: 293–94).

26. Ar. *DC*, 2.2(285a15f.). Cf. A., *DCM* 2.1.4.

suited for motion. Thus he does not press his right foot to the earth so much as the left.

2. The answer to the second argument is clear in the same way. For the left foot is much more stable, and for that reason it supports him more, because it moves more slowly, unless affected by some accident.

Question 8: Whether the human body should be less hairy than the bodies of other animals.

Further one inquires whether the human body should be less hairy than the bodies of other animals.

1. It seems not. Hairs serve as a covering for animals, but a human needs a covering just as other animals do. Therefore, he ought not to have less hair.

2. Likewise, hairs arise from many dry and earthy superfluities. But such superfluities do not occur any less in humans than in other animals; therefore, etc.[27]

The Philosopher says the opposite.[28]

One should say that the human body ought not to be hairy on two accounts. One reason is to enable good sensation, because a human has a better sense of touch than all the other animals.[29] But we perceive something by touch better in an immediate rather than a meditated fashion. If, then, the human body were hairy, the human would only perceive touch through the medium of the hairs. But hairs are without sensation on account of their earthiness, and, as a result, the human would not then have the best sense of touch.

Another reason is owing to the complexion. For the human especially abounds in heat and moisture, through which his life continues. Thus his heat consumes a great deal of the dry and earthy superfluities, which are the material for hair. And for this reason the hairs are only thin and fine except on the

27. See A., *DA* 12.3.3.192–94 (*SZ* 2: 968–69).

28. Ar., *HA* 2.1 (498b16f.); cf. Avic., *DA* 1 (fol. 29vbG); A., *DA* 2.1.2.20–21, 12.3.3.192 (*SZ* 1: 294–95; 2: 968).

29. See A., *DA* 21.1.1.5 (*SZ* 2: 1412).

parts where there is a superabundance of these [dry and earthy superfluities], for example, under the arms and on the groin and on the rear part of the head, because the vapors rise to the brain and the posterior part is drier than the anterior part, and for that reason hairs are more abundant on the posterior than the anterior part of the head and face, etc.

1. On to the arguments. To the first, one should say that a human is ruled by reason. Thus what is given to other animals by nature the human can prepare and furnish for himself with his hands, because he has the organ of organs, namely, hands, and for this reason he does not need hairs for a covering because he can make a covering for himself.[30]

2. To the second argument one should say that the dry and earthy superfluities are distributed, thinned, and exhaled in the human imperceptibly, and little of them remains. This is why the human body has only thin and fine hairs.

Questions 9–12: Whether every animal ought to be horned.

Why does the stag have solid and not hollow horns?

Why does he alone cast off his horns and acquire new ones?

Why does an animal that is without teeth on the upper jaw ruminate?

Further, one inquires whether every animal ought to be horned.[31]

1. It seems so, because horns are given for defense. But every animal—for example, a horse—requires defense, just as do the others. Therefore, etc.

2. Likewise, horns are generated from smoky and gross superfluities. But these exist in every animal.

30. See A., *DA* 14.2.2.31, 21.1.1.6 (*SZ* 2: 1058–59; 1412–13); Ar., *Part. An.* 4.10 (687a7f.).

31. See Ar., *HA* 2.1 (499b15f.); Avic., *DA* 1 (fol. 30raH); A., *DA* 2.1.2.27 (*SZ* 1: 297).

Sense indicates just the opposite.

After this, one asks why the stag has solid and not hollow horns.[32] And, why does it alone, among the other animals, cast off its horns and acquire new ones and, when it has cast them away, why does it hide them? And why does an animal without teeth on its upper jaw ruminate, and why does it have several stomachs and twisted intestines and horns that are not twisted?[33] Examples include the cow and lamb.

1. To the first, one must respond that the superfluity in the human as well as in any animal is moist and dry. The moist one is eliminated through the urine or through perspiration. But the dry superfluity, which is left behind after the last digestion, is of three types. From the superfluity of the first digestion, because it is the dregs, there is no power at present. Of these three, one is subtle, one is gross, and one is in-between. From the subtle superfluity hairs are generated; from the gross one, horns; from the one in-between, the teeth are generated. Now, however, it is the case that in certain animals the heat is sufficient to disarrange and scatter the gross fumes, which are the material for the horns. And for this reason not every animal has horns. Similarly nature provides horns for defense. But some animals defend themselves with their feet and some with their teeth—examples being the falcon, the lion, and the dog—and for this reason not every animal needs horns. Similarly there is another cause why those animals that do not have molars have horns and ruminate, like the sheep, because the material for the teeth is converted into horn. But those that have all the teeth, like the horse, the human, and the ass, lack horns, because the material for the horns is converted into teeth, etc.

2. From this the response to the second argument is clear.

To the first question one must respond that some animals defend themselves with their feet and teeth, and so they do not need horns. And, moreover, not all have the gross fumes that

32. See Ar., *HA* 2.1 (500a6f.); Avic., *DA* 1 (fol. 30raH); A., *DA* 2.1.3.34–35, 3.2.2.77–78 (*SZ* 1: 301–2, 382–83).

33. See Ar., *HA* 2.17 (507a34f.); Avic., *DA* 1 (fol. 30vbB); A., *DA* 2.2.1.93–95 (*SZ* 1: 330–31).

are the material for the horns, or it is because these fumes are converted in these animals into teeth, as has been shown.

To the second question one must respond that the stag is a very melancholy animal.[34] Thus the stag has a lot of gross, earthy fumes that are converted into horns, and for that reason too it has the biggest horns. And because it belongs to the male to defend the female, as the Philosopher says below,[35] nature therefore provides horns to males rather than to females. Nevertheless, horns are given to both male and female in some animals, as in the case of the oxen and goats. This is because they are not as swift of foot as the stag. Now, although the doe does not have horns, she nevertheless has swift feet so that she can flee attackers. But this is not the case in cows and she-goats, and for that reason nature has bestowed upon them horns so that they may defend themselves when they cannot flee.[36]

Furthermore, all marrow is naturally warm and moist, except the marrow of the stag, because black bile [*melancholia*] is especially abundant in the horns of the stag.[37] For this reason there is very little marrow in them, but instead they are solid and not hollow.

Besides, the horns grow so much that they put a burden on nature, and it is especially onerous for nature to carry and bear them, and for that reason—because they are beyond the order of nature—nature allows for them to be cast off or removed, just as in the bellwether ram.[38] And this is also why the stag casts off its horns. But because it is instinctive for the stag to defend itself from this or that animal, it perceives through its particular estimation that if rapacious animals, like a bear, lion, dog, or some other of this sort, see that it has cast off its horns, they will quickly attack it as if the stag were unarmed. This is why it naturally hides its horns as soon as it has cast them off, owing to the wisdom of nature, lest they be discovered by other animals.

34. See A., *DA* 12.3.7.226 (*SZ* 2: 982–83).

35. See Ar., *Part. An.* 3.1 (661b28f.).

36. See A., *DA* 12.3.7.222–23 (*SZ* 2: 980–81).

37. See A., *DA* 3.2.7.149 (1: 416–17), 12.1.7.89 (*SZ* 2: 927).

38. The text here actually reads "just as in the wisent [*in wisente*]" but the *corrigenda* provide *in vervice* as a better reading. For the *vervix,* see *DA* 22.2.1.128(88) (*SZ* 2: 1528–29).

To the last point one must reply that animals lacking teeth in the upper jaw cannot digest or masticate their food properly in their mouths owing to its grossness, since they live on the things born from the earth, and for that reason they immediately send it to the inner stomach in which it is partly softened and ground up. But then it sends it to the mouth a second time in order to chew it up completely, and that is when such animals ruminate and masticate such food completely. Once they have chewed the cud, it is sent again to another stomach, and, because that nutriment is gross, such animals have very twisted intestines so that it will remain in them a very long time and be digested better. And because the gross and poorly digested foods produce vapors and they produce gross fumes on their own as well, this is the reason why such fumes abound in these animals and why they ascend to the head, from which horns are generated.

And this is why these five things accompany one another: to wit, ruminating, having several stomachs, twisted intestines, a lack of upper teeth on the upper jaw, and having horns that are not twisted on its forehead, unless the animal is warm and can consume such fumes, like the camel and strong he-goats and oxen and others of this sort.

Questions 13–16: Whether breasts ought to be located on the chest in a woman.

Why it is that in elephants and apes the breasts are located on the chest, just as in the human.

Why there are more breasts in some animals, and fewer in others.

Why there are breasts even on some males, as, for example, sometimes on the he-goat.

Further it is inquired whether breasts ought to be located on the chest in a woman.

1. It seems not. For Avicenna says that where the digestive function ends, there begins the nutritive function.[39] But the di-

39. Avic., *De anima* p. 1 c.5 (fol. 4vb).

gestive function ends at the lower part of the belly. Therefore, the nutritive function begins there and, consequently, that is where breasts ought to be located.

2. Likewise, among all the parts of the body there is a special congruity between the vessels for the milk and the vessels for the sperm. Therefore, they ought to be located in parts next to one another. But the seminal vessels are on the lower parts; therefore, the milk vessels will be there too.

The Philosopher says the opposite, and it is evident to the senses.[40]

After this one asks why it is that in elephants and apes the breasts are located on the chest, just as in the human.[41] And one asks why there are more breasts in some animals, and fewer in others.[42] And one asks why even on some males there are breasts, as for example on the human or the pig, even though they do not nourish a fetus, and sometimes on the he-goat.[43]

To the first question one must reply that the nobler an animal is, the more it needs a nobler and better digested nourishment. But according to the Philosopher in book two of *On the Soul* and according to the medical authorities, the digestion of nourishment occurs through the mediation of heat.[44] Since the human is the noblest animal, it therefore demands that its nourishment be situated quite near to the source of heat. This source is the heart, and this is the reason why the breasts are on the chest in a woman. Moreover, it would be disgraceful and embarrassing for a woman always to uncover her shameful parts, if her breasts were located there. For that reason nature located them in a decent place and not between the feet.[45]

40. Ar., *HA* 2.1 (500a13f.). Cf. Avic., *DA* 1 (fol. 30raH); A., *DA* 2.1.3.36 (*SZ* 1: 302).

41. See Ar., *HA* 2.1 (500a18f.); A., *DA* 1.1.2.18, 2.1.3.36, 2.1.4.58–60 (*SZ* 1: 51, 302, 312–13), 14.2.3.41–44 (*SZ* 2: 1063–64).

42. See Ar., *HA* 2.1 (500a21f.); A., *DA* 2.1.3.36–37 (*SZ* 1: 302–3).

43. See Ar., *HA* 2.1 (500a30f.); A., *DA* 1.2.23.441, 2.1.3.36 (*SZ* 1: 213, 302).

44. Ar., *De anima* 2.4 (416b28f.).

45. While strange on the surface, this statement does, in fact, give us an insight into what was considered "acceptable" in Albert's time, indicating that breast-feeding must have been widely accepted in public.

1. To the first argument one should say that nothing nourishes before it is in a disposition near to conversion. But this only occurs when digestion precedes it. Thus, nutrition begins when digestion is completed. Moreover, Avicenna understands this statement concerning digestion and nutrition as it is in relation to one and the same person. But there is no nourishment in the breast for the one who digests the milk, she being the person in whom the digestion occurs, but rather nutrition is present in the breast for the nourishment of someone else.

2. To the second argument one must reply that because milk and semen are closely related, an abundance of the one impedes the abundance of the other.[46] So if the breasts were on the lower part of the human, there would be less milk owing to the abundance of the semen, when, nevertheless, a larger amount of milk is required because the human nurses on breast milk longer than other animals do.

To the second question one must reply that the elephant is a melancholic animal that has little blood. The length and hardness of its bones attest to its size. And likewise the ape has but little blood. An indication of this is that it has very poor estimation and good motion, and for that reason there is in them naturally but a modest abundance of milk. This is why their breasts are near the source of blood and heat, to enable the milk to be more abundant there and to be digested.

To the third question one must respond that some mothers (that is, female animals), such as the dog, sow, and sheep, have and give birth to several young, and this is why they have several breasts, so that all the young may suck the milk and be fed at the same time. If there were fewer breasts than there are young, then it could happen that a given young one would be deprived of nourishment, which is not seen in nature, and for that reason, etc. Therefore, young are most frequently generated in proportion to the number of breasts, etc.

46. The relationship of breast milk and sperm or semen is intelligible only because women were commonly understood by the followers of Galen to have semen or sperm in addition to menses. See *QDA* 10.3;5; *DA* 9.2.3.99–103, 10.1.1.7, 16.2.1.92–93 (*SZ* 1: 813, 829–30; 2: 1207).

To the fourth, one should say that not only is the milk conserved in the breasts, but also the spermatic blood. An indication of this is that when females enter adolescence, their breasts begin to itch a great deal due to these things [i.e., milk and spermatic blood]. Thus in those animals in which the breasts are located near the seminal vessels, there are no breasts on the males. But in the human the breasts are located at a distance from the location of the seminal vessels or the testicles, and the same is true for the pig.[47] And for this reason there can be breasts in these males, but in others this is not the case due to their location and proximity to the testicles, unless in terms of their complexion and disposition they very much resemble the mothers, etc.

Questions 17–19: Whether growth is necessary for living things. At what age is there more growth? Why in the human, when he is fully grown, are the upper parts smaller than the lower ones?

One inquires now into the growth of living things. First, whether growth is necessary for living things.

1. It seems not because the operation of a nobler form is itself nobler. But the form of something living is nobler than the form of something not living. But a non-living thing, generating by means of its mediating form, simultaneously bestows being and perfected or completed size. This is clear from fire, and is therefore all the more clear in a living thing that generates.

2. Likewise, the form of a non-living thing is nearer to matter than is the form of a living thing. If then the form of the non-living receives being and perfected size at the same time, so much the more does the form of the living do so.

Second, one asks at what age more growth occurs.

1. And it seems that more growth occurs during adolescence than in childhood. Growth is a movement from incomplete

47. See Ar., *HA* 2.1 (500a30f.).

to complete. Therefore, the more something approaches full growth [*completum*] and its natural end, the more it grows, etc.

2. Furthermore, natural motion is directed to an end and is weakened in the beginning, although violent motion is the opposite. But growth is a natural motion, and is therefore directed to an end so that it can grow; therefore, etc. Now adolescents are nearer to the end of that which completes them than are children; therefore, etc.

The Philosopher says the opposite in this book.[48] For he says that a person grows more in the first seven years than in the second, and more in the second seven years than in the third, and grows less in the fourth seven years because then an equilibrium is achieved.

Third, one asks why in the human, when he is fully grown, the upper parts are smaller than the lower parts whereas, in the beginning, it is the other way around and they are larger. This is the opposite of what happens in all the other animals. Now it seems that the upper parts are larger in all cases, because according to the Philosopher the upper parts extend from the head to the bottom of the belly, and the lower parts extend from there to the feet, and the head with the trunk is larger than the feet.[49]

To the first, one should say that growth is necessary to a living thing because it is impossible that a part should be equal to the whole at the same time in the same subject. But living things are generated from seed, which is a part cut off from the one generating. Which is why it is impossible that an animal or a living thing generated from a seed should be equal in the beginning to the one doing the generating. And for that reason living things have three powers: the nutritive, through which it is conserved in its acquired being; the augmentative, through which it proceeds, with nourishment as its mediating agent, from the

48. Ar. *HA* 2.1 (500b26f.); cf. Avic., *DA* 1 (fol. 30raH), 12.3 (fol. 45rb); A., *DA* 2.1.4.43–45 (*SZ* 1: 305–6), 12.1.5.67–69 (*SZ* 2: 918–19).

49. See Ar., *HA* 2.1 (500a26f.). Recall that the term "feet" can include the entire lower body. As A. points out in *DA* 1.496, he uses *pes*, "foot," for the Arabic *rigil*, the entire lower walking assemblage. Cf. *DA* 1.2.1.110 (*SZ* 1: 87).

imperfect to the perfect; and the third is the generative power, through which it produces one like itself.

1. To the first argument one should reply that in non-living things the matter of the one generated did not exist earlier in the one generating, but rather in its opposite. Thus, the one generating introduces form in the matter that is externally determined with respect to its quantity. But in living things the matter of the one generated did exist earlier in the one doing the generating. And this same matter is augmented by the reception of nourishment, and for this reason the form is introduced earlier under one quantity, and then the quantity and the augmented matter are increased. This is why this is attributed to the nobility of the form itself and not to its ignobility.

2. To the second, one must say that non-living things are so near to matter that they have a determinate matter, as is evident in a rock, and for that reason it receives being and a perfected quantity at the same time. But animated things do not receive their entire matter determined at one and the same time, and for that reason they do not receive a perfected quantity [full-grown size] all at once.

To the second inquiry one must reply that two things are required for growth, namely, an agent and one that is acted on. The augmentative power requires heat that digests the food, and requires a nutrimental moisture through which growth might occur. Therefore, more growth occurs at the age when these things are more abundant. But this is during childhood, although sometimes it seems to the senses that there is more growth during adolescence, unless the natural heat is weakened by excessive moisture during childhood. Therefore, as long as an animal can convert nourishment into its own substance, so long as it is sufficient to replace what is lost and more, just so long will it grow. And when it can only convert an amount sufficient to replace what is lost, then it is in stasis, and when it can no longer do this, then it is a time of decreasing size.

1. To the first argument one should reply that the more a living thing approaches full growth, so much more will it grow, in order to achieve larger size, yet in the near term it does not pro-

portionally acquire more size *de novo* than it had acquired in an earlier period, because the more it approaches full growth, the more its power is weakened owing to the reaction of contraries.

2. To the second argument one must reply that a natural motion is of two types. One is according to its approach to its natural and proper end and occurs with the strengthening of its cause, and such is the motion of the light and the heavy, which inclines to an end. The other is a natural motion resulting from a withdrawing from its cause and accompanied by the weakening of the cause, and such is the motion of growth. And it is weakened for the same reason as the illumination of a medium, for the further something is from a luminous body, like the sun or a candle, the weaker it is on account of the distance from the cause.

To the other, the third, question one must reply that a woman's womb comes near to being round, and for that reason long members, like the tibia, thigh, and leg bones, are not located as conveniently as a round member, like the head and the trunk, and for that reason the head and the trunk are proportionally larger in the womb than smaller. But in other animals the wombs are more stretched out, and for that reason their long members enjoy more solidity. Thus the human, in the beginning of his life, has upper parts that are broad and large in comparison to the lower ones, but as his age increases the upper parts become dried out and as a result they are proportionally reduced.

One must reply to the argument that in every animal the upper parts are larger in size, but are not proportionally larger, and for the reason that, etc.

Questions 20–26: On the nature of teeth.

Further one asks about the nature of teeth.[50]

Why do some animals have teeth set in order like a saw, and are called the *karcharodonta,* like the dog, wolf, and lion, but others do not have teeth that intersect one another?[51]

50. See Ar., *HA* 2.1–4 (501a8f.); Avic., *DA* 1 (fol. 30raH–rb1); A., *DA* 2.1.4.46–57 (*SZ* 1: 307–12).

51. Cf. in general *DA* 2.1.4.46–57 (*SZ* 1: 307–12). *Karcharodonta* is a perfectly

Second, one asks why some teeth are pointed and some are broad.

Third, one asks why the front teeth are replaced, but not the molars.

Fourth, one asks why there are more teeth in males than in females.

And fifth, why a large number of teeth and their closeness indicate a long life, and few teeth that are spaced out indicates a brief life.

And sixth, one asks why teeth grow in width but not in number.

And seventh, one asks why in the human and in other animals in their youth the teeth are white, but in old age they are dark, except in the horse.[52]

To the first, one should answer that in some the teeth are arranged only for chewing and grinding food. And in other animals they are arranged for chewing and for defending, as in the case of the dog, wolf, and lion, and therefore the teeth in these latter animals are arranged in more of a saw-like manner, and this is not so in the case of the former, for example, the horse, ox, and human. But the first animals, like the dog and lion, because they are rapacious and predators, seize their prey as a hook and a gaff seize a fish, and as a saw cuts wood better when it is toothed. For that reason nature has taken pity on them and endowed them with such teeth, etc.

To the second question one must say that food is received and divided by the front teeth and ground up by the molars, and this is why the front teeth are sharp, for cutting and separating, and the molars are broad, for grinding and crushing, like millstones, etc.[53]

To the third question one must answer that the molars are more necessary than front teeth and that is why they are replaced less often, because the molars are produced more from

---

good Greek word meaning "saw-toothed" and is used in this context by Ar. at *HA* 2.1 (501a16–17). But it does not seem to be used in the *DA*.

52. A., *DA* 2.1.4.52 (*SZ* 1: 309–10).

53. "Millstones": *lapides molares.* A. continues the word play immediately below by calling molars *dentes molares.* See also *QDA* 19.12–13.

the spermatic and solid material than are the front teeth, and for that reason nature has strengthened the molars more, as they cling to the gums with four roots. But the front teeth are fashioned more from the flowing spermatic blood, and for that reason they are more easily replaced, because their material flows and if they are pulled out during an age near to the sperm they grow back—not, however, in old age.

To the fourth question one should respond that strong heat or the power of heat and sufficient and proportionate matter are required for the generation of teeth, and because the power is stronger in males and the spermatic blood is more abundant, teeth are more abundant in them than they are in females, for the opposite reasons.

To the fifth question one should say that a large number of teeth and their close spacing are witness to strength of power and a large quantity of spermatic material, and these two things are necessary for a long life, and for that reason, etc. And furthermore, the more teeth there are and the more closely spaced they are, the better the digestion and chewing in the mouth will be, and as a result the digestion in the stomach and in the liver will be better, and, further, the conversion of the food into the substance of the members will be better. This is why people who eat very quickly and gulp down their food like gluttons have a shorter lifespan, because the food, which is not well chewed, putrefies and generates blockages, and makes the body hectic owing to the lack of nourishment.[54] Thus death occurs more quickly in ones such as these, etc. Teeth that are few and spaced further apart are worse, for the opposite reasons.

To the sixth question one should say that a large number of teeth occurs from the matter, but their size and their breadth can be due to the strength of the agent. Since in the human the spermatic material plays the dominant role with respect to the teeth, they therefore do not grow in number, but because the

54. I.e., suffering from hectic fever, from a Greek term meaning "recurrent." These fevers attacked the body's solids, and hence are often rendered as "consumptive" and can be associated with tuberculosis. Cf. Hall (1971), 7f., with Galenic references. Cf. DA 20.1.4.21 (*SZ* 2: 1369–70).

teeth are solid and less corruptible and since a potent power for converting some kind of food into the substance of the one nourished is always present in a person or in a living thing, it is for that reason that the teeth always grow in length and width. And, moreover, this is also due to the fact that they are constantly involved in the motion of chewing, and would be worn down if they did not grow continuously, for that reason nature causes them to increase continuously, etc.

To the last question one must reply that the closer animals are to the spermatic age, the whiter their teeth will be on account of the sperm, which is white. But in old age the fumes—earthy, melancholy, and dark—multiply owing to a deficiency of heat, and as a result there ensues a corruption in the blood, which is the nourishment for the teeth, and this is why, on account of the dark blood, teeth turn dark in old age. But in the horse a great deal of moisture abounds, and its heat is greater in youth than in old age. This is why, if it is not reined in, it often may kill itself from an excessive rage of wantonness. This is why in its youth (for it is an especially wanton animal in youth) its teeth grow dark on account of the abundance of heat, because the overpowering heat at this time burns and darkens the teeth. But its heat is reined in and tempered in old age, and the teeth are nourished then with a nourishment more appropriate to them and are not corrupted or burned by the overpowering heat, and they therefore whiten at this time, and the desire for intercourse ceases in them.

Question 27: Whether the human ought to lack a tail naturally.

Next one asks whether a human ought to lack a tail naturally.

1. It seems not. Tails are conferred upon animals for covering the anus, which is a nerve-filled and delicate member, in order to cover and safeguard it from injuries, like cold and other such things. Since the human, however, needs to safeguard the anus no less than other animals, therefore he will no less lack a

tail. Avicenna reveals the fact that there is a need to safeguard the anus when he discusses the things it undergoes, etc.[55]

2. Moreover, animals' tails contribute to motion. For if the tail is removed from a bird, a fish, or a dog, its motion will be retarded. This is why wise hunters never shorten the tails or the ears on their dogs, because this would clearly harm their hunting, because they are directed through the powerful movement of these members to go after prey, just as is a ship by oars. Otherwise, their movement would be hesitant. And a little below the Philosopher will say that a dog that loses its tail will not dare to cross a bridge because it may fall, just as a ship sinks without an oar.[56] Since a human needs to be made fit for motion no less than other animals, so he ought no less to be deprived of a tail.

The Philosopher says the opposite, and it is clear to the senses.[57]

One must respond that tails are given to animals for two reasons: as a covering for the anus, and as an aid to motion, just as the two arguments above conclude. But the human does not need a tail as a covering. For he has reason and the intellect, through whose intervention he can prepare for himself a covering and other things that are given naturally to others. Nor does he need a tail for the sake of motion, because ease of locomotion in the human depends on the arrangement of the hip, the leg, and the foot, to which a tail adds nothing.

And, moreover, the human sits and rests on the posterior part and the anus, and a tail would prove an impediment to this rest and sitting.

Furthermore, many nerves extend through the vertebrae of the back in the posterior part toward the anus, and in the human these nerves are bent back toward the penis. So the nerves, which in other animals extend toward the tail, proceed in the human to the penis, and for that reason the human has, among

55. Avic., *Can. med.* 3.17.1.1–22.

56. See A., *DA* 1.1.7.89, 22.2.1.28 (*SZ* 1: 78–79; 2: 1458–59); A., *De motibus animalium* 2.2.3.

57. Ar., *HA* 2.1 (498a11f.), 2.8 (502a16f.); *Part. An.* 4.10 (689b2f.). Cf. Avic., *DA* 1 (fol. 30rbI–K), 14.7 (fol. 58ra–b); A., *DA* 2.1.4.56–60, 14.2.4.47–48 (*SZ* 1: 311–13; 2: 1066–67).

all the animals, the greatest and strongest penis for its size, erect as a spear. Thus it penetrates a virginal vulva, which is harder and stronger and more difficult to penetrate and open up with the penis than is the same one in other animals, owing to the excessive narrowness of the womb. It is clear on account of this.

On to the arguments. One should say that although in other animals the tail contributes to covering and motion, this is not the case in the human. And the reasons have been discussed.

Question 28: Why birds imitate the human in speech more than the other animals.

Further one asks why birds imitate the human in speech more than the other animals.[58]

1. And the contrary seems the case. For those parts of birds that conform more closely to the parts of the human ought to imitate him more in their operations. But the parts of walkers are of this type, because at least their tongues conform more to the human tongue than does the bird's; therefore, etc.

2. Furthermore, the animals that have greater discretion ought to participate more in the principle of discretion. But walkers, like the dog and the ape, participate more in discretion or in a likeness of discretion than does the bird.[59] This is evident, because the ape learns to stitch cloth together and the dog learns to serve, which is a sign of great discretion.[60] Therefore, such animals as these should imitate the human more, so it seems, in speaking than does the bird.

The opposite is evident to the senses, because the parrot, the magpie [*pica*], and the jackdaw [*monedula*] imitate the human in speaking, and one never sees a walker speak.[61]

58. See Ar., *HA* 2.12 (504a35f.); Avic., *DA* 1 (fol. 30rbK–vaL); A., *DA* 2.1.6.74, 4.2.2.97–98, 21.1.5.26–27 (*SZ* 1: 320, 481; 2: 1425–26).

59. See A., *DA* 21.1.2.10, 21.1.3.15–18, 22.2.1.30 (*SZ* 2: 1415, 1419–21, 1460).

60. Presumably tricks taught to the animals. For discretion among quadrupeds, cf. *DA* 21.1.4.20–25 (*SZ* 2: 1422–25).

61. See A., *DA* 2.1.6.74, 4.2.2.95, 12.3.5.204, 23.1.24.129 (*SZ* 1: 320, 480; 2: 973, 1638–39).

To this, one should respond that birds, especially, imitate the human in speaking, and birds having a broad tongue do this, because two things are required for speaking: apprehension of something heard and the suitability of the instrument. But now it is the case that some birds apprehend quite easily and their tongues have quite suitable nerves and can move flexibly toward their palate and they have a broad, wide windpipe [*canna pulmonis*],[62] well disposed in the throat. This is why such birds especially can imitate the human in giving voice. And this happens especially among birds that make a sound [*vox*] in the throat, for those forming a sound at the tip of the beak, like the sparrow, nightingale, and similar ones, cannot perform this operation. Walking animals have more earthy vocal instruments, and the nerves of their tongues are not as well suited as are those of the aforementioned birds, and that is why they cannot adapt their instruments to form a voice.[63]

1. To the first argument one should reply that although parts of other walking animals may conform better to human parts in terms of shape and motion, they conform less well in the formative principles of voice.

2. To the second argument one must reply that although the dog and the ape and such animals are more teachable in certain areas, nevertheless their organs of voice are not so well formed as those of birds, as has been said, and for that reason, etc.[64]

Question 29: Whether the motion of birds, which is flight, is natural.

Further one asks whether the motion of birds, which is flight, is natural.[65]

62. The term is synonymous with trachea, but, as this is also a term A. employs, "windpipe" is used to keep the two distinct.

63. Note that A. carefully uses *vox* throughout this passage both to mean human articulate sound, "voice," and formed sound such as that of a bird's chirp.

64. It is interesting that before scientists hit on using sign language in apes, many an experiment had failed trying to teach a chimpanzee to form simple words since an ape's vocal equipment is simply not up to the task.

65. See A., *DA* 2.1.6.69–71 (*SZ* 1: 318–20).

1. It seems not, because every mixed thing is moved according to the nature of the element that is dominant in it.[66] But water and earth dominate in every mixed thing and their natural motion is downward. Therefore, the ascent of a mixed thing is contrary to nature.

2. Besides, if the flight of birds were natural, it would be so either by reason of the body or by reason of the soul. Now, not by the body, because once the soul has withdrawn from the bird, its body, placed in the air, will descend. And it is proved in the same way that it is not by reason of the feathers. Neither is it by reason of the soul because the operation of a nobler form is itself nobler; but the human's soul is nobler than the soul of any other animal. Thus flight will be much more natural to the human than to the bird, but this is false, and therefore the premise is false.

To speak to the opposite, natural motion is one whose principle is from within *per se* and not *per accidens,* but flight in birds is from their intrinsic principle, because it is by their power; therefore, etc.

To this question one should reply that the motion of flight can be related to a bird in two ways. It can be by reason of the body, just as the Philosopher responds in the fourth book of the *Physics,* and thus the flight of a bird is an unnatural motion because the motion of the mixed body follows the nature of the element that is dominant in it.[67] But in every mixed body the heavy elements are dominant. Or it can be related to the bird by reason of the soul, and in this way flight is a natural motion. Nevertheless, it is not natural absolutely, like the motion of the heavy and the light, but it is the motion of an animal even with apprehension, because, simply put, a natural motion is toward the one end in which the movable [*mobile*] naturally rests, like the heavy going down and the light rising up. And on account of this the Philosopher said in the second book of *On Interpretation* [*Peri Hermeneias*] that a natural power does not relate to op-

66. See A., *DA* 12.1.1.1–3 (*SZ* 2: 894–95).
67. Ar., *Phys.* 8.4 (254b7f.); cf. A. *Phys.* 8.2.3.

posites. But animal motion can be toward several ends owing to the difference of its apprehension.[68]

1. On to the arguments. To the first, one should reply that flight is not a natural motion for a bird with respect to its body, and reason proves this, but rather it is natural with respect to and by reason of its soul.

2. And you prove that it is not by reason of the soul because it is of a nobler form, etc. One must say that a nobler operation is appropriate to a nobler form. Thus the proper operation of the human soul—which is to understand— is nobler than the operation of another animal as far as this is concerned. Nevertheless nothing prevents some brute beast from surpassing the human in certain operations.[69] For example, the eagle and the lynx surpass the human in sight, the ape in taste, the dog and the vulture in smell, the wild boar in hearing, the spider in touch, and so on for others.[70] Whence the verse:

> The wild boar surpasses us in hearing, the spider by touch,
> The vulture in smell, the lynx in sight, and the ape in taste.[71]

Question 30: Whether any animal ought to lack a spleen, a gall bladder, which is its reservoir [cista], *and lungs.*

Further one inquires whether any animal ought to lack a spleen, a gall bladder, that is, its reservoir [*cista*], and lungs.[72]

68. Ar., *Peri Hermeneias* 13 (22b39f.), trans. Boethius (ed. Charles Meiser, [Leipzig, 1877–80]), vol. 1, p. 24, v. 5–6.

69. See A., *DA* 19.1.5.24 (*SZ* 2: 1344).

70. See A., *DA* 22.2.1.113, 23.1.7, 23.1.24.144 (*SZ* 2: 1518–19, 1547, 1653–54). A. repudiates the claim attributed to the ancients that the spider surpasses all other animals with its sense of touch. See *De anima* 2.3.23 (133, 55–57).

71. Cf. John of Salisbury, *Policraticus*, 7.2, ed. C. C. I. Webb, CC CM 118 (Brepols, 1993), p. 96.

72. The vocabulary here and throughout this section is difficult since the words can often mean several things. Thus, *fel* can indicate either "bile" or "gall bladder" whereas *cistis fellis* is used just below for "gall bladder." Cf. *DA* 1.3.5.605–6, 611 (*SZ* 1: 277, 279–80) and 12.1.6.84 (*SZ* 2: 925). Likewise, words such as *cholera* can seem at one point to indicate a specific type of bile, at other times to be used generically. Throughout this section, then, the Latin is often left in place to assist the reader in attempting to evaluate A.'s intent.

1. It seems not, because the heart is a more principal member than the brain.[73] But no animal having a brain lacks protection for the brain, which is afforded it by the cranium, its webs, and its tunics. Therefore, by analogy, no animal that has a heart will lack a protector for the heart. But the lungs are the protector of the heart, because the heat of the heart is tempered through the air or through the cold of the air that has been breathed in; and therefore, etc.

2. Moreover, the gall bladder [*cistis fellis*] is the receptacle for bile [*cholera*], in order to aid in digestion and elimination, and it is something purified from the blood.[74] Therefore, similarly, every animal that has blood requires the gall bladder [or "bile," *fel*] and something else for the sake of digestion and the elimination of feces. Therefore, etc.

3. The same can be argued with respect to the spleen since it is the receptacle of black bile [*melancholia*] and there is no animal without black bile.[75] Therefore, etc.

But the Philosopher says the opposite.[76]

One must reply that the lungs are a necessary assistant and fan for the heart.[77] For by expanding they draw in air for tempering the heat of the heart, and by contracting they expel air that was drawn in and warmed. In some there is therefore an abundance of heat in the heart, and for these the lungs are even more necessary. But in those lacking blood, heat does not abound in the heart owing to the coldness of their complexion. Bloodless creatures are like this and neither does heat abound in fish, because they continuously live and are nourished in a cold element, namely, water, and such animals therefore lack lungs.

Similarly bile is something purged from the blood, as was argued. For the gall bladder is a receptacle for bitter, red bile [*cholera rubea*]. But in some animals the whole nourishment

73. See A., *DA* 1.1.5.73, 12.1.1.15 (*SZ* 1: 72; 2: 899).

74. See A., *DA* 13.2.4.99 (*SZ* 2: 1031–32).

75. See A., *DA* 2.2.1.86–87 (*SZ* 1: 326–27).

76. Ar., *HA* 2.15 (506a1f.); cf. Avic., *DA* 1 (fol. 30vaA); A., *DA* 2.2.1.85–92 (*SZ* 1: 326–30).

77. See A., *DA* 1.1.5.75, 13.1.1.9, 13.1.6.46 (*SZ* 1: 72–73; 2: 988, 1005).

[*nutrimentum*] gives way and is converted over to the nourishment [*fomentum*] of the animal and to the restoration of what has been lost. This is the case in birds on account of the abundance of their heat and their dryness. And this is why there is no gall bladder in them distinct from other parts, but rather [the bile] is scattered among various parts along with the nourishment.[78] And if some bile [*cholera*] remains in the liver, it immediately sends it to the stomach in order to strengthen their digestion. And because they use little food in comparison with quadrupeds, their elimination therefore obeys the urge for expulsion without the need of bile [*cholera*] as a stimulant. And there is still another reason for this: not all birds drink, neither do they emit urine, as is apparent in birds with curved talons,[79] and this is why they do need some moisture to enable the food to penetrate the narrow vessels of the members better. For this reason both *cholera* and *melancholia* are mixed into their blood in the vessels, so that the nourishment might have easier transit to the individual parts of the members, because *cholera* can penetrate by virtue of its heat and *melancholia* does so by virtue of its acidity, because it is acidic. This is why the Philosopher says that every animal has a bladder, but according to more and less.[80] For some have a proper receptacle for the bile, like the human, the cow, the dog, and such animals, and some do not have their bile in a dedicated receptacle but rather dispersed among various parts.[81] So, in the stag, the bile is dispersed through the intestines, and this is why the intestines are bitter tasting and dogs will not eat them unless they are starving. And this is why hunters hunt them at a very fast pace, because otherwise it would be impossible to capture them using dogs. For the *cholera* is very much aroused in their intestines by all the motion, and *cholera* flees there from other parts of the body and corrodes and ul-

78. See A., *DA* 2.2.1.89 (*SZ* 1: 328).

79. Although the text here reads *cansonitis,* the *corrigenda* provide *gampsonicis,* which is derived from the Gr. *Gampsonyx.* For other appearances in *QDA* of *gampsonica,* see 8. 8 (175, 73), 8.12–16 (177, 5), and 8.15–20 (195, 51). For the absence of a bladder in birds, see A., *DA* 1.1.5.60 (*SZ* 1: 68).

80. See Ar., *HA* 2.15 (506a20f.).

81. See Ar., *HA* 2.15 (506a31f.); A., *DA* 2.2.1.88, 22.2.1.45 (*SZ* 1: 328; 2: 1473).

cerates them. As a result of the excessive pain they will not run further and thus are detained by the dogs, because they cramp from the pain in the intestines just as a human does suffering from colic [*cholerica passio*]. You can assign the same reasons concerning the spleen and the *melancholia* in such animals.

1. On to the arguments. To the first, one must say that just as the brain needs an exterior body as a protector, so too does the heart need one, but nevertheless it needs lungs only if there is an excess of heat in it. This is why the relationship of the cranium to the brain is not the same as that of the lungs to the heart.

2. To the second argument one should say that although every animal has what is purged from the *cholera* and *melancholia* of the blood, nevertheless not all have a proper receptacle for them. Instead, in some animals these are dispersed through various parts. Now because the spleen is the proper receptacle for *melancholia,* and the gall bladder for *cholera,* and these are distributed among various parts in some animals, thus not every one has a spleen or a gall bladder.

Question 31: Whether it is necessary for every animal to have members specially designated for generation.

One asks further whether it is necessary for every animal to have members specially designated for generation, namely, testicles.

1. It seems not. A power that resides in every part of the body does not need a determinate part. But the generative power exists in every part of the body; otherwise, the one generated would not be assimilated to the one generating both *in toto* and in respect to its individual parts. For unless the generative power existed in the hand of the one generating, and so too in other parts, the one that is generated would not have a hand; therefore, etc.

2. Moreover, the generative power is no less in the plant than in the animal, which is evident from the plurality of parts in the plant. But there is no part specially designated for generation in the plant, and therefore not in the animal either.

The opposite is evident from the determination made by the Philosopher.[82]

One must respond that some animals are generated from propagation, and some from putrefaction. In those generated from putrefaction there are no members designated for generation, because they are not generated from semen. But in those generated through propagation this is necessary, because in such as these the male emits semen outside itself, which would not happen if the semen were not collected and digested in a determinate part.[83] And this is why males have testicles, in which the semen is digested and fermented, and a penis, through which it is emitted at the time of generation. But the female has a womb that receives and informs it, and a mouth of the womb (which is the vulva), through which it receives it when it is spewed forth by the penis.

1. On to the arguments. To the first, one must reply that semen arises from the superfluity of the nourishment that has been prepared almost to the point of conversion into a member, but because there is so much of it, some is not converted. This is why it has the disposition of every member potentially, because, if it had it in act, it would have been converted into those members. So the semen, when it receives the form of the semen or the sperm or of the one doing the generating (which are one and the same thing), receives a power for producing one like the one whose semen it is, and does so by virtue of the power of the one in which it exists. And although semen is digested or collected in a determinate part, like the testicles, it is nevertheless a superfluity of food that is potentially like the whole, and for that reason not only is the one produced from semen similar when born to the part, but also to the entirety of the one doing the generating.[84]

2. To the second argument one must respond that the more imperfect the form is, the less variety there will be in its matter. For there is less variety in the matter of fire or earth than in that

82. Ar., *HA* 3.1 (509a30f.). Cf. Avic., *DA* 1.1 (fol. 31raD); A., *DA* 2.2.3.108 (*SZ* 1: 336).

83. See A., *DA* 1.1.5.61–63 (*SZ* 1: 68–69).

84. See A., *DA* 18.1.3.23 (*SZ* 2: 1291–92).

of a plant, and less in the plant than in the animal, and this is why it is not necessary that the generative power of a plant be in a determinate part. Instead, there can be a principle sufficient to produce another plant like itself and it can be present in any part. And this is owing to the imperfection of the plant, because there is little variation in its material. But this is not the case in animals generated through propagation; and for that reason, etc.

And once more the plant is not generated by propagation, but rather through sprouting, because it is evident to the senses among plants that sprout from some part of themselves, that the power of sprouting exists in every part of the plant unless a hard covering should prevent it, as is the case on its trunk. But it is not this way on animals, and for that reason, etc. Hereby the solution to the whole question is clear.

BOOK THREE

Question 1: Whether veins are necessary in an animal body.

"WE HAVE ALREADY discussed the disposition of the members,"[1] etc. With respect to this third book we will first ask whether veins are necessary in an animal body.

1. It seems not. The natural power [*vis naturalis*] is in charge of providing nourishment for plants and in animals.[2] But there are no veins in plants delegated for nourishment. But a vein does not exist for another purpose, and therefore veins are not necessary.

2. Moreover, bones are nourished just like other members, but there are no veins in bones.[3] It is therefore not necessary to posit the existence of veins in other parts for the sake of the distribution [*delegatio*] of nourishment.

3. Moreover, the conversion of the nutriment is almost completed in the third digestion, but veins are not required during the third digestion, and are therefore not required in the other digestions either.

To the contrary: every bodily operation occurs by means of an intermediate body. But the distribution of the nutriment is a bodily operation, as is clearly evident in its own right [*de se*]. Therefore, it occurs by means of a bodily medium. But such a medium is nothing other than a vein; therefore, etc.

One must reply that the word "vein" is an equivocal noun which has two meanings.[4] These are "pulsating," which is properly called an artery, and "non-pulsating," which is proper-

1. Ar., *HA* 3.1 (509a27–30); cf. A., *DA* 3.1.1.1 (*SZ* 1: 344).

2. A., *DA* 20.2.3.73 (*SZ* 2: 1399).

3. A., *DA* 16.2.6.124 (*SZ* 2: 1223).

4. Simplifying the unnecessarily duplicative *vena nomen aequivocum est ad duo vel duplex est . . .*

ly called a vein in the human.[5] Now, a pulsating vein (that is, an artery) is necessary for the sake of sustaining the heat or to bear the spirit to individual members. Since these members are like a medium for life, their arteries arise on the left side of the heart, where the heart's heat and the spirit are especially abundant. But a non-pulsating vein is necessary for the distribution of the nutriment, that is, the humors, to individual members and for the sake of sustaining the heat and the spirits and to restore what has been lost, because when aliment [*alimentum*] is received into a particular part, it will be unable to restore whatever has been lost in these individual members unless it has access to individual members.

Moreover, everything that has a part in which it is generated, has also a part in which it is kept; otherwise, it would be generated in vain. But the blood is generated in a given part in the body such as in the liver and the heart.[6] It therefore requires a part in which it is kept. But this part is the vein, and this is why, etc.

Nevertheless, one must observe that sometimes the blood is in a remote power, and sometimes it is in a power that is close to conversion, and sometimes it is in a middle power.[7] When it is in the stomach it is in a remote power. From there it has a pathway through which it comes to the stomach (like the esophagus) and a pathway through which it is distributed from the stomach (like the delicate or mesenteric pathways through which it is sent to the liver). It is in a power close [to conversion] when it is dispatched to a given member, and then it requires no connection beyond the member. But it exists in a middle power in the liver where the second digestion occurs, and this is why the liver requires veins through which the blood may be dispatched to other members and through which it can receive the moistures not converted into blood as a sediment [*hypostasis*].[8] And this is why the veins are necessary.

5. A., *DA* 1.2.20.381 (*SZ* 1: 189).

6. Cf. A., *DA* 1.3.5.601 (*SZ* 1: 275).

7. A., *DA* 3.1.1.1. On a remote or proximate power, see also *QDA* 9.18 (211, 1–5). The former can result in diverse effects; the latter seems determined to a single effect.

8. A., *DA* 3.1.3.100 (*SZ* 1: 393–94). *Hypostasis* usually means "sediment" or "dregs," as at *QDA* 15.16 (269, 72).

1. On to the arguments. To the first, one must reply that the nutritive power in the animal exists under the natural and animal regimen. But in a plant it is not ruled by the animal regimen, and this is why the power requires more in the animal than in the plant.[9] And in addition, the plant has a marrow in the middle of every part in place of a vein. But these "veins" extend from the root to the extremities. But in animals the nutriment is taken into the middle, as it were, and the veins extend out from here just as if from the center to the circumference.

Moreover, plants always have their aliment at hand, and they therefore do not absorb more than they can digest. But animals do not have their aliment always at hand, and as a result they frequently eat in a single day more than they can digest over several days, and this is why certain parts are necessary in an animal where this undigested or unconverted nutriment may be stored. Veins are parts of this sort.

2. To the second argument one must reply that every animal part, although it may appear solid, like bone, is nevertheless porous and capable of receiving moisture. Thus blood is conveyed to the extremities through the veins, and in the extremities the blood seeps into the parts connected to them. Therefore, although bones do not receive nutriment through external veins, they nevertheless do receive it from internal veins through internal pores.

3. To the third argument one must reply that the digestive power is weak in the third digestion, and this is why it should not receive much at one and the same time, because if it did receive a lot, and more than it could digest, then some disability would occur in it. This is why there are veins in which the blood is stored and gradually and successively sent to the members. And this is why there are no veins in the first or third digestion, but only in the second.

9. Cf. A., *DA* 12.2.1.99 (*SZ* 2: 930–31).

Question 2: Whether veins exist in every part of the body.

Further one asks whether veins exist in every part of the body or are extended throughout the entire body.[10]

1. It seems so. A vein is the site of blood, but there is blood in every part of the body. Therefore, veins are also in every part of the body.

2. Moreover, a fever generated in the blood is only generated from the blood within the veins. But this would not be so if there were blood somewhere other than in the veins. Therefore, everywhere there is blood, there is a vein. But blood exits from any part of the body that has been punctured. Therefore, there is a vein in every part of the body.

3. Besides, according to the physicians there are four humors in addition to the veins. The first moisture[11] is at the ends of the veins. The second is in the members but is not disposed for conversion and is called "dew" [*ros*], because it is white and it moistens the members. The third is in a disposition close to conversion and is called *cambium*.[12] And the fourth is called gluten [*glutinum*]. But three arise from the first. Therefore, where the first one exists, so do the other three. But the first one is in the veins, and therefore, etc.

The opposite is evident to the senses.

To this one must reply that the veins are not extended throughout the entire body because in everything having diverse parts one finds one part different from another; otherwise, the whole would be homogenous. But an animal is just such a totality, and therefore one finds in it diverse parts with distinct locations. But if the veins were extended throughout the entire body, one would find simple flesh in every part of the body. But this is not true, and therefore, etc.

Again, according to the Philosopher in the third book of *On*

10. Ar., *HA* 3.2–4 (511b1f.); A., *DA* 3.1.1–2.1–28 (*SZ* 1: 344–58).

11. To assist the reader in tracking A.'s use of language, "humor" is reserved for the Latin *humor* in this passage and *humiditas* is rendered "moisture."

12. Latham (s.v. *cambium*) identifies a medical usage for *cambium*, stemming from the thirteenth century, meaning "transformable matter."

Animals, in the chapter on the veins, the (larger) veins extend neither to the womb nor to the bladder, and therefore, etc.[13]

1. On to the arguments. One must reply that the veins are said to be the site of the blood because the blood is distributed through the veins to individual members. Nevertheless, the blood seeps from the extremities of the veins to the fleshy exterior parts (that is, externally), is drunk into their pores, and nourishes them later. And this is why it is unnecessary that there be a vein everywhere there is blood, but rather that wherever there is blood it is either in a vein or seeps from a vein.

2. To the second, one must reply that a fever generated from blood is generated within veins, because, if that blood is corrupt, some injury or corruption will immediately befall the other parts since it is the nutriment for the individual members.[14] Abscesses [*apostemata*] occur outside the veins as a result of corrupt blood, as do skin eruptions [*impetigo*] and other illnesses, although not fevers. So it does not follow from this that there be blood outside the veins just as there is inside.

3. To the third argument one must reply that although three moistures arise from the first, it is nevertheless not necessary that they exist in the same location.

Question 3: Whether the veins arise from the heart or the liver.

Next one asks whether the veins arise from the heart or the liver.

1. And it seems that they arise from the liver. Because the first principle of a thing is that by which, once it is first set in place, the thing is set in place, and upon the removal of which the thing is removed. But once the liver is set in place, the blood is set in place, and, once it is removed, so too is the blood removed; therefore, etc.

2. Again, the goodness of the branch follows upon the goodness of the root, and an injury appears in the branch following

13. The word "larger" has been added by the editor of the text to make sense of the passage, based on Ar., *HA* 3.4 (514a23f.).

14. A., *DA* 3.2.3.105, 3.2.6.146 (*SZ* 1: 396, 415).

upon some bad quality in the root. And conversely, goodness follows not from some bad quality, but from the good. But whatever injury or weakness arises from the blood, the liver and not the heart always provides a remedy. But this would not be so were the liver not the root of the blood and consequently also of the veins, in which the blood is contained.

3. Moreover, the branches are moved by the movement of the root, and not contrariwise. If, then, the veins should have their origin in the heart, then all the veins would be pulsating veins.

4. Moreover, the nearer the root in a plant is to its principle, the thinner it is, and the more remote it is, the thicker it is. But veins in animals are analogous to roots in plants, because, just as aliment is conveyed to the members through the veins, so too is it conveyed to the branches through the roots. But the veins near the liver are thinner than those near the heart, and therefore, etc.

The Philosopher says the opposite.[15]

And it appears by this argument that the more something participates in something else the more it approaches the nature of that same thing's principle. But the heart participates more in blood than does the liver. Therefore, the heart is more the principle of blood than the liver is, and, as a consequence, it is more the principle of the veins.

Moreover, that at which a thing terminates and which is not penetrated by it is a thing's principle. But the veins terminate at the heart, and it is not penetrated by them. Nevertheless, the liver is penetrated by them, that is, by the veins, and therefore, etc.

There is a great controversy between the physicians and the Philosopher or philosophers surrounding this question.[16] Some of the physicians, like Galen and his followers, posit that the veins arise at the liver because the generation of the blood occurs in the liver and, as a result, the liver is the principle of the veins through which the blood is borne.

15. Ar., *HA* 3.4 (514a23f.); *Part. An.* 2.9 (654b11), 3.4 (666b33f.), 3.5 (667b16f.); *GA* 4.8 (776b12f.).

16. Galen, *De usu part.* 4.12 (Kühn 3: 297); Avic., *Can. med.* 1.1.5.1; *DA* 3.1 (fol. 31raA-vaC); A., *DA* 3.1.5-6.43-65; 16.1.16.84 (*SZ* 1: 365-76; 2: 202).

Moreover, it is very apparent in plants that the roots are thinner and more subtle at the base or trunk. But the veins that penetrate the liver are subtle and thin, and the vein connected to the heart is large in the manner of a trunk. Therefore, the liver is the veins' root.[17] And in addition to this the liver has only coagulated blood, which is why it has the color of blood. But the heart is just like coagulated melancholy [*melancholia*], whereby it is hard and earthy.

Nevertheless, this is contrary to the view of the Philosopher and Avicenna.[18] For Avicenna says that one ought to hold to the opinion of the first philosopher and the prince of philosophers (namely, Aristotle) regarding this part, who claims that the heart is the first principle of the veins, as does the physician, Master Nicholas of Poland, as well.[19] And the reason is that in an animal, since there is one substance that has several parts which have various powers, it is necessary that there be one power that is the principle of the others and from which the others receive some influence. This, however, is the vital power, because without the vital power the nutritive power does not operate in an animal. Therefore, it is necessary that the organ for this power have a basis for beginning, that is, have the basis for a principle, higher than the organs for the other powers. But the organ for the vital power is the heart, and therefore, etc.

Moreover, the veins are capable of containing and bearing the blood. But although the blood is generated in the liver, nevertheless this does not occur without the assistance of heat, and it can descend to the members only with heat mediating. But the principle of heat is in the heart;[20] therefore, in the liver there is no power that can nourish through the mediation of

17. Cf. A., *DA* 13.1.7.55 (*SZ* 2: 1009).

18. Avic., *DA* 3.1 (fol. 31rbB).

19. The identity of Nicholas of Poland remains uncertain. The editor notes as one possibility a Dominican, Nicholas, who was named prior provincial of Poland in 1249. Another and perhaps better possibility, however, is the physician Nicholas of Poland, author of two works: *Experimenta* and *Antipocras*. This Nicholas composed these works in the first half of the fourteenth century, however. If our text refers to him, then it must be a later interpolation. See the prolegomena to Filthaut's edition of *QDA*, xlvii, 34-46.

20. Cf. A., *DA* 13.1.1.9 (*SZ* 2: 988).

the blood that has been generated in it except through the influence it receives from the heart.

And this is why the Philosopher claims that the heart is the principle of the veins.[21] For he claims that it is the first part in the animal's generation upon which all the other parts are based and without which no other part can perform its proper operation. Therefore, so that the view of all those discussing this may be reconciled, "origin" can be distinguished into two types: one is virtual and radical, and the other is corporeal. In the first way the veins arise from the heart, but they arise from the liver in the second way. Thus the branching of the veins clearly and in an evident way begins at the liver, just as the appearance of branches from the trunk of a tree is evident, although in reality it proceeds from the root, but nevertheless appears clearly to us to be from the trunk. So too for the heart, that the veins proceed from it in reality, although they appear to do so from the liver, but because physicians are [dabblers in] sensibles, this is why, etc. The Philosopher speaks in the first way; the physicians in the second.

In this way one can disclose the arguments. The arguments to the first part prove that the veins arise corporeally and immediately and in an evident way from the liver. Yet the arguments to the contrary prove that the veins arise virtually or radically or with respect to origin from the heart.

Question 4: Whether the arteries are necessary to an animal.

Again one asks whether the arteries are necessary to an animal.

1. It seems not. For the arteries exist for the sake of conveying life and the natural heat.[22] But natural heat and life exist in every part of the body. Therefore, since these exist in the animal without the arteries, the arteries will not be necessary.

21. Ar., *Part. An.* 3.4 (666a10f.); *GA* 2.4 (740a17f.); A., *DA* 3.2.6.145, 10.2.3.61, 13.1.4.25, 16.2.3.112–13 (*SZ* 1: 415, 851; 2: 995–96, 1217–18).
22. A., *DA* 1.1.5.75 (*SZ* 1: 72–73).

2. Moreover, according to the Philosopher in *On the Difference between the Soul and the Spirit,* the soul is united to the body by a mediating spirit.[23] But the soul is united to every part of the body; therefore, there is spirit in every part. But it is not in the arteries, and therefore the arteries are not necessary for the sake of the spirits. But they are posited for the sake of the spirits; therefore, etc.

3. Besides, life and heat exist in a plant as well as in an animal. But arteries are not necessary for the plant, and are therefore not necessary for the animal either.

To the contrary. An animal cannot exist without life and heat. But the vital power and the natural heat are borne from the heart to the individual members by the arteries. Therefore, the arteries are necessary.

One must reply that the arteries are necessary because two things are required for the animal's preservation: namely, the restoration of what has been lost and the maintenance or preservation of what has been acquired. But the veins, which are quiet and non-pulsating, are posited for the sake of the restoration of what has been lost, so that the blood—that is, nutriment—may be conveyed through them to the individual members.[24] But the arteries, which are pulsating veins, are posited for the preservation of what has been acquired, so that through them the vital power and the natural heat and spirits, without which no operation occurs, descend from the heart to the individual members. Thus, just as the <lungs'> arteries are necessary to attract air to cool and temper the heart, so too the arteries proceeding from the heart are necessary for [delivering] the power of its influence to the other parts.[25] Thus the Philosopher says, in book fourteen,[26] that every corporeal operation (sensation and life are corporeal operations) occurs with the mediation of the heart and the heart's influence on the other parts. This therefore occurs through a corporeal medium, and that

23. Although ascribed here to Ar., the work is actually by Costa ben Luca, *Excerpta e libro Alfredi Anglici* De motu cordis, *item* De differentia animae et spiritus liber, trans. Johanne Hispalensi, c. 4, ed. C. S. Barach (1878), 136.

24. A., *DA* 1.1.5.75, 1.2.20.381, 1.2.21.397 (*SZ* 1: 72–73, 189, 195–96).

25. "Lungs'": added by the editor.

26. Ar., *Part. An.* 3.4 (666a11f.).

medium is the artery that is filled with spirit and natural heat. And because of the continuous pulsing of the spirit in the heart and in the arteries, every artery has two tunics.[27] But a vein has only one tunic, except for that one that is directed from the heart to the lungs. For it contains not only blood but also spirit, and this is why it has two, just as an artery does.

Another reason is that it conveys nutriment. And it must be firm on account of this, lest it burst on the way from the weight of the nutriment that travels through it and lest, as a result, the entire animal fail, etc.

1. On to the arguments. To the first, one must reply that although the natural heat and vital power are in every part, these nevertheless are connected only through the influence they receive from the heart. For heat is flowing from the heart and the vital power flows from the heart through the arteries, and this is why, etc.

2. To the second argument one must reply that the soul is not said to be united to the body by spirit, as if the spirit were the medium, but rather that it operates with the mediation of the spirit. But spirit alone is conveyed through the arteries. For even if other parts operate by means of spirit, these spirits nevertheless flow through the arteries, and this is why, etc.

3. To the third argument one must reply that animal life has a more perfect regimen than plant life, and this is why many things are necessary for the animal that are not necessary for the plant.

Question 5: Whether the arteries take their origin from the heart or the brain.

One asks whether the arteries take their origin from the heart or the brain.

1. It seems it is from the brain.[28] An artery is a vehicle for spirits. Therefore arteries naturally arise from a place where spirit

27. "Tunic": *tunica,* a type of membrane whose name suggests a resemblance to woven cloth.

28. Cf. A., *DA* 1.2.20.381–83 (*SZ* 1: 189–90).

abounds. But spirit naturally arises abundantly from the area of the brain, as is said in the book *On the Motion of the Heart*;[29] therefore, etc.

2. And the argument is confirmed in this way. The lesser world imitates the greater world,[30] but transparent and gleaming bodies (such as air, fire, and supercelestial bodies) are especially abundant in the superior part of the world. Therefore, it will be this way in the human, who is a lesser world. Now the human is called a microcosm, and microcosm is said from *micros,* which means "lesser," and *cosmos,* which means "world"; it is a sort of a lesser world, just as the "macrocosm" is the greater world.

3. Besides, the arteries arise from that part in which they [the spirits] especially abound, but this is from the area of the brain. For its one web, which is called the "marvelous net" [*rete mirabile*], is made from arteries.

To the contrary. That which comes from a source attests to its source. But the arteries have a tough substance and composition, whereas the brain is soft. They therefore cannot arise from it, since like arises from its like.

One must reply that both virtually and corporeally the arteries have their origin in the heart. This is clear first from the fact that all the powers that exist in an animal are rooted in the vital power, which is in the heart. Second it is evident that the artery is the vehicle for the natural heat, the spirits, and the vital power. An artery draws its origin corporeally from that area in which these are generated. But these are all generated in the heart, and this is why, etc. Thus, just as the vein takes its radical and virtual origin from the heart (because this is the first part in an animal) but takes its corporeal origin from the liver (in which is generated the blood for which veins provide a vehicle), so too does an artery take both its corporeal and virtual origin from the heart: the virtual because the heart is the first member, and the corporeal because the generation of those things for which the artery is the vehicle lies in the heart.

29. Alfred of Sareshel, *De motu cordis* 10.14, ed. C. Baeumker (1923), 45.
30. A., *DA* 16.1.11.60 (*SZ* 2: 1186–87).

1. On to the arguments. To the first, one must reply that spirit is of two types, namely, the vital and the animal, which is itself divided into sensible and motive. Animal spirit abounds in the brain, and the author indicates this in the book *On the Motion of the Heart.*[31] But the vital spirit abounds in the heart.

2. To the second argument one must reply that the heart in an animal is analogous to superior or supracelestial bodies,[32] as the Philosopher indicates in book two of *On Heaven and Earth.*[33] For although the solar body is not the center of the world in the way that the heart is located in the animal, it is nevertheless the center of nature.

3. To the third argument one must reply that some of the arteries are just like roots and these are located near the left ventricle of the heart. Other arteries are like branches, and the web in the head is covered with these. In this way a solution to the whole is apparent.

Question 6: Whether the nerves are necessary.

Next one asks about the nerves. And first, whether the nerves are necessary.[34]

1. It seems not. Because a power that exists indiscriminately in every part does not require a determinate part as its bearer. But sensation exists in every part of an animal. Therefore, it does not require a part bearing it but only one for sustaining it. But nerves are only posited for the sake of bearing sensation and are therefore posited superfluously.

2. Besides, every part necessary to an animal is ordered to sense and exists for the sake of sense. But we do not experience [*sentio*] by means of hairs and nerves, according to the Philosopher in the third book of *On the Soul,* and therefore, etc.[35]

31. Alfred of Sareshel, *De motu cordis* 10.15, ed. C. Baeumker (1923), 46.

32. A. changes from his previous *supercaelistia* to *supracaelistia.* The meaning would appear not to change.

33. Ar., *DC* 2.2 (284b6f.).

34. Ar., *HA* 3.5 (515a27f.); Avic., *Can. med.* 1.1.5.1–6; *DA* 2.8 (fol. 47vb–48va); A., *DA* 3.1.7.66–70 (*SZ* 1: 376–78).

35. Ar., *De anima* 1.5 (410a30f.), 3.13 (435a24f.).

To the contrary. "Nature does nothing in vain."[36] But nerves exist in every animal; therefore, they do not exist in vain, but are necessary.

To this, one must reply that there are four powers in a human: the nutritive or natural, which is in the liver; the vital, which is in the heart; the animal, which is in the brain; and the generative, which is in the parts designated to generation (specifically the testicles in men and women).[37] In just the way an artery corresponds to the vital power (for the power flows through the artery to the individual members), and a vein corresponds to the natural power, so too does a nerve correspond to the animal power. And because the animal power consists in sense and motion—for by these we distinguish an animal from a non-animal, as the Philosopher says in the first book of *On the Soul*— this is the reason why some nerves are sensitive and some are motive nerves. Therefore, just as sense and motion are necessary to an animal, so too are the nerves.

Moreover, bone is not directly connected to bone in an animal because, if it were, then no flexion could occur between the bones. This is why they are positioned appropriately next to one another and are joined by nerves.[38]

1. On to the arguments. To the first, one must reply that although sense exists in every part in terms of participation, it nevertheless exists in one part in terms of its roots [*radicaliter*], for it flows from here to the other parts by means of specific members. For example, the power of sight flows to the eye through the optic nerve. Thus sense exists in one part only for bestowing [a power], as in it does in the heart, and in another part only for receiving [a power], and in still another part it exists both for giving and receiving.

2. To the second argument one must reply that some nerves carry sense and motion, and these are very sensitive [*sensibiles*].[39]

36. Cf. Ar., *De anima* 2.2 (413b2), 3.12 (434a30).

37. On women's "testicles" see also *DA* 9.1.1.1, 9.2.2.96, 9.2.3.109, 15.1.6.31, 15.1.8.52, 15.2.7.117 (*SZ* 1: 774, 810, 816; and, 2: 1098, 1107 and 1138).

38. *Nervus* can mean both sensory nerve and "ligament" or "sinew." A. comments on this immediately below.

39. Cf. A., *DA* 1.2.1.113, 124, 12.1.2.26 (*SZ* 1: 88, 92; 2: 904).

Thus an injury to the arm muscles is especially onerous owing to the branching of the nerves. Other nerves only join bones together, but these are not properly called nerves, as the Philosopher notes.[40] Nerves like this are hardly sensitive at all, so that the bones can be moved without pain in the joints, for otherwise intolerable pain would be caused by their continuous motion. This is why they do not have sensation.

Question 7: Whether the nerves take their origin from the brain or the heart.

One asks next whether the nerves take their origin from the brain or the heart.

1. It seems, from the heart. For "flesh is the medium of touch," according to the Philosopher in the second book of *On the Soul,*[41] and an organ is something internal, "near the heart," as the Philosopher says in *On Sense and the Sensed.*[42] But a nerve is the organ of touch according to Avicenna.[43] Therefore, the nerves take their origin from the heart.

2. Besides, the nerves are very sensitive; therefore, they take their origin from a very sensitive part, and the heart, but not the brain, is such a part. Therefore, etc.

To the contrary. The nerves arise from that part from which sense and motion flow, and they carry sense and motion as well. But this is immediately from the brain; therefore, etc.

Moreover, the nerves arise from that part at whose injury the operation of sense and motion cease. But this occurs following injury to the brain; therefore, etc.

One must reply, as was done earlier regarding the veins, that origin is spoken of in two ways: one is virtual and radical, and this is how the nerves and all the official parts of an animal arise from the heart; another is corporeal and immediate, and this is

40. Ar., *HA* 3.5 (515b10f.); cf. A., *DA* 3.1.7.67 (*SZ* 1: 377).

41. Ar., *De anima* 2.11 (423b26); A., *De anima* 2.3.31–34.

42. Ar., *De sensu et sens.* 2 (439a1f.).

43. Avic., *Can. med.* 1.1.5.1; *De anima* 2.3, 5.8.

how the nerves arise from the web of the brain itself and from its *nucha.*[44]

The first two arguments proceed according to the first way, and the arguments to the contrary proceed according to the second. And thus the solution is clear.

Question 8: Whether the nerves are of a melancholic or phlegmatic complexion.

One asks next about the complexion of the nerves—whether they are of a melancholic or phlegmatic complexion.[45]

1. It seems that they are of a phlegmatic complexion. For it is said in the text that a watery moisture surrounds the nerves.[46] But phlegm is harmful owing to its viscosity, blood is harmful owing to its abundance, bile [*choler*] owing to its sharpness, and black bile [*melancholia*] owing to its harshness. Since viscosity and wateriness are attributed to phlegm, the nerves will have just this sort of complexion.[47]

2. Moreover, a thing attests to its origin. But the nerves arise immediately from the brain, which is cold and moist.[48] The nerves will therefore be of the same sort.

To the contrary. Whatever is made of a hard substance has a melancholy complexion. But the nerves are of this sort, and therefore, etc.[49]

One must reply that the nerves can be considered in two ways: either in terms of the matter from which they arise or in terms of the type of complexion they have. If in terms of the matter, then the nerves are cold and moist, because if they were cold and dry they would be dark and hard in the manner of earth or rock.[50] Nerve, however, is transparent and clear, which

44. *Nucha*: Either the marrowy matter of the brain itself or, more likely, the spinal chord. Cf. *DA* 1.2.11.257 (*SZ* 1: 141).

45. Cf. A., *DA* 12.1.2.26 (*SZ* 2: 904); Avic., *DA* 3.1 (fol. 31rbB).

46. Ar., *HA* 3.5 (515b16f.).

47. Cf. A., *DA* 3.2.3.105–11 (*SZ* 1: 396–99).

48. Cf. A., *DA* 1.2.18.355–69 (*SZ* 1: 178–84).

49. A., *DA* 12.2.3.113 (*SZ* 2: 936).

50. Cf. A., *DA* 1.2.18.357, 12.2.5.136 (*SZ* 1: 179; 2: 944–45).

attests to the nature and matter of water and, as a consequence, to the nature and matter of phlegm. If, however, they are considered in terms of the type of their complexion, they are cold and dry, and this is why they suffer injury from excessive moisture, as is clear in the case of paralysis and spasm.[51] According to the physicians,[52] this is also very clear among animals, because goose flesh, in terms of the matter, is cold and dry, and it therefore generates a sickness on the basis of the cold and dry matter—quartan fever, for example. But in terms of its composition it is moist and warm, and is the opposite of chicken flesh.[53]

With this a response to the arguments is clear. The first ones prove that the nerves are cold and moist, and this is true in terms of the matter. But reason proves the opposite, that they are cold and dry, and this is true in terms of the manner of their composition or complexion.

Question 9: Whether the bones are necessary.

One asks about the bones. First, one asks whether the bones are necessary.[54]

1. It seems not. For the entire body and every single part of the body is ordered toward sensation, but bones are insensible, according to the Philosopher in the third book of *On the Soul*.[55] Therefore, bones are unnecessary.

2. Besides, every part of the body is capable of receiving aliment and the vital spirit. But the bones, since they are solid, are not capable of receiving aliment or spirit, because they are opaque bodies. Therefore, etc.

The opposite is apparent to the senses. For nature would not always produce bones in an animal unless they were necessary, and therefore, etc.

One must reply that the bones are necessary for three rea-

51. Perhaps: "convulsion," *spasmus*. 52. Cf. A., *DA* 23.1.24 (*SZ* 2: 1558).

53. Cf. A., *DA* 23.1.24.116 (*SZ* 2: 1628).

54. Cf. Avic., *DA* 12.9 (fol. 48va–vb); A., *DA* 3.2.1.71–78, 12.2.6.137–44 (*SZ* 1: 379–83; 2: 945–49).

55. Ar., *De anima* 3.13 (435a24f.).

sons:[56] one is that they are the foundation for the entire body. For everything that flows requires something solid, which is its boundary, so that it will have a determinate shape, and this is why flesh and fat and like things have the bones as their foundation.

Another reason is that they will be the armor and defenses for the principal parts. Now the skull is the protector for the brain, and the pectoral bone is the protector of the heart, and so on for the others. For just as a wise king, when he builds a city or a fortress, not only fashions them out of things from which a fortress or city is constructed, but actually even makes defenses to protect what he has built, so too nature—wise in that very wisdom through which it was born to obey its Creator and which is the overseer and the artisan of all things—fashions bones in the animal as a defense for the body, etc.

The third reason is that they are capable of supporting the body. For if there were no bones in the feet and in the thighs and legs, they would not support the body.

1. On to the arguments. To the first, one must reply that some parts of the body, like the organs, are ordered to sense, and others are ordered to be capable of sustaining and supporting the organs, and this is why, although the bones are insensible, they nevertheless are not superfluous, because they support the body.

2. To the second argument one must reply that, although the bones are solid when compared to the flesh and similar parts, bones are nevertheless somewhat porous, and as a result are capable of receiving the nutrimental moisture, as much as is sufficient for them, and this is why, etc. In this way a solution is evident.

Question 10: Whether the bones have sensation.

One asks next, based on this, whether bones have sensation.

1. It seems so. Sensation is present in anything that has a power that can sense what is agreeable or injurious to it. But a power of this sort is in bones, and therefore, etc.

56. Cf. A., *DA* 1.2.1.118 (*SZ* 1: 90).

2. In addition, pain is a sensation of something that is contrary, according to the second book of *On the Soul*[57] as well as according to Galen. But sometimes the bones experience pain. Therefore, there is sensation in them.

3. Moreover, the same argument applies to both bones and teeth since, according to Avicenna, the teeth share the nature of bones. But we experience sensation through the teeth, and therefore also through the bones.[58]

The Philosopher says the opposite in the third book of *On the Soul*[59] as well as in this book.[60]

One must reply that bones do not experience sensation in their own right [*per se*]. And the reason for this is that sense is receptive of sensible species apart from matter, and this is why it is required that there exist a mean proportion of sensibles in sensibility. For which reason "plants do not have sensation," according to the Philosopher in the third book of *On the Soul*,[61] because earthiness abounds in them and they do not receive [sensible] species apart from matter but are rather altered by matter. But bones are likened to plants by the Philosopher in the book *On Plants*,[62] and they do not exist in a medium conducive to tangibles.[63] They therefore receive [sensible] species in their own right [*per se*] in a material fashion, and this is why they do not have sensation. And an indication of this is that if the web surrounding the bones is removed, and they are struck with a painful blow, they do not feel it. Yet because the bones are often joined to a nerve and fleshy parts, this is why it appears to us that sensation occurs in bones, because it occurs through the sensation of the other parts. Thus they have sensation by vir-

57. Ar., *De anima* 2.2 (413b23f.); Galen, *De elementis* 1.2.3; *De locis affectis* 2.1.

58. Avic., *Can. med.* 1.2.2.19; *DA* 12.13 (fol. 50rb). Cf. Galen, *De ossibus* 5.

59. Ar., *De anima* 3.13 (435a24–25).

60. Ar., *HA* 3.3 (514a21f.).

61. Ar., *De anima* 3.13 (435b1–2).

62. No clear analogue can be found in this Ps. Aristotelian work, although the reader may want to examine *De plantis* 1.3 (313a18–22; Meyer, 1841).

63. "In a medium conducive to tangibles": a troubled phrase, *in medio tangibilium.* Recall that touch consists in a mean proportion of sensible or tangible qualities. See *QDA* 1.17 (92, 39–40).

tue of the nerve-filled parts and the pannicular-membranes attached to them, but not in their own right [*per se*].

1. On to the arguments. To the first, one must reply that the first act of actual sensation exists in the bones, because, just as life is diffused through all the parts of the body, so too is sensation. But the second act of sensation only exists in an organic part attendant upon the sense itself. This is why, owing to the first act of actual sense and of life, it has a power of discriminating between what is suitable and of repelling what is harmful, but that sensation nevertheless does not exist there. And beyond this one can say that a power receptive to what is suitable and that repels what is harmful exists in a plant, and that sensation nevertheless does not exist there either.

2. To the second argument one must reply that bones do not feel pain *per se* but only *per accidens* as a result of the pain in other parts, just as blood does not exist in the brain in its own right [*per se*] and yet nevertheless, as a result of a lesion in other, adjoining parts, swelling and a flux of blood can occur in parts like these which do not have blood in their own right [*per se*].

3. To the third argument one must reply that teeth adhere firmly to fleshy parts with certain nerves in between them [*mediantibus*], and this is why the teeth know pain—because the pain in the nerve travels to the teeth, but this occurs *per accidens,* specifically, by virtue of the nerves. And because dry things greatly retain or incorporate impressions received, pain in the teeth is therefore very strong, because they are dry and hard. And this is why the pain imprinted on them lasts a long time.

Question 11: Whether cartilage is necessary.

Further one asks whether cartilage is necessary.[64]

1. It seems not. Whatever serves neither motion nor sense is unnecessary to an animal. But cartilage is of this sort, and therefore, etc.

64. Ar., *HA* 3.8 (516b31f.); Avic., *DA* 12.9 (fol. 48vb); A., *DA* 1.2.1.120–23, 3.2.1.76–77 (*SZ* 1: 90–92, 383).

2. Moreover, "nature does nothing in vain and does not lack what is necessary"; but in many animals cartilage is absent and yet nature does not lack what is necessary. Therefore cartilage is not necessary.

As previously, the opposite is evident.

One must reply that cartilage is necessary. For some parts need to flex and to be stabilized. But these two occur together in neither flesh nor bone. For there is no stability in flesh and no flexion in bone, and this is why it is necessary that there exist some part in which the two occur together, and cartilage is like this, and this is why there is cartilage in the nose and in the ear. Therefore, cartilage takes the place of bone for the purpose of stability and takes the place of nerve in them for the purpose of flexion. Similarly, it takes the place of flesh for the purpose of softness.

1. On to the arguments. To the first, one must reply that although cartilage is not an organ of sense or motion, it nevertheless supports organs of sense and motion.

2. To the second argument one must reply that some parts are intended to be so solid that they are inflexible; otherwise, the leg and the hip bone would not support the body, and this is why cartilage does not exist in all the parts.

Questions 12–16: On hair.

One next inquires about hair: why it exists on brute beasts from birth, but on a human is present only on certain determinate parts, like the head, eyebrows, and eyelashes.[65]

Second, one asks why the hairs on the head do not always grow, namely, in discrete number, whereas the hairs of the beard grow and increase in number.

Third, one asks why women do not have a beard, as do men.

Fourth, one asks why hair color changes more during old age than it does during youth.

65. Cf. Ar., *HA* 3.10–12 (517b3f.); Avic., *DA* 4.3 (fol. 31vaA–32raA); A., *DA* 3.2.2.79–99 (*SZ* 1: 383–93), and *QDA* 1.25–28.

Fifth, one asks why hairs that have been cut off grow back but feathers do not.

To the first, one must reply that of all the animals the human has parts that are moister at the beginning of his life. Hairs are actually caused by hard, dry, and solid matter, and this is why there is more hair on other animals at birth than there is on a human, since they have more dry matter which provides the material for hair, whereas in a human more hair exists on hard and more solid parts, like the skin and the head's skull and the eyelashes and eyebrows.

In addition, the eye is the most fragile of the external members, and this is why nature provides hairs on the eyebrows from the beginning—as a protection for the eyes in order to defend them better from external injury such as a bit of straw falling into the eye or the like, etc.

To the second question one must reply that heat is required for the generation of hair, a heat that expels the fumes that form the material for hair and are a material suited to the generation of hair. Yet we know a great number of hairs is caused by the quantitative division of matter because, as the Philosopher says in the third book of the *Physics,* number is caused by the division of a continuous [quantity].[66] But the growth in quantity is caused by growth in power. Now, then, heat is powerful in youths and the matter for hair is abundant, namely, the dry fumes, and this is why the hairs on the head increase in number all the way through youth.[67] But in older people the heat is weakened, and a dry and earthy superfluity are more abundant. Thus the weak heat cannot propel the matter for hair as far as the head. Rather, the heat is itself turned back as far as the cheeks and the beard, and the matter, which is heavy, descends that far because the weak heat cannot raise it up; nevertheless it accomplishes what it can and raises the heat to the beard as well as to a nearer location.

And if one objects to this, that therefore the chest and the back will be hairier than the beard because they are nearer, etc., one must reply that this is not correct because although the

66. Ar., *Phys.* 3.7 (207a33f.); A., *Phys.* 3.2.12.
67. Cf. A., *DA* 12.1.5.68–74 (*SZ* 2: 919–22).

matter goes there, much of it is not retained but escapes due to the looseness and openness of the pores there, but the beard, which is bony[68] and cold, retains the fumes well until they are converted into hairs, etc. And, moreover, because a warm fume always moves upwards, as it were, this is why more of it goes to the chin than to the chest. Actually it cannot go to the head because it does not have enough power. And this is why the hairs of the beard multiply among old people, whereas in youths they multiply on the head, and the reason is evident.

There is another reason, according to the physicians, which is that semen is drawn from the brain to the seminal vessels, through the cheeks and the beard. And this is why when the generative power begins to dominate and the semen increases in a person, the hairs then begin to grow on the beard, because the heat is rising from the genital members and from the heart, and the spermatic spirit turns back to the brain as far as the beard, and there it is strengthened and generates hair. Thus, when the hairs of the beard are greatly increased in number, then this is an indication that old age is quickly approaching, because the heat is lessening, etc.

To the third question one must reply that moisture, which blocks heat, abounds in women and this is why they do not grow beards or hair on their other parts, because the matter cannot be raised up owing to the coldness of their complexion, except among those in whom heat greatly abounds. Sometimes women like this produce a beard[69] and, similarly, hair on the other parts, like in the armpits, the groin, and around the mouth or the opening of the vulva, owing to the strong heat resulting from the striking of the penis against the vulva during intercourse etc.

To the fourth question one must reply that the cold and moist are the cause of whiteness.[70] Thus congealed water turns white as do snow, hail, and crystal, which is nothing but congealed water.[71] But in old age the heat fades, and as a result the cold is increased, so that the heat does not have the power to expel all

68. Bony: *ossuosa,* apparently dense, like bone.

69. See A., *DA* 1.2.9.227 (*SZ* 1: 131).

70. Cf. A., *DA* 19.1.7.34–37 (*SZ* 2: 1349–51).

71. Crystal: *crystallus,* more commonly *crystallum.*

the matter of the hair but only that matter that is thinner and more watery. And this is why hairs in old age turn white, owing to the weakness of the heat and the wateriness of the moisture, because this is when excessive cold tinges and whitens them.

To the last question one must reply that feathers are created from a more phlegmatic moisture than that which gives rise to hair. An indication of this is that they resemble organic parts.[72] Now the feathers do not have the same nature in every part, and thus, just as an organic part, like a finger that has been cut off, does not grow back, neither does a feather that has been cut off return, because there is not sufficient power in the remainder of the feather still affixed to the body to produce a part similar to the one cut off. If, however, a feather is completely removed, another can grow in its place because the power capable of producing a whole feather still remains in the entire body, although the power capable of producing another one does not exist in a single part. Therefore, when the feathers are removed at the root they return, as is evident in geese that have been plucked to make pillows, etc. And the reason is evident.

Question 17: Whether skin and web [tela] *are necessary to an animal.*

Further one asks whether skin and web [*tela*] are necessary to an animal.[73]

1. It seems not. Whatever impedes sensation is unnecessary for an animal. But the skin impedes sensation, for by interposing itself it keeps the object at a distance from the organ of sense, and therefore, etc.

2. Besides, the skin is at an even greater distance from the principal parts than is the flesh. But flesh is not necessary because it ebbs and flows, and in some parts there is no flesh at all; therefore, etc.

72. Cf. A., *DA* 7.3.4.169, 19.1.6.31 (*SZ* 1: 663; 2: 1347).

73. Cf. Ar., *HA* 3.13–16 (519a30f.); A., *DA* 3.1.1.2, 3.2.2.84 (*SZ* 1: 345, 385–86). "Web" seems to be interchangeable with "membrane" at this juncture of the *QDA,* although *tela* is not as commonly used as *membranum.*

3. The same seems to hold for a web, for "nature does nothing in vain." But the interior members are adequately preserved by the exterior members, and therefore the web is not needed.

To the contrary: these are necessary because these both exist in animals, which would not be so if they were unnecessary.

One must reply that skin and web are necessary. And the reason is that every moisture not contained in something solid flows and is not bounded. Because there are fluids [*humiditates*] arising throughout an animal's body from the nutriment, it is necessary for the animal's body to be wrapped with something solid, so that the fluids do not flow out of it. That which contains them is the skin. In addition, the soft members of an animal are prone to injury, and therefore they need some things that are solid to protect them from injuries. The skin is such a solid, and therefore, etc.

One can be persuaded by the same argument for the web, because the principal members, like the heart and the brain, are enclosed by hard parts—the heart by the bones of the chest and the brain by the skull. Therefore, these members need some parts thinner than bone so that neither the principal members nor, as a result, all the others, will be harmed by their hardness, because once a principal member suffers something, all the rest suffer as well. The webs, however, are parts just like this, and this is why they are surrounded by a web, that is, by the tunics of webs or pannicular-membranes.

1. On to the arguments. To the first, one must reply that something can be a medium between a sense and its sensible in two ways. Either it is there in such a way that it will be susceptible to the sensible species by means of a medium in the sense. Such a medium does not impede sensation but is rather necessarily required by it, because "a sensible placed upon a sense is not perceived," according to the Philosopher's way of thinking.[74] Thus a medium is necessarily required so that, if something were in the heaven and the medium were a vacuum, it would not be seen. And the Philosopher proves this. [Or] some-

74. Ar., *De anima* 2.11 (423b6f.).

thing else can be a medium which is not susceptible to the species but conveys it to the sense, like rock or wood, and such a thing impedes [sensation]. Skin is a medium in the first way, keeping the object at a distance from the sense, and this is why it does not impede sensation but strengthens it.

2. To the second argument one must reply that something is at a distance from a principal part in two ways: either according to its location or according to its nature. According to location, the skin is more distant from the principal parts than is the flesh, but not according to nature, because skin is made from parts that are more spermatic than is flesh, and this is why flesh that is cut away grows again, but skin does not, except when it is closely joined to fleshy parts, because facial skin and the skin of the prepuce or the labia does not grow back when it has been cut away.

3. To the third argument one must reply that an impediment can arise from the principal parts in two ways: either from outside—and the skin and the bones are opposed to such an impediment—or an impediment can arise from the hardness of the adjacent members, and the webs and the pannicular membranes are opposed to this impediment. Again, the webs exist for the sake of something else, namely, so that the interior fluids will not flow out too much, that is, due to either some intrinsic or extrinsic cause.

Question 18: Whether blood is necessary for an animal.

One asks further about the blood. And first, whether blood is necessary for an animal.[75]

1. It seems not, because "we exist from and we are nourished by the same things."[76] But blood is posited as necessary only for the sake of the nutriment and its preservation. Therefore it is not necessary [to the animal].

2. Moreover, in the seventeenth book of this work the Philos-

75. Ar., *HA* 3.19 (520b10–521b3); Avic., *DA* 4.3 (fol. 32raA); A., *DA* 3.2.6.142–48 (*SZ* 1: 413–16).

76. Cf. Ar., *DG* 2.8 (335a10–11).

opher argues in this way: whatever is not present in all animals is not necessary to an animal.[77] But the testicles are not present in all animals, and therefore, etc. The same argument can be made about the blood, since it is not present in every animal, and therefore, etc.

3. Moreover, one and the same thing cannot be the cause of contraries. But blood is the cause of an animal's corruption, as is said in the text. Therefore it is not necessary to it.

To the contrary: but that thing is necessary to an animal which, when absent, means the animal cannot survive. But blood is something like this, as is said in the text,[78] and therefore, etc.

One must reply to this that both order and degree exist in animals. Now the human is the noblest animal. But some animals are closer to plants, like the immobile animals, and some are closer to the human, like the horse and cow and ones similar to these. Some occupy the middle ground, like fish and serpents. Therefore, those that are nearer to plants lack blood, and there are four genuses of these, as is said in the beginning of book four.[79] But those that are nearer to the human have a great deal of blood, while those that occupy the middle ground have blood but have less of it. Therefore, blood is not necessary to all animals, but only to those that have blood, because their life depends on warmth and moisture, and blood is naturally warm and moist. And this is why it acts especially to preserve life, since preservation stems from a like thing just as corruption results from a contrary.[80] For blood is said to be the seat of life according to the physicians and the Peripatetics and especially according to Plato.[81]

1. On to the arguments. To the first, one must reply that "we exist from and we are nourished on the same things" in a mediated but not in an immediate sense. For we exist from mixed things, since we are nourished by semen and mixed things. This

77. Ar., *GA* 1.4 (717a9f.); A., *DA* 15.1.3.13 (*SZ* 2: 1090).
78. Ar., *HA* 3.19 (521a1f.); A., *DA* 3.2.6.143–45 (*SZ* 1: 414–15).
79. Ar., *HA* 4.1 (523a31f.); A., *DA* 4.1.1.3–4 (*SZ* 1: 433–34).
80. For the remainder of this question, and for q. 19, cf. *QDA* 7.27.
81. See Ar., *HA* 3.19 (521a9f.); A., *DA* 1.2.2.128–29 (*SZ* 1: 94).

is why everything that nourishes is mixed. Thus certain birds eat iron and other birds eat poison, like flies, just as the quail eats henbane [*iusquiamum*] and the she-goat and the stag eat serpents.[82] Nevertheless not every mixed thing nourishes, owing to a defect in the heat and a disproportion in those things by which nourishment occurs. Nevertheless, the proximate principles of generation and nutrition are not the same, since semen is a principle of generation and blood is a principle of nutrition.

2. To the second argument one must reply that blood is not necessary to an animal insofar as it is an animal, because it is not present in every animal; nevertheless, it is necessary for some animals, as was said.

3. To the third argument one must reply that the same thing is not a cause of contraries uniformly and in the same way or according to one and the same reason, but it is entirely possible in another way. Thus it is when proportionally disposed blood is the cause of preservation. If, however, the blood is superfluous or diminished, it can be a source of corruption.

Question 19: Whether blood is a principle of the animal's preservation or is a natural source of corruption.

One asks whether blood is a principle of the animal's preservation or a natural source of corruption.[83]

1. And it seems that it may be the occasion or principle of corruption. Corruption is of two types. One is combustion, and this terminates in ashes; the other is putrefaction, and this is a natural and proper corruption, as is said at the beginning of the fourth book of the *On Meteorology*.[84] But putrefaction arises from heat and moisture, and therefore, since the blood of all the parts is especially warm and moist, it will especially be the principle of putrefaction, so it seems.

82. Cf. A., *DA* 23.1.40 (*SZ* 2: 1570); Isid., *Etym.* 12.1.41.18 (s.v., *cervi*), (PL 82: 427).

83. Ar., *HA* 3.19 (521a1f.); A., *DA* 3.2.6.144–45 (*SZ* 1: 415).

84. Ar., *Meteora* 4.1 (378b31f.).

2. Moreover, of all the animals, those having blood are more prone to corruption, according to the physicians. But this would not be so were the blood not the principle of corruption, and therefore, etc.

To the contrary. The philosophers chose to study in areas near to the sea, which are warm and moist. But this would not be so unless heat and moisture were the principle of preservation. Therefore, since the blood is particularly hot and moist, it is especially the principle of preservation, so it seems.

One must reply that blood can be considered in two ways. One way is that which falls under nature's regimen, and in this way blood is especially the principle of preservation, because life depends on heat and moisture, which are found in blood, and this is why a sanguineous complexion, according to which the individual is under nature's regimen and under the requisite temperament, is the best and suited to long life.

Blood can be understood in another way, to the extent that it is beyond nature's control on account of some defect. But this defect can also be of two types: either specifically, such as blood corrupted in an abscess, or generally, as in a fever, which arises from the blood. And this latter defect causes more corruption than does the first. But this defect can still be understood with respect to the substance and to the quantity and quality [of blood]. For if blood is considered according to its substance, then it is hardly corruptible at all and hardly passes over into the matter or into the nature of poison. This is not the case for the other humors because a potent poison arises from a quartan fever, which results from a hard, earthy, dry melancholy.

But if one considers blood in terms of its quantity, then it is especially corruptible, because it is very abundant in the body. And in like manner if one considers it with respect to its quality, because it contains within itself the father of putrefaction, which is heat, and its mother, which is moisture. But in terms of its own substance, since it is very well digested and very pure, it does not putrefy, etc.

Thus one must reply to the question that blood that falls under nature's regimen is not a principle of corruption but of preservation. If, however, it falls outside nature's control, then it can be a principle of corruption, not, however, by reason of its substance, but by reason of its quantity and quality, as was seen.

1. On to the arguments. To the first, one must reply that heat is of two types: natural and accidental. Putrefaction arises from accidental heat, which is external and foreign. Because external heat opens the body's pores and causes the internal, natural heat to escape and as a result it releases the moisture that is the subject of the natural heat. Thus one must reply to the form of the argument that heat and moisture are natural things in the blood, but putrefaction does not arise from the natural heat and moisture but from a heat and moisture accidentally acquired in the blood itself.

2. To the second argument one must reply that corruption is of two types: namely, natural and accidental. The more sanguineous bodies are, the more slowly are they naturally corrupted because the heat and airy moisture that are in the blood are the cause of long life. Nevertheless, such things are more quickly corrupted accidentally because the complexion of bodies of this sort is more temperate and therefore more easily moves away from the mean and from a temperate condition, and this is why the complexion of such bodies can easily become distempered, etc.

Question 20: Whether of all the humors blood alone nourishes.

Further one asks whether of all the humors blood alone nourishes.[85]

1. It seems so, because there are four powers in every animated being: the attractive, retentive, digestive, and expulsive. But the attractive only attracts the sweet, as the Philosopher implies in the second book of *On the Soul*,[86] because only the sweet nourishes. Therefore, the digestive power will digest only the sweet.

85. Cf. Ar., *HA* 3.19 (520b18f.); Cf. A., *DA* 3.2.3.100–121, 3.2.6.142, 12.1.6.82, 12.2.1.101 (*SZ* 1: 393–404, 413; 2: 924, 931–32).

86. Ar., *De anima* 2.10 (422a11f.).

But only what is digested nourishes. Therefore, since only the blood is sweet, only the blood nourishes.

2. The same view seems defended by the authority of the physicians, who posit that only the blood nourishes.[87]

3. Moreover, the other humors have their own proper receptacle, like yellow bile [*cholera*] which has the gall bladder, and melancholy the spleen, which would not be the case if they contributed something to the animal's nutriment.

To the contrary. Nature does nothing in vain. But all the humors can be generated from any part of the nutriment, according to its diverse disposition. Therefore, for the same reason, if one nourishes, so too will another.

To this, when one asks whether the blood alone nourishes, one must reply that blood can be considered in three ways. In one way, such that blood alone nourishes and no other humor does so. But this understanding is false, because nutrition is the complete assimilation of the nutriment to the one nourished.[88] It is therefore necessary that whatever is lost be restored by the nutriment. But what is lost can be choleric or phlegmatic or melancholic or sanguineous, because the member that has to be nourished may be of this type. Therefore, one must say that all the humors can enter into the nutriment for a complete restoration and perfect assimilation to occur.

That blood alone nourishes can be understood in another way, because it nourishes in a solitary manner without another humor present, and this understanding is true, because blood can nourish though not conjoined with or mixed with other humors, and this is not the case for the other humors.

It can be understood in a third way, that blood alone nourishes effectively and the other humors nourish only materially, because they nourish only with the blood mediating their activity, just as one may say that only medicine heals because it acts effectively, even though food and drink heal materially because they generate the material for health.

87. Galen, *De venaesectione* (Kühn 9: 262).

88. Cf. A., *DA* 12.1.2.25 (*SZ* 2: 903–4).

Therefore, one must reply to the question that the other humors nourish along with the blood, because there is no mixture so choleric or so melancholy in and of itself that it cannot be nourishment for certain birds like the ostrich. Nevertheless, that they do not nourish all animals stems from the fact that the complexion of an animated being is not always proportioned to the mixture.

1. On to the arguments. To the first, one must reply that in the nutriment and in the attractive power, that by reason of which something is attracted is sweet. But nothing prevents that which is attracted from being bitter, sharp, or pontic, or having other tastes.

2. To the second, one must reply that the physicians say that only the blood nourishes because it alone nourishes effectively and in a solitary manner, but nevertheless it is not the only one that nourishes; rather, the other humors nourish as well.

3. To the third argument one must reply that choler exists in an animal's body in two ways: one is necessary, and the other is helpful. The choler that is necessary flows with the blood and makes a path for the blood, which is thick, by penetrating and opening the narrow veins of the body, because choler is very penetrating. The other, helpful choler is received into the gall bladder for three reasons: (1) so that it may arouse the expulsive power by stimulating the intestines; (2) so that it may wash away the oily phlegm and scrape it from the wrinkles of the stomach; (3) to strengthen the digestive power with its heat. Therefore, the choler that is necessary can contribute to the nutriment of any part, and the one that is helpful acts as a nutriment for the gall bladder [*bursiculae fellis*] and for those three forms of assistance that have been mentioned. And the same must be said for the melancholy, with respect to the spleen and the attractive power, which it strengthens.

Question 21: Whether blood is more abundant in thin animals than in fat ones.

One asks whether blood is more abundant in thin animals than in fat ones.

1. It seems, in fat ones. For fat is caused by what remains from blood; thus, where the blood is abundant, so too will be the fat, and conversely.

2. In addition, fat is warm and moist, and this is why it is placed around cold parts. But such is the nature of blood, and therefore, etc.

The Philosopher says the opposite.[89] For he says that "the more the fat abounds, the more the blood is diminished." Therefore, etc.

One must reply that fat is of two types. There is one that is caused by the strength of the heat digesting a large quantity of aliment, and this one tends to be yellowish in color, and the fat that is in a goose or a hen is of this type, and this type of fat is attended by an abundance of blood. The other type is the fat generated from moisture that is not completely digested owing to the weakness of the heat, and this one tends to be white in color and argues for a paucity of blood, as in the pig. The Philosopher is thinking of this one, whereas the arguments are based on the first type. It is true of this one that such fat argues for blood, but where the other fat is present, the blood is, as a result, more abundant in thin than in fat animals.

Question 22: Whether blood is more abundant in men than in women.

One asks further about another property, whether blood is more abundant in men than in women.[90]

1. And it seems to be more abundant in women. Because the more form something possesses, the less matter it has. But the male has more form and less matter, and therefore as a consequence he has less blood.

2. Moreover, a woman is a flawed man [*vir occasionatus*], as will be stated below.[91] Therefore, a woman will have more blood. Proof for the conclusion: a woman is only produced owing to a

89. Ar., *HA* 3.19 (520b29–521a2).
90. Cf. A., *DA* 3.2.6.147 (*SZ* 1: 415–16).
91. See *QDA* 15.2. Cf. A., *DA* 9.2.6.134, 16.1.14.73 (*SZ* 1: 826; 2: 1195).

defect of the heat acting on the semen and to an abundance of matter. Since, then, blood has the nature of matter, there will be more blood in a woman.

3. Moreover, where something overflows then it is more abundant there. But in a woman the blood overflows, an indication of which is the frequent flow of her menses. Therefore, etc.

The Philosopher says the opposite in the text.[92] For he says that there is more blood in men than in women.

One must say that just as semen is of two types, so blood is of two types. For one semen is well-digested, and this is the sort found in men. The other is poorly digested, and this is the sort found in women, for example, the menses itself. And the same is true of the blood. For one type is well-digested and dense, and this sort is in men. The other is crude and very fluid and phlegmatic, and this is the one in women, since in men the heat is more powerful and, as a result, is sufficient as a principle for digestion. But in women the heat is weak and impeded, and this is why their blood is crude and fluid and, as a result, phlegmatic. On account of this they often suffer the flux of menses, since nature rejects this phlegmatic blood and purges it through menstruation.

1. On to the arguments. To the first, one must reply that if one should compare different species, then certainly the more form one has the less matter it has. If, however, one should compare the same species, then it is not necessary that the one that has more form will have less matter. But the male and the female are in the same species, and therefore, etc.

2. To the second argument one must reply that <because> a woman is a flawed man, for this reason she has less blood. As proof one must say that because a woman is produced from a defect in the heat in the semen, it follows from this that there will be more fluid and undigested blood in women.

3. One must reply in the same way to the third argument. Because there is more fluid and crude blood in women, this is why

92. Ar., *HA* 3.19 (521a21f.). Cf. Avic., *DA* 4.3 (fol. 32raA); A., *DA* 3.2.6.147 (*SZ* 1: 415–16).

their menstrual flow is very frequent. But this does not apply to digested blood, which is in men.

Question 23: Whether the first generation of the blood occurs in the heart or in the liver.

Again, one asks whether the first generation of the blood occurs in the heart or in the liver.

1. And it seems to be in the liver. Because the first generation of blood occurs in that member in which the nutritive power is present, but this is in the liver. Therefore, etc.

3. Moreover, blood is generated from chyle. The place where chyle is first found is the same place in which the generation of blood occurs. But chyle is found first in the liver, after it has been separated off from the stomach, and therefore, etc.

The Philosopher says the opposite in the text.[93] For he says that "the blood is first generated in the heart in the fetus when it is in the womb." And he proves this with an argument. That in which blood participates well, being well disposed to it, is the principle of blood. But the heart is just like this, because the blood is warmed in the heart by its superabundant heat and is tempered by its temperament; therefore, etc.

One must reply that we speak of the first generation of blood in two ways: either in the womb in the first stages of the production of the fetus, or in the completed animal, after birth, from the aliment it absorbs. If we speak in the first way, then the first generation of the blood occurs in the heart, because the heart is the first member generated in the embryo, and this is why the ventricle that can contain the blood is in the heart itself. Thus the blood is contained as if in a vessel in no member save the heart, as is said in the text.[94] If, however, we are speaking about the generation of blood in the second way, then its first generation occurs in the liver because when the nutriment is digested in the stomach, whatever is purified is sent to the liver through the mesenteric veins, and there it is converted into blood. Nev-

93. Ar., *HA* 3.19 (521a9–10); cf. Avic., *DA* 3.3 (fol. 32raA).
94. Ar., *HA* 3.19 (520b18f.). Avic., *DA* 4.3 (fol. 32raA–B).

ertheless, this only occurs by means of the mediating heat of the heart. Nor does the blood generated there have the power and aptitude for nourishing the individual members except through the influence that it receives from the heart. Now the blood receives a power from the heart, and it is moved by this power everywhere, up and down, back and forth, in every different direction, being moved by that basis [*ratio*] with which it is informed by the soul and by the influence of the heart, which power also moves in every different direction.

Replying to the arguments, one must reply that the first two arguments concern the generation of the blood from the nutriment in a completed animal, and this occurs in the liver.

The first argument actually applies to the blood generated in the womb for the existing animal, and the second argument shows that the blood's generation does not occur without the heart's power. And this we concede, etc.

Questions 24–26: Why the blood in the heart is liquid and is coagulated outside of it.

Why the blood of some animals coagulates.

Why the blood putrefies near the bones.

After this one asks why the blood in the heart is liquid and is coagulated outside of it, since the semen or sperm is, to the contrary, well coagulated and thickened in the body but liquid outside of it.

Second, one asks why the blood of some animals, like the human and the cow, coagulates, as is evident during phlebotomy, whereas the blood of other animals, and especially of melancholic animals like the stag and the hare, does not coagulate.

Third, one asks why the blood putrefies near the bones.

To the first, one must reply that the first digestion proceeds by thinning, whereas the second and the third proceed by thickening. Thus the blood in the body always has a moisture flowing along with it, and likewise the heat and spirit move the blood itself to the individual members. This is why it is liquid and flows in the body, but when it is outside the body the wateriness is separated from it and the heat and spirit escape through evaporation of the heat containing it or through the mortification of the cold containing it, and they ebb, and this is why it thickens quickly. But the semen is generated during the last digestion, which proceeds by thickening. Thus the semen exists in a proximate disposition toward conversion, as it were, and this is why it is coagulated in the body, but, according to the fourth book of Aristotle's *On Meteorology*, everything that is coagulated by heat is dissolved by cold and vice versa.[95] Since, then, semen is coagulated by heat in the body, this is why it flows when it is outside the body since, owing to its ability to be acted upon [*passibilitatem*], it is dissolved and liquefied by cold.

To the second question one must reply that coagulation is of two types: one occurs from compression, occupying a smaller space, and every wet thing can be coagulated in this way by cold. Another coagulation occurs from the isolation of two natures, namely, of the aqueous and earthy. Such is the case in milk, for when the whey is separated from the curd then the curd gels. And coagulation of this sort does not occur in every blood. The reason for this is that some animals are particularly melancholic, and they are nourished on a coarse and melancholic nutriment, and this is why all the earthiness and melancholy that is in their nutriment is converted into their substance. But some animals, like the cow and the human, are nourished by what is refined [*subtile*] in the nutriment, and what is thick remains in their blood. And because what is thick like this becomes the material for hair and threads, this is why the blood of animals such as these contains hairs and threads, as it were, whereas the blood of the first ones does not.[96] The blood of melancholic animals does not coagulate, therefore, because everything that provides the material for the threads that cause coagulation passes over into their nutriment and the blood remains fluid. But the blood of the other animals which has threads and *ines*,[97] that is,

95. Ar., *Meteora* 4.2 (379b10f.).

96. Cf. A., *DA* 12.2.1.102 (*SZ* 2: 932).

97. The Latin *ines* is a plural form transliterated from the Greek and means "strength, nerve, or sinew."

certain pellicles, coagulates because this sort of thing contributes greatly to compression and coagulation.

To the third question one must reply that heat abounds near the bones, and for this reason the blood there cannot resist putrefaction. In addition, the blood near the bones is thinner, because the bones are nourished by what is thicker in the blood. For this reason the flesh nearer the bones is softer, sweeter, and more flavorful. Therefore the abundant heat more quickly overcomes it, owing to its thinness, and as a result it putrefies.

Question 27: Whether the marrow is necessary for the bones' nourishment.

Further one asks whether the marrow is necessary for the bones' nourishment.

1. It seems not. Those things that are not well compacted and coagulated are continually being lost, and this is why they need nourishment to restore what is lost. But the bones are well compacted, solid, and coagulated. Therefore, they do not require marrow for nourishment.

2. In addition, the brain is filled with marrow and yet it does not provide nourishment for any bone, because it is cold and moist, and therefore, etc.[98]

The Philosopher says the opposite in the text.[99]

One must reply that marrow is necessary for the sake of nourishment and for the fomentation of the bones, because some parts in the body are solid, like bone, and others are permeable, like flesh, and some occupy a middle ground. But all these parts exist in flux and endure continuous loss. Now the Philosopher says in the first book of *On Generation* [*and Corruption*],[100] that for all those possessing a species (that is, form) in a material way, in one case it is present with respect to matter and is in flux, and in another case it is present according to form [*species*], and en-

98. Cf. A., *DA* 12.2.3.113 (*SZ* 2: 936).

99. Ar., *HA* 3.20 (521b4f.); cf. Avic., *DA* 12.19 (fol. 48va–vb); A., *DA* 3.2.1.74, 3.2.7.149–50, 12.1.7.89 (*SZ* 1: 381, 416–17; 2: 927).

100. Ar., *DG*, 1.3 (318a9f.); cf. A., *DG*, 1.1.22.

dures.[101] Therefore just as the permeable parts need a thin nutriment, and the parts occupying the middle ground—like the nerves and veins—require a thicker nutriment, so the bones need something very thick with which they may be nourished. And an indication of this is that bones, when they are outside the body, more quickly tend toward corruption and turn to dust than when in the body, which would not be the case were they not supported in the body by something which does not support them when they are outside the body. This, however, is the marrow, and therefore, etc.

On to the arguments. To the first, one must reply that this argument has to do with things—inanimate and animated—that are compacted, because in the animated ones heat does not cease to act on the moisture, consuming it. And on account of this some parts are moved by the power of the soul against their own nature, and every unnatural motion causes weakness and some imbalance in the thing to which it belongs. But this is not found in inanimate things, and this is why the extent to which things in animated beings are compacted is the extent to which they require nutriment. But those that are inanimate do not. The milk and sperm will be discussed below, and this is why, etc.[102]

101. This passage seems troubled by interpolations that attempted to comment on the original.

102. See *QDA* 15.11–*QDA* 16.1.

BOOK FOUR

Questions 1–2: Whether bloodless animals are naturally cold. Whether bloodless animals living in the water are edible, but not those that live in air.

“WE HAVE ALREADY said above,”[1] etc. With respect to this fourth book one may first ask about bloodless animals. And first, whether bloodless animals are naturally hot or cold.

1. And it seems that they are hot. This is because the container draws the content to its own complexion. But the sea is salty and, consequently, hot. Therefore, marine animals and those lacking blood are hot.

2. Moreover, flyers are hot and dry. But some bloodless animals are flyers, like bees and wasps. Therefore, such creatures are hot. Therefore, conversely, bloodless creatures will <not> be naturally cold. The proof of the consequence or of the converse follows this rule of the *Topics:* “When you have two contraries, if some property is in the one, the contrary property will be in the other.”[2]

3. Moreover, an animal has a certain complexion and it tolerates badly weather similar to its complexion, and tolerates well weather unlike its complexion. This is why a choleric creature tolerates summer badly but winter well, and a phlegmatic one tolerates winter badly but summer well, and similarly for the

1. Ar., *HA* 4.1 (523a31–b2).

2. This is a loose translation of the exceedingly terse and cryptic *Si oppositum in opposito, et propositum in proposito,* which seems to reproduce Boethius’s rendering of Ar., *Topica* 7.3 (153a33ff.)—*Nam si opposita oppositae, et eam (quae dicta est) propositi necesse est esse.* (PL 64: 990D.) Cf. A., *Topica* 7.2.1. The phrase is invoked again at *QDA* 13.1, to show that if sickness is present in one who is ill, so health will be present in one who is healthy.

others. But bloodless animals, like flies and bees and crabs, are generated in summer and tolerate it well, but they are corrupted in winter. Therefore, they are naturally cold.

Related to this, one asks why bloodless animals living in water are edible, like oysters and crabs and marine shellfish, but those that live in the air are not, like bees and flies, because bees and wasps are poisonous and hot, just as scorpions are poisonous and cold.

To the first, one must reply that complexion can be considered three ways: either according to genus (and thus every living thing is said to be hot and moist, because life endures on account of these); or, in a second way, according to species, and thus a human is said to be hot and moist, but the ass is hot and dry, as is the lion. There is a third way, according to the individual, and this can be considered in two ways: either comparatively or absolutely: comparatively, just as one woman is said to be choleric in comparison with another, even though all women are naturally phlegmatic; if absolutely, then this person is said to be choleric, and that one sanguine.

When, then, one asks whether bloodless creatures are cold or hot, one must respond that according to the genus, i.e., in general,[3] they are hot and moist, insofar as they are alive. With respect to the species, they are all cold, inasmuch as they are bloodless, because when the heat in the heart of an animal is adequate, it converts food into blood, but when the heat is weak, food cannot be adequately digested or converted into blood, but instead is converted into slime [*virus*] and wateriness. If, however, we speak according to the individual and comparatively, then some are warm in comparison to others, because flyers are warm in comparison to those that exist in water and when compared to themselves, when they are dead. Thus one must reply to the question briefly by saying that they are naturally cold.

1. On to the arguments. To the first, one should say that complexion is twofold: natural and accidental. Now the sea, or the complexion of the sea, since it is water, is naturally cold and

3. *Secundum genus, idest* [*sic*] *in genere.*

moist, but accidentally it is warm and dry on account of the saltiness and earthiness with which it is mixed, and the container draws the content to its natural, and not accidental, complexion. And besides this, although such animals live in a salty, bitter, or warm sea, they are nevertheless nourished by that which is sweet and cold and moist in it.

2. To the second argument one must reply that flyers lacking blood are naturally cold; nevertheless, in comparison with others, they can be called hot (or they are hot) with an elemental but not with a life-giving heat.

To another matter, which is asked in relation to this, one must say that in the human the complexion abounds with blood, and bloodless animals thus are far removed from the human complexion, but in aquatic animals, that which is gross and poisonous is converted into shells and hard skins, and that which remains of their substance is cleansed through the salt of the sea. Thus cleansing with seawater is effective against itching and mange. This is why the skin of a crab is red, because whatever in it stems from slime (or poison) and heat goes into the shell, and reddens it. This is because it is in the nature of heat to tinge by reddening, and when cooked they are redder, because the black earthiness and muddiness darkening their skin is washed away by cooking. But in flying animals this cause is absent (that is, the washing-away of poison through seawater or through something else, like a shell). For this reason all bloodless flyers are poisonous to humans and are not a food to any extent [*secundum magis et minus*].

And if it is objected that such animals are nourished by sweet and suitable food, as bees are (and because flies are nourished by flesh, milk, and the like), and that as a result it would seem that they are not poisonous to humans, it must be said that the sweet nutriment that is received from flowers is converted into the substance of such animals and it is infected and altered in them, and for that reason, although they are nourished by agreeable foods, nevertheless they themselves are not suitable food.

Moreover, although that which is made by them, like honey, is nourishment for a human, it still does not follow that these in themselves and in their substance are edible, because a honeycomb does not come from the substance of such animals, but is rather fashioned with skill, using a kind of prudence and a kind of natural instinct or skill they possess, and it is not generated from their substance but comes from somewhere else. One might imagine that it is gathered by them from flowers and other things and this is why it does not receive infection from their complexion, and although the honey may receive a bit of sharpness from them, it is nevertheless slight.

Questions 3–4: Whether an animal having a shell must naturally be enclosed inside the shell.

Whether animals having shells differ in shape, number, and substance.

Further, one asks whether shellfish-animals (or those having shells) must naturally be enclosed inside their shells.[4]

1. It seems not. A shell in bloodless creatures is just like bone in those having blood. But bone lies within the flesh and not outside of it. Therefore, this will have to be so for a shell.

2. Moreover, every fluid naturally seeks an external location. But the flesh of such animals is fluid and watery and will naturally seek an external position; therefore, etc.

Related to this, one asks about the difference in shells with respect to shape, number, and substance.

To the first, one must reply that moisture is, naturally, poorly bounded by its own boundary and well bounded by another's. But the flesh of some animals is very moist, and this is why they naturally require something solid on the outside as their boundary and to retain and preserve them. But the flesh of some animals is sufficiently bounded by a fairly hard, viscous skin, and this is why something solid that is internal is adequate for them upon which their flesh is established and through which it is sustained. Examples include perfected animals such as the cow,

4. Cf. Ar., *HA* 4.2–4 (525a30f.). Avic., *DA* 4.1 (fols. 32rbA–33rbC); A., *DA* 4.1.1–4.1–48 (*SZ* 1: 433–57). Here, as in *DA*, A. struggles with terminology for shellfish. He uses *conchilia* and then defines it by saying *vel habentes testas*.

the human, and others of this sort, and this is why the crab, oysters, and others of this type have external shells, because of their softness. But the genuses of cephalopods [*malakie*][5] have something that is hard or shell-like inside for its hardness, and are fleshy on the outside.

1. On to the arguments. To the first, one must reply that flesh has a different disposition in various animals, and that is why it is sufficient for some to have something solid that is internal while in others something solid that is external is required.

2. To the other argument one must reply that because moisture naturally seeks an exterior location, animals that have soft flesh necessarily require a hard exterior by which they may be bounded and conserved.

To that which is asked in relation to this, one must reply that the material of the shell is a viscous humor and the agent is hot, dry, and combustive. And when this humor is very viscous and cannot be easily divided, but rather it is uniformly dense, muddy, and *mulcidus* throughout, then a shell is generated for the whole animal, as is seen in the tortoise and snails owing to the uniform character of the material.[6] When, however, the viscous humor is not so uniform but can be divided, two or more shells are generated through the heat, depending on the requirements and species of the animal and on the divisibility of the material. If the heat is very sharp and penetrating and the humor is very viscous with some earthiness mixed in, then lines or bumps or rough places appear on the shell because, in a place or part to which the heat and the humor are driven, there appears a bumpiness and a roughness, and in a place from which

5. A. uses *malakie,* in a variety of forms, as a generic term for the cephalopod mollusks, such as the squid, cuttlefish, and octopus. A transliteration of the Gr. *malakia,* "soft-bodied ones." See *GF,* 155f.

6. The word *mulcidus* is clearly a corruption as it appears in no lexica and an electronic search of the PL revealed no allied words. The Latin words most likely to yield a *mulc-* stem (or a form easily corrupted into *mulci-*) yield meanings with the senses "to stroke," "to milk," and "to beat." None is especially apt for this context. The reading seems secure from the editor's apparatus (though one manuscript omits the offending passage) and the corrigenda offer no assistance in this regard.

it is driven, there appears a concavity. But when the heat is weak and the humor is oily, then the shell is soft; but when the humor is very earthy with a burning heat, then the shell is as hard as a rock, as in the case of the snail, which is poisonous, because it always lives in a marsh and in mud and not in the sea.[7] Thus all of the differences in shells are caused either because of the humor or vapor, or because of the material, or because of the heat acting in them, and because the humor of the sea is very viscous owing to its saltiness and is bitter from its brackishness. This is why shells or shelled animals are more often generated in the sea than in other places. They are generated less often in other places because the salt content of the sea makes for the dryness of their shells, etc.

Question 5: Whether every animal has every sense.

"The disposition of the instruments," etc.[8] First one asks whether every animal has every sense.

1. It seems so. The human is like plants in the vegetative power, and like brute beasts in the sensitive power. But every plant participates in all the parts and potencies of the vegetative power with the human. Therefore, by the same argument, every brute beast participates in all the parts of the sensitive power with him.

2. Moreover, the whole intellective power is present to each one participating in it, unless it is impeded accidentally. Therefore, for the same reason, each part of the sensitive power is present to one participating in it unless it is impeded accidentally.

The Philosopher says the opposite in the third book of *On the Soul*[9] and argues from reason that hearing, sight, and smell are

7. Just above A. used *limax* for snail. Here he uses *testudo*. The terms are interchangeable. Cf. *DA* 26.1. 17(22) (*SZ* 2: 1750).

8. Ar. *HA* 4.8 (532b29–33); cf. Avic., *DA* 4.2 (fol. 33rb); A., *DA* 4.2.1.75–76 (*SZ* 1: 470–71).

9. Ar., *De anima* 3.1 (425a8f.); 3.11 (433b31f.); 3.12 (434a27). Cf. A., *De anima* 3.5.1.

situated in the head; but not every animal has a head, as is clear in the same chapter. Therefore, etc.

To this, one must reply that not every animal has all the senses. And the reason is this: Touch is present in an animal in order to protect it from injurious things outside itself and so that it might cleave to things that are suited to it. Taste is present, moreover, so that it may distinguish between nourishment that is proper for it and nourishment that does not suit it, when it is nearby. Smell is given to it for the sake of aliment that is somewhat distant from it, so that it will seek out that which is suitable and avoid that which is harmful. And in a similar way sight and hearing are given to animals so that they will seek out and distinguish suitable things from harmful things, when they are at a distance. Now, however, there are some animals that are immobile, and these do not have organs for progressive motion, and they therefore do not need sight or hearing or smell because if, by some of these senses, they did perceive something at a distance that is suitable for them, still they would be unable to seek it out or, if they should perceive something unsuitable or harmful, to avoid it. And this is why these three senses are present only in perfect animals having the power of progressive motion but not in those that are immobile, because these do not seek their aliment from a distance but rather it is carried to them through the water, specifically to those clinging to the sea shores, and this is why all immobile animals are aquatic and not land animals.[10]

1. On to the arguments. To the first, one must reply that the vegetative power is not as perfect as the sensitive, and this is why it demands less diversity in the subject. On account of this in every part there can be a part of the threefold vegetative power: to wit, the nutritive, the augmentative, or the generative. But the sensitive power is more perfect and demands greater diversity in the subject, because every part requires a different organ, and this is why, etc.

Or, in another way one can say that not every plant has all the parts of the vegetative power, because some do not generate

10. See also *QDA* 1.7.

others like themselves, but are only generated from the earth, etc.

2. To the second argument one must say that intellect is immaterial and thus does not use an organ. But the sensitive power is organic and uses an organ. And this is why the intellective power is not diversified as the sensitive power is among the several and diverse organs which it uses, etc.

Question 6: Whether fish have hearing.

Further one inquires whether fish have hearing.

1. It seems not. Sound is the object of hearing. But sound is not a motive force for hearing unless air that is continuous all the way to the organ of hearing acts as a medium. Since air is not continuous for fish owing to the medium of water, fish cannot hear.

2. Moreover, all things that are mute are deaf, as is said in the text. But fish are mute, therefore deaf.[11]

The Philosopher proves the opposite through many experiments.[12]

One must reply that fish have hearing, which the Philosopher sufficiently proves by various experiments. But their hearing is not as perfect as that of animals that live in air. The reason for this is that sound is multiplied by the mediating motion of that in which it is generated. Therefore, the most suitable medium for sound is one that is more mobile, but according to the Philosopher in the second book of *On Meteorology*[13] and according to the Commentator in book three of *On Heaven and Earth,* air is just this sort of medium.[14] This is why air is a more suitable medium for sound than water, although sound can be made in both. Thus, fish do not perceive sounds that are as faint as do animals that hear in air.

11. Ar. *HA* 4.9 (535b14f., 536b2f.). Cf. A. *DA* 4.2.2.96 (*SZ* 1: 480–81).

12. Ar. *HA* 4.9 (533b4f.). Cf. Avic., *DA* 4.2 (fol. 33rb), A., *DA* 4.2.1.77–82 (*SZ* 1: 471–73). *Experimenta:* experiments, tests, or, in general, experience.

13. Cf. Ar., *Meteora* 2.8 (368a16f.).

14. Averroes, *DCM* 3, 28; *De anima* 2, 79.

1. On to the arguments. To the first, one must say that sound is caused in air and moves the continuous air as far as the hearing either through itself or through another. Now if the air extends continuously as far as the ear, then the species of the sound is multiplied in each part of the air. If the air, however, ends at the water, a certain tremor is caused by a tremor of the air that is touching the water, and the water trembles continuously as far as the hearing of the fish living in the water, and in such a case the sound is moving the continuous air through the water as far as the ear.

2. To the second argument one must say that not all things that are mute are deaf, but rather the contrary. Now we acquire variations of speech by being taught to do so. Thus a person who is deaf from birth cannot be instructed in any language [*idioma*], and this is why he can speak no language in an ordered way. But there is nothing that prevents there being some impediment in the nerves of the tongue while the nerves of the hearing remain well disposed.

Thus one must say with regard to the Philosopher's statement that he does not understand that all who are mute are deaf, but rather he understands that those who are mute because of deafness do not have the power of speech, etc.

Question 7: Whether voice is natural or voluntary or artificial.

"Voice," etc.[15] Here one asks about voice. And first, one asks whether voice, or the generation of voice, is natural or voluntary or artificial.

1. It seems that it is not natural. We cannot accelerate a natural operation, but we can accelerate or retard the formation of voice; therefore, this is not natural.

2. Moreover, "what is natural is the same for all," according to Boethius's *On Division*.[16] But the formation of voice is not the same in all, and therefore, etc.

15. Ar. *HA* 4.9 (535a27f.). Cf. A. *DA* 4.2.2.90 (*SZ* 1: 476–77); Avic. *DA* 4.2 (f.33rb).

16. Boethius, *De divisione* (PL 64: 879B).

3. Moreover, a man and a boy have the same nature. If, therefore, the formation of voice were natural, then boys would give voice in the same way as men, which is false, and therefore, etc.

To the contrary. The effect of a natural potency is a natural effect. But the vocative potency is natural, seeing that its instruments are natural.

Moreover, the tongue is suited for two tasks according to the Philosopher in the second book of *On the Soul:* to wit, taste and speech.[17] But taste is a natural operation, and therefore, so too is speech.

One must reply to this that the formation of voice can be related to the instruments by whose mediation it occurs (think of the power of the heart and the lungs and the arteries and the tongue and the palate and the teeth), and thus it is itself natural because a natural operation proceeds with a natural instrument mediating.

In another way it can be related to the animal powers commanding the aforementioned powers, for example, to the will (whether to the rational appetite in a human and the concupiscent appetite in a brute beast) and in this way the formation of voice can be called voluntary, for it is in the human to form voice. For even though he has the instruments or principles of voice, he can nevertheless desist from the formation of voice.

Third, the formation of voice can be related to an accidental variation in the voice. And in this way the formation of voice is artificial insofar as it is up to the will of different individuals who, for the sake of expressing different concepts, form different voices. Seen this way, the Philosopher speaks for the first [solution when he says] that voice is by nature and is shared by all. But the Philosopher speaks for the last [solution when he says] that speech is not natural because speech properly adds articulation and order to voice.[18]

Thus, briefly one can say that insofar as voice is a sound emitted by an animal in order to express a concept, namely, of joy or sadness, it is natural. Insofar as it is an articulate and ordered

17. Ar., *De anima* 2.8 (420b16f.).
18. Ar., *HA* 4.9 (536b19f.).

sound, it is artificial and conventional. And for this reason one is accustomed to say that speech is from nature, but speaking in this or that manner is conventional.

1. Through this one can respond to the arguments. To the first, one should reply that because the operations of vital parts obey the appetite with respect to movement, and since the formation of voice occurs in many with the mediation of appetite, then an animal can retard or speed its voice according to what it desires.

2. To the second argument one must reply that although an ordered, articulate, and signifying voice is not the same among all, the initial formation is nevertheless the same among all as far as the natural instruments are concerned. Thus all people give voice in the same way although they do not speak in the same way.

3. To the third argument one must reply that the vocative power is active, and in children the active power is weak and the operative instruments are weak, and this is why a boy cannot give voice like a man.

Arguments to the contrary proceed in the same fashion.

Question 8: Whether every voice is significative.

Next one asks whether every voice is significative.

1. And it seems so. Voice is the striking upon the windpipe [*vocalis arteria*] by air that has been breathed in by the soul or by a power that is in these parts, and accompanied by an image being signified.[19] But every utterance with an image being signified is significative; therefore, etc.

2. Moreover, just as speech is varied in humans, so also there is a voice that is varied in animals, like mooing in a cow and a horse's whinny and a sheep's bleating or baaing and a dog's barking, etc. But every speech is significative. For Plato says that words were invented just for this purpose, to give an immediate

19. On *vocalis arteria* as "windpipe," cf. *DA* 9.1.1.2 (*SZ* 1: 774–75).

indication of a shared desire. Therefore, every voice is significative.

3. Moreover, voice is something sensible to the hearing that is present in animals, so that they may apprehend shared concepts. But this cannot occur unless voice is significative. Therefore, since art, like nature, does nothing in vain, every voice—whether natural or artificial—will be significative.

To the contrary. Every voice signifies either by imposition or by nature. But there is a certain voice that is neither imposed significatively on a thing nor represents something naturally. Therefore, not every voice is significative.

Besides, the thing's intention is in the soul. If then the voice is imposed on it, then the intention is either wholly in the soul or wholly in the imposed voice, or is in both. If it is wholly in the one so that it is not in the other, this [intention] cannot be in the voice as long as it is not in the soul. It is therefore necessary that what is not in the voice be in the soul. It cannot be in both, because then the same thing would be in different things.

It must be said that voice can be informed in two ways, because there is a vocative power in its formation which can be informed by the estimative or imaginative power, just as occurs in clever animals, or it can be by means of reason, as is the case for the human, and this is for the sake of expressing something. And the voice formed in this way is significative because such voice is a sign of an internal concept.

It can also be considered in another way, insofar as the vocative power is not governed by an animal power, and in this way not every voice is significative because a human can form many things and vocalize them by accident, not for the sake of expressing anything at all. He rather forms or fashions various voices in response to a joke or jest, and voices of this sort are not significative. Nevertheless, something like this is not a voice without being able to be imposed on a thing to be signified.

1. On to the arguments. To the first, one must respond that the Philosopher speaks of a voice which is made by nature and is governed by imagination, and such voice is significative. For

although it is not significative of something outside the sign, it is nevertheless a sign of its own previously imagined thing, and voice is nevertheless a sign in the utterance because it was imagined thus.

2. To the second argument one must say that although words were invented for the sake of giving expression to concepts, not all words express determinate concepts, and this is why they lack a proper end.

3. One must respond to the third argument in the same way.

Question 9: Whether sleep is necessary for an animal.

"Let us discuss sleep," etc.[20] First, one asks whether sleep is necessary for an animal.

1. It seems not. Act occurs by virtue of form and affect [*passio*] by virtue of matter. Therefore, to the extent that one has more form, then to that extent one can act more. But an animal has more of form than fire does. Therefore, since fire can act continuously and does not require rest, an animal can perceive all the more without an intervening period of rest.

2. Moreover, the nobler the form is, the more it is capable of numerous operations. And the more superior a power is, the less is it given to fatigue. But an animal power is higher than a vital or natural power. Since then a vital or natural power does not require rest, like sleep, then neither does the animal power. But sleep is only necessary to provide rest for an animal power; therefore, etc.

3. Furthermore, sleep occurs in an animal, according to Avicenna, on account of nourishment, because during sleep the powers turn inward so that the nourishment can be digested better. And this is why the exterior parts are insensible, as it were, during sleep. But plants receive nourishment just as animals do. If sleep is necessary for animals on account of this cause, then it will be necessary too for plants, which is false. Therefore, the first is false; therefore, etc.

20. Ar., *HA* 4.10 (536b24f.). Cf. Avic., *DA* 4.3 (f.33va) and A., *DA* 4.2.3.99–101, 19.1.2.5–7 (*SZ* 1: 482–83; 2: 1334–36), *De somno et vigilia* 1.2.2.6.

The opposite is evident to sense, because something is necessary if a thing cannot exist without it. But an animal cannot exist without sleep. Therefore, etc.

One must respond that sleep is necessary for an animal. And there are many causes of this. One stems from the matter, because after nourishment has been received the vapors ascend to the brain and there they are thickened and condensed by the brain's coldness. This is just as we see in the larger world where vapors that rise up from the earth and water ascend by means of the sun's power to the middle layer of the air. In the brain, then, these vapors are next condensed and block the paths of the animal spirits that minister to sense and motion and prevent the animal power from reaching the exterior senses (this is just as clouds placed between the sun and the moon prevent their light from reaching us), and then the exterior senses must desist from their operation. Sleep then occurs, because "sleep is the restraint of the senses," as is said in *On Sleep and Waking*.[21]

Another cause stems from the end, because every mover, in moving the moved, is fatigued by the continuous motion. But the soul is the principle of moving the body, and, while moving, it is moved in an accidental manner and thus is fatigued. For this reason, rest is necessary to it, and this comes upon it as sleep.

A third cause can be that when two powers are in something, if one ceases then the other is strengthened. Thus, if a human having two eyes loses one, he will see better with the one left to him than he did previously with the other. But an animal that receives a great deal of nourishment, not continuously but all at once, needs, then, its natural power strengthened after the reception of nourishment. But this can only occur when the other powers are at rest, and this is why sleep is necessary. Thus digestion is performed better during sleep than during waking.

1. On to the arguments. To the first, one must respond that power differs in an animal and in an inanimate thing because the former is much nobler. But now it is the case that there are few operations in inanimate things. For example, in fire there

21. See Ar., *De somno et vigilia* 1 (454b10f.).

is an operation that follows from its substantial form, and that is why it can operate as long as its form remains. But in animate beings there are some operations that relate to well-being that are not consequent on the potency of its form, but come forth from the form itself, with the mediation of its powers. And thus, because they do not follow the form immediately as do the operations of inanimate things, therefore, etc.

Or, it can be said in another way that although the animal may have more of form, it is nevertheless not necessary that it operate this way continuously.

2. To the second argument one must respond that an animal power is one thing, and both the vital and the natural powers are something else. For the animal power is present in a human for the sake of well-being [*bene esse*], and this is true at least of the external senses. But the natural and vital powers are for the sake of being [*esse*], since the natural power exists for the restoration of the moisture that is lost whereas the vital power serves to regenerate the heat and spirit that have been consumed, and since an animal cannot exist without these and since they are nevertheless being lost continually, this is why the vital power (which is in the heart) and the natural power (which is in the liver and in other members) act continuously, whereas the animal power does not.

Another cause can be that the natural power and the vital power are strengthened by their objects, but that this is not the case for the animal power. For the visual power is not strengthened by color, and hearing is not strengthened by sound.

3. To the third argument one must reply that sleep is in the animal not only for the sake of the digestion of nourishment but also for the sake of rest for the senses, which are continually fatigued. And, furthermore, a plant continuously takes in nourishment but does not take in more than it can digest, and this is why sleep is not necessary for a plant. But the animal takes in its nourishment all at once and this is why it needs the digestive power to be strengthened more after receiving nourishment than before doing so, which cannot occur unless some other powers cease to act. This is why, etc.

Question 10: Whether sleep is present in every animal.

Next one asks whether sleep is present in every animal.

1. It seems not. For many animals live the life of a plant, as will be said below. But sleep is not present in plants and therefore is not present in such animals either.[22]

2. Furthermore, sleep is a restraint of the exterior senses. It prevents the influence of the interior powers on the exterior senses. But there are many animals that lack exterior senses, and therefore, etc.

3. Moreover, when the agent is near to the object acted on, it is necessary that this one act and that one be acted upon, just as, when sight is near to a visible object, vision necessarily occurs. But there are many animals that never close their eyes, like fish and the hare and other animals with a ringed body. They therefore see all the time, and as a result they never sleep.

The Philosopher proves the opposite in the text.[23] For he implies that every animal occasionally ceases from its sensory operations; but such a cessation occurs by means of sleep, and therefore, etc.

One must say that sleep and waking are present in every animal. For sleep is the restraint of the sensitive part. And this occurs at some time for every animal because the operation of sense occurs with a corporeal organ mediating it, and the exertion of one operation or the operation of one power yields to the operation of another. This is evident because when someone intensely imagines something, he does not perceive the things which are offered to sense.[24] This is clear in very studious people who do not show interest in corporeal pleasures. But the sensitive part is very active in a waking state. At such a time, therefore, the natural power yields, and this is why it is necessary at other times that the sensitive operation should yield to some extent and the natural operation be intensified. And sleep

22. Cf. Ar., *GA* 5.1 (779a1); A., *DA* 19.1.2.5–7 (*SZ* 2: 1334–36); *De somno et vigilia* 1.1.1–6.

23. Ar. *HA* 4.10 (536b24f.).

24. Cf. A., *De natura et origine animae* 2.16 (p. 42, 85–89).

therefore comes upon every animal so that the sensitive powers might rest and the nutritive powers work more freely.

1. On to the arguments. To the first, one must respond that some animals are said to live the life of a plant because they are immobile after the manner of plants and they share much in common with plants. But this sharing exists among the powers of the vegetative part, for which sleep is not suitable. But every animal above the plant level has sensation, and this is why sleep comes upon animals, although not upon plants.

2. To the second, one must respond that sleep is the restraint not only of the exterior sense but also of the common sense that is generally found in every animal.

3. To the third argument one must reply that some animals live in a dense, fluid element, like water, and they therefore have hard eyes, so that they will not be harmed, and on account of this these animals do not need eyelids or a covering for the eyes. As a result, they sleep with their eyes open. Thus, although the eyes are not covered, nevertheless the visual power can be withdrawn or directed away from such eyes. There is another reason regarding the hare, namely, that nature is very solicitous toward the preservation of the individual. This is why when someone approaches a terrible place or a place in which something terrible dwells, nature frequently causes a numbness and a shiver and this is so that the individual will be protected from the terror. This is just what frequently happens when someone encounters his enemy or a wolf. On, then, to the proposition. Because the hare is a melancholy animal and very timid,[25] nature permits its eyes to remain open so that it will perceive better if it encounters something harmful.

Or one can say that this is due to the shortness of its eyelids, because it naturally has short ones for the reason already cited, namely, on account of impediments, so that it may perceive those things even while sleeping and waking, etc.

25. A., *DA* 22.2.1.110 (*SZ* 2: 1515).

Question 11: Whether sleep is an affect [passio] *of the heart or the brain.*

Further one asks whether sleep is an affect of the heart or the brain.[26]

1. It seems it is an affect of the heart. For the same thing is the principle of motion and rest. But the heart is the principle of motion, and therefore also of rest and, consequently, of sleep.

2. Furthermore, motion follows sense. Therefore, when sense is restrained, motion is prevented. But when motion is prevented, an impediment occurs or its principle is impeded, and this is the heart; therefore, etc.

3. Moreover, during sleep heat and the spirit turn back to the interior parts. Therefore, sleep is an affect of that part from which the heat and the spirit flow. But this part is the heart; therefore, etc.

To the contrary. Sleep is not an affect of that part whose power is not restrained during sleep. But the heart is not restrained in sleep, for it infuses heat and spirit and life, which is evident from the pulse that moves during sleep just as it does during waking.

Moreover, sleep is a fetter for sensation as far as its operation is concerned. But the brain is the principle of sensation so far as its operation is concerned. Therefore, sleep is an affect of the brain.

One must respond that sleep is an affect of the brain. For sleep is caused by a multiplying of the vapor all the way to the brain. For when the vapors that have risen are condensed by the cold and moisture of the brain, the external senses are clouded and the animal sleeps. And this is why animals with small brains, like flyers, or with dry brains, like the dog, sleep less than a human, whose brain is large, cold, and moist. Thus in a small brain, as in a flyer, the vapors are not so condensed, and this is why it does not sleep as much. And, once again, children sleep more

26. Cf. A., *De somno et vigilia* 1.2.2–4.

than old people owing to the abundance of moisture that is in children. It is clear from this that the brain is the principle of sense and motion with respect to their operation. This is so even though the heart is their principle with respect to their first origin, and during sleep sense and motion are restrained with respect to their operations and not with respect to their first act, in which respect they are referred to the heart. This then is a sign that sleep is an affect of the brain and not of the heart.

1. On to the arguments. To the first, one must say that the principle of sense is threefold. The first exists with respect to its origin; the second, with respect to its operation; and the third with respect to its turning back internally. The heart is the principle in the first and third modes. The brain is the principle in the second, and this is why, just as the brain is the principle of sensitive motion, so too is it the principle of rest. And this rest is sleep, because opposites have to concern the same thing.

2. To the second, one must say that the heart is the immediate principle of heat and of the spirits, but not of sensitive motion, and this is why, although sensitive motion may be impeded, it is not necessary that the heart be impeded.

3. To the third argument one must reply that heat exists for the sake of three things: namely, for life, for nourishment, and for the operations of the senses, because without heat there can be no operation of sense. And during sleep a turning-back of heat occurs with respect to the third instance, namely, the heat of the sensitive operations, so much so that it abandons the external senses altogether, but does not do so in the first and second instances. And this is why it is not necessary for sleep to be an affect of the heart.

Question 12: Whether sleep is an affect of the common or a particular sense.

Further one asks of which power is sleep an affect [*passio*], namely, of the common sense or of a particular sense.

1. And it seems to be an affect of a particular sense. For sleep is a restraint and an affect of the first sense, as *On Sleep and*

Waking maintains.[27] But the first sense is touch; therefore, etc.

2. Moreover, all the members are restrained during sleep. Therefore, it belongs to that power that extends through all the members. But such is touch. Therefore, etc.

3. Moreover, sleep is the result of that power whose nature is to be acted on by things that cause sleep.[28] But sleep is caused by the action of sensible qualities. Therefore, it is an affect [*passio*] of touch since qualities are the object of touch.

4. Besides, an interior power is strengthened during sleep, and an external power is weakened. But the common sense is an internal power. Therefore, sleep is not its affect, but rather the affect of something external.

The Philosopher suggests the contrary in *On Sleep and Waking*, for he says that "sleep fetters the first sensitive."[29] But the first sensitive is the common sense; therefore, etc.

It must be said that sleep is an affect [*passio*] of the common sense. For no one particular sense is restrained during sleep, but each one is. It is then necessary that something be restrained first on which all the particular senses are based, and such is the common sense. For common sense is related to the particular senses just as the center is related to the lines stretching out to the circumference, as the Philosopher has it in the second book of *On the Soul*.[30] The sensitive power is diffused by the common sense to each of the particular senses, and all the variations of all the senses are terminated at that one, and this is why the particular senses are restrained when the common sense is restrained.

Moreover, sleep is an affect of the brain, as has been said. Therefore, it is an affect of some power whose organ is the brain. But the brain is not an organ of one particular sense.

27. Ar., *De somno et vigilia* 2 (455b10); 3 (458a28–29). Cf. A., *De somno et vigilia* 1.2.1.

28. We have emended the unintelligible *virtutis cuius natum* to read *virtutis cuius natura*.

29. Ar., *De somno et vigilia* 2 (455b10); 3 (458a28f.).

30. Ar., *De anima* 3.2 (which in Albert is 2.2), (427a10f.). Cf. A., *De anima* 2.4.11; *De somno et vigilia* 1.1.9.

1. On to the arguments. To the first, one must reply that something can be called a principle with respect to something else in four ways: in one way, with reference to its origin; in another way, with respect to the immediate influence of a power; in a third way with respect to a necessity of life; and in a fourth way with respect to cognition. In the first way the vital power is a principle of sense; in the second way, the common sense is a principle of the particular senses; in the third way, touch is the principle of the other senses; but in the fourth way it is sight, because it reveals to us the many differences among things, as the Philosopher says in the first book of the *Metaphysics*.[31] Thus when the Philosopher says that "sleep is the affect of the first sensitive," one must say that he is speaking of the first sensitive in the second way, and not in the third way.

2. To the second argument one must reply that during sleep restraint can exist in individual members in two ways: either because that which is diffused throughout all the members is restrained first, or because something is restrained earlier because it is diffused in this way. Thus, in sleep, there is a stupor in all the members, not because touch is restrained first but because the common sense, from which the power of touch flows, is restrained.

3. To the third argument one must reply that something that is acted on [*pati*] by those things that cause sleep can be acted on in two ways: either through a corporeal impression alone, or with the judgment [*discretio*] of such an impression. Touch is thus acted on by tangible objects, because it judges among them. But the common sense is not acted on during sleep in this way; rather, the organ of the common sense receives a corporeal impression from the tangibles themselves, and that is why its power is dulled. Thus, although sleep is caused by the action of tangibles, nevertheless it does not belong to its affect [*passio*] <neither is it the result of its power>[32] that it is acted on by them with judgment, but belongs to that which is acted on through the mediation of a corporeal impression.

31. Ar., *Metaph.*, 1.1 (980a24f.).

32. The material within <> represents the editor's addition.

4. To the last, one must respond that an interior power is multiple: one is natural, vital, and another animal. In sleep the vital power is strengthened, to wit the natural power of the heart and the lungs and the like. An indication of this is that digestion is performed better during sleep than during waking. This is why sweat is expelled more easily in sleep than while awake, because sweat arises from the work of the natural power. But the interior animal power is not strengthened [during sleep]. Still it can be said that one interior animal power is only receptive, like common sense, and another is retentive, like the estimative and imaginative. And in sleep the interior sense is restrained with respect to apprehension but not with respect to the retention of things apprehended earlier, and not with respect to the comparison of these things to one another. And this is why many people while sleeping imagine many things which they do not imagine while awake, and they recollect and dream. Nevertheless the interior sense is sometimes restrained absolutely with respect to all its powers and then nothing whatsoever appears to the one asleep.

Additional Questions for the Fourth Book
[Quaestiones . . . annexae]

Question 1: Why salt placed in water liquefies, but it does not liquefy when placed on a burning tile or in oil (or boiling honey).[33]

One asks why it is that salt placed in water liquefies, but it does not liquefy when placed on a burning tile, but hardens instead. Similarly, when placed in oil or boiling honey, it does not liquefy.

In order to see this better, one must know that the generation of salt occurs from heat just as if that were its craftsman and from salt water as its matter. For the heat of the sun acting on the water consumes the more liquid and more subtle parts of it, and, once these are consumed, a thickening and harden-

33. These questions are missing from book four in MS *T*.

ing results and a transmutation into the essence of salt from the remaining thick and earthy parts.

Therefore, we say that when salt is placed on a hot tile with the heat acting on it, the more liquid parts are consumed and imperceptibly escape, and there occurs a greater thickening and hardening in the remaining thick parts. We perceive this on a newly made tile that is placed in the sun or on the fire; if it is left there for a long time, the action of the heat changes it. Since it does not encounter liquid and subtle parts on which to act, it acts on the solid parts and consumes and annihilates them, and thus destruction occurs. Similarly, salt placed in honey or oil does not liquefy, and this is because a blockage occurs from the viscosity and unctuousness of the oil. As a result, the liquid and subtle parts of the oil cannot penetrate [the salt] so as to effect a separation and liquefaction of the same [salt] by means of interspersing [the oil] between one part and another of the salt.

Similarly it does not liquefy in honey on account of its viscosity causing blockage in the pores, and this is why it is hardened in these two by their heat and dryness, whereas placed in water it liquefies. And this occurs because the fluid and more liquid parts of the water enter into the substance of the salt and separate part from part, and once separated these are softened; as a result of such separation and softening there occurs a liquefaction and a reduction to the proper matter.

Question 2: Why musk placed among fragrant things loses its aroma and later, when placed in a foul-smelling place, it recovers its odor.

One asks why it is that when musk is placed among fragrant things it almost loses its aroma and later, when placed in a foul-smelling place (for example, when it is put into a latrine) it recovers its odor.[34]

In order to see this better, one must know that an aromatic

34. Cf. A., *DA* 22.2.1.120(75) (*SZ* 2: 1552), where this *muscus* is defined as the exudation of a musk deer that is found on trees. It was probably some sort of resinous sap.

odor has to occur from a warm and dry complexion and from a subtle and pure substance, and it is necessary for every aromatic to be hot, both partially and totally, and to have a subtle substance. When, therefore, musk is placed among odoriferous things that are themselves warm and dry, and with the warmth and dryness of these others acting on it, that which is more subtle and pure in the musk is released, and thus the aroma that was contained in the musk but has been released is attracted by the odoriferous materials surrounding it, owing to the likeness that they bear toward the musk, and it is then consumed. That, however, which is thicker and more impure (the type of things that have no aroma in them) remains in the musk, and thus the odor of the musk is reduced.

Another reason that the odor of the musk is not perceived so well is because when a change occurs in the odors and properties of the other aromatic materials, then once the instrument [of smell] is placed nearby, the spirit of smell is changed by all of these others. Thus, when the soul is intent upon perceiving all the other properties, it cannot as fully perceive the odor of the musk. Thus, "intent on many, there is less for an individual of the senses."[35]

But when it is put in a foul-smelling place, it recovers its own odor. Let us see why this is. A foul smell results from an accidental heat and a great deal of moisture and a crude substance. In the musk the fetid thing is preserved as a result of the moisture. The heat of the fetid thing acting on the musk resolves the heat of the musk, and, when it finds something contrary to itself, it struggles against it and gathers strength within itself, and, strengthened by this struggle and acting on the cruder parts of the musk, it resolves them into fiery parts, and purifies and refines them into aromatic parts. Another reason is that nature delights more in an odoriferous thing than a fetid one, and the musk, infected by the property and odor of the fetid thing, when it is placed near the nose, causes the soul, on account of its horror at the fetid odor, to perceive more fully and more avidly the odor of the musk, which is more delightful and friendly to it.

35. See Wilhelm Binder (1861), 287, n. 2512.

Question 3: Why charcoal is kindled by a crystal when it is placed opposite the sun's rays.

One asks why it is that if a crystal is placed opposite the sun's rays entering through a window of a house or through some opening and if a bit of charcoal or some little piece of food is held behind the crystal from the rear, it is kindled. I have heard that the same thing happens when a bottle filled with water is exposed to the sun.

For this one must know that the crystal has to be made by a congealing coldness. For it is by means of cold that water is congealed into snow, and by a more intense cold it is congealed into hail, and by an even more intense cold it is congealed into a crystal.

Again, one should know that a crystal is naturally moist and contains in itself a certain moisture that is a fomentation [*fomentum*] for the heat.

And again a crystal is a radiant body that is transparent to the rays of the sun. When, then, it is placed opposite the sun's rays, these rays, penetrating through the most narrow pores into the substance of the crystal, are then made very subtle, and in the crystal they strike one another, and, because of this thinning and striking of the rays, and because there is moisture there, which is a fomentation for the heat, <they ignite>, and when the rays exit from the other side of the crystal they are thinned even more and strike one another and are ignited. Even the air between the crystal and the food is ignited—that is, it is combustible—and this is why the food itself is ignited. For air is a material that is close to fire, and a very rarified air is changed into fire, and the fire is condensed into heat.

Note that the lion is a choleric animal, namely, very hot and dry, but the burned bile [*cholera adusta*] is burned in it, and therefore it continuously suffers a quartan fever from the natural melancholy.

Note too that every poison is very dry either in act or in power.

Question 4: Why snow is preserved in barley chaff.

One asks why it is that snow is preserved in barley chaff.

We say to this that chaff is cold and dry and both barley and snow are very cold. Therefore, since that which is cold has its movement toward the center and since that which is cold and dry also has its movement toward the center, the snow is not dissolved but is preserved undissolved.

Question 5: Why pears placed in barley chaff become ripe.

Again one inquires why it is that pears placed in barley chaff become ripe.[36]

We say it is because generation has to occur from the heat of the pears. Now there is a certain earthy and crude substance in pears, and a certain moisture exists in pears. Moreover, in pears there is also a certain spiritual and smoky substance. When, then, they are put in the cold and dry chaff of the barley, their pores are constricted, and, once they are constricted, the airy and spiritual substance is retained so that a particular heat is born whose efficacy is increased in the pears. Acting then on the substance, it dissolves and softens it, and so ripening occurs. But if the pears are left for a long time, a greater dissolution occurs than the consumption and purification of the things dissolved can bring about, which is why they rot and sometimes putrefy.

Question 6: Why women have such large breasts, and men do not.

One asks why it is that women have such large breasts, and men do not.

One must reply that there is a threefold cause, namely, a final, a material, and an efficient cause. The final cause is this: infants do not have instruments suitable for chewing and grinding up hard and solid foods, and this is why provident and pious

36. Cf. A., *De veg.* 7.2.3.

nature established certain instruments, namely, loose-textured and spongy breasts, within which a blood-filled mass might be received and bubble up. And by means of this bubbling action a certain white sweetness, set free from the substance of the breasts, whitens it, and so too by this bubbling action it is converted into the milk on which the fetus is nourished.

Another reason is that some veins proceed from the liver to the womb and on the way they split into two. One branch extends to the breasts and the other to the womb.

Again, the womb is joined to certain veins that intersect the middle of the breasts. Therefore, the bloody mass and superfluity contained in the womb—and especially a mass of melancholy blood—are sent through the branches and through these veins in large quantity, and in a quantity even greater than is needed or which suffices for the nourishment of the breasts. By means of the heat and motion of the heart, which is placed under it, the mass bubbles up, and, once digested and dried out by this bubbling action, it is changed into the essence or substance of the breasts. And this is why the breasts of women become large and grow.

The material cause is the great moisture that is in women. For a woman is naturally conceived from the matter of the [moist] humor. Then, the heat acts on such great moisture present in the breasts, digests it, decocts it, and changes it into the essence of the breasts. Therefore, the breasts become large, and moisture is the material cause.

Heat coming from the heart is the efficient cause. Thus the breasts are naturally conceived from a drier material.

Know too that the breasts grow more from the fourteenth year, when women begin to menstruate, because then the superfluities of all of the humors and of the melancholy blood flow together to the womb, and these, when turned back in large amounts to the breasts, are the cause of large breasts. But before the time of menstruation the superfluities do not flow so much to the womb. The breasts increase a great deal in pregnant women, because the superfluities, which are customarily cleansed by menstruation, are then retained in large quantity and sent to the breasts and are a cause of large breasts.

Question 7: Why the members are discovered to be cold in one who is dying but they are warm in one already dead.

One asks why the members are discovered to be cold in one who is dying but they are warm in one already dead.

It is on account of this: that when someone is dying, the blood and the spirits are recalled from the exterior members to the interior ones to assist nature, and this is why the exterior ones are cold. But once the person dies and the war of illness versus nature has finished, nature succumbs and the blood is then redirected to the exterior members along with the spirit, and this is why they are warm. Some, after the soul's exit, become engorged when earlier they were not swollen, because at that time a great deal of windiness is generated in them from the mortifying cold, on account of which the belly swells and becomes hard.

Question 8: Why marine fish are very bland although they are nourished on salt water.

Again one asks why marine fish are very bland, although they are nourished on salt water.

Response. Seawater consists of three elements. Although it is salty, there is still something contained in it that is not salty but is sweet and likeable and similar to the fish's substance, namely, a certain moisture. The members then attract it to themselves for their nourishment by a natural sensation.[37] That is why fish are not salty.

Or, in another way. Although they are nourished by salt water, on account of the digesting and decocting heat of the members they lose the saltiness and are despoiled of it.

37. Lit., "a judgment that is natural to sense," *sensu naturali sententia*. This redundancy seems to be an interpolation.

Question 9: Why insanity follows after being bitten by a rabid dog.

Again one asks why insanity follows after being bitten by a rabid dog and why the man who has been injured spews forth poison just like one who is driven by an evil spirit.

Response. While the wound is open, the poison is not perceived and it evaporates. After the wound has closed, the poison ascends to the brain and infects it.

Cure: Place a cautery on the head and use theriac.[38]

Question 10: Why a person who passes by a corpse suffers goosebumps.

Again one asks why a person who passes by a corpse, or if there is someone hidden, suffers goose bumps [*horripilatio*].

Response: the hidden person emits an exhalation that touches the one passing while infecting the air, and thus goose bumps occur, or he imagines that there is something fearful there, and this is why he is afraid.[39]

It is the same for a corpse, because it infects the air, etc.

Question 11: Why madmen have so much strength.

<Again, one asks why madmen have so much strength.

Response.> Madmen have such great strength because, although they have an infected brain, they are not afraid of anything. Thus the brain is not constrained or straitened. A large quantity of the motive spirit distributed throughout the members produces great strength.

Another reason is that although the brain is infected, they do not think about the future nor that anything bad can happen to them, and this is why they dare anything.

38. On the famous poison antidote called *theriaca,* see *DA* 1.1.8.102, 2.2.4.119 (*SZ* 1: 84, 342–43). Yet another valid meaning is "treacle" (Latham, s.v. *theriaca*) which might better suit a cure of this sort. Cf. *De veg.* 6.455.

39. Cf. A., *De causis propr. elem.* 2.2.1. *Horripilatio* also implies that the person's hair stands up on end.

Another reason is that although the spirit in them is not intent upon producing imagination, reason, or memory, it is nonetheless diffused in large quantity throughout the members and produces great strength on behalf of the amount of fumes generated in them, which, wandering among the members, produces great strength.

Question 12: Why the blood of the one slain flows while the murderer is present.

One asks why the blood of the one slain flows while the murderer is present.

Response. It is by divine vengeance, so that he would be moved to repent. Or the killer is moved himself in his conscience, and thus the spiritual organs[40] are moved, and when these are moved, they emit a spirit and fume from themselves. This then moves the air, and with the mediation of the moved air the breath is borne away to the body of the one killed and this warms the body, because it is still fresh. Thus the humors are moved, and thereby a discharge of blood occurs.

Question 13: Why a morsel of bread placed in a very strong wine increases the sharpness of the wine.

One asks why it is that a morsel [*mica*] of bread placed in a very strong wine increases the sharpness of the wine in strength and violence.

We say that although there are four elements in wine, nevertheless it gets a hot sharpness and strength only from the fiery parts, which are warmer and subtler. Thus the stronger the wine, the warmer it is.

We say that since the substance of bread is easily penetrable, the fiery parts, which are more subtle, sharper, and more penetrable, enter into it. Therefore, with the gross and less warm remaining behind, the strength and heat of the wine are diminished.

40. "Spiritual organs": i.e., organs having to do with respiration and the movement of spirit.

Moreover, we say that the morsel [*offa*] consumes the stomach's heat, which acts on it and releases in it the more subtle parts of the wine first, which, as the more suitable parts, had entered the bread's substance. Once these subtle parts have been resolved and are acting on the morsel and even on the stomach's heat, from there a great vapor [*fumus*] rises up, which penetrates the head with its power and presents an injury to its accidental moisture and, since the heat is being blocked by its movement, indigestion occurs. An indication of this is that the ancients gave to those suffering a fever some water in which there was a morsel of bread and they did this so that the warmer (or less cold) and subtler parts of the water might enter the substance of the bread while the cruder and colder parts of the water might remain behind, and thus the water was cold and became still colder. Moreover, they gave the water for another reason as well, since with the admixture of the bread the water developed a certain affinity and inclination [*amicabilitas*] for the members, which is why it is better received by the members and remains longer in the digestive process, and the unmixed can alter and change something better in its own right.

Question 14: Why vision takes pleasure more in the color green than in any other color.

Again one asks why it is that vision takes pleasure more in the color green than in any other color.

One must know that in order to be seen better, some of the colors are median and some are extremes. The extremes are like black and white, whereas the median are red and all the rest. Again, some colors are simple, and some are composite. The simples are those that are present as a result of individual elements, as red is present from fire, white from water, and black from earth. The composites are those that are present from various elements, like green, which results from fire and earth, and is a composite resulting from the color red and black. Again, some colors break up the visible spirit, like white and red, and some unify it, like black.

Question 15: Why roasted meat strengthens more than does boiled meat.

One asks why roasted meat strengthens more than does boiled meat.

I answer that one must reply that roasted meat resembles the heart's heat once its viscosity has been consumed, and since it is most like [the heart], it strengthens it more than does boiled meat.

Question 16: Why one who has been wounded in the arm's muscle loses the power of speech.

Again one inquires why one who has been wounded in the arm's muscle loses the power of speech—whether it is due to fear or weakness.

One must reply that it is on account of both. For on account of fear the spirit suddenly withdraws to the interior parts, and since it is confined there and is not diffused to the nerves and muscles, this is the cause why speech ceases. It can also be owing to weakness, because a great deal of blood has flowed out and the spirits evaporate. Thus, once the body has been weakened by a loss of blood, the spirits' instruments have been rendered incapable of forming speech.

Question 17: Why members that are weakened by hunger are immediately strengthened by food that has been received, though not yet incorporated, in the stomach.

Again, when all the members have been weakened by hunger, how is it that, once food has been received in the stomach, even though nothing is incorporated yet, all the members are strengthened and they are made stronger?

We say: Desiccated on account of the preceding hunger, they absorb food avidly, and as a result of this the released moisture represses the dryness a bit, and a strengthening thus occurs. But when a little bread is received, all the members are strengthened. Thus a great deal of moisture is released from it and diffused through the members.

We say otherwise that although all the members are weakened, nevertheless the stomach is weakened more. For the remaining members draw from it, but it draws almost nothing from the others. When it happens that the food is consumed, those parts through which it passes and in which it is found are strengthened by the food's moisture. Thus the body's weakness is strengthened to some degree.

Or it can be referred to the soul. For when the body is afflicted with hunger, the person (that is, the soul) is afraid that it may suffer some greater affliction. When, then, it eats, it begins to be strengthened, observing that it has escaped some future evil. We see somehow that if someone is condemned and despairs of life, and if one should say to him: "Do not be afraid," he hopes on the one hand and despairs on the other.

Question 18: Why does a dead person staying in the water rise to the top, and a living one sink to the bottom?

Again one asks why a dead person staying in water rises to the top, and a living one sinks to the bottom.

Response. We know that the spirit is the bond between the body and the soul, and once it is released, the soul withdraws from the body. But because nothing can exist in a vacuum, air, which is subtler, moves in to where the spirit was earlier. And just as what is heavier seeks the lower, what is lighter seeks the higher.

One objects: does not water enter into a dead body and not a living body? Therefore, the dead body ought to be heavier.

Response: that water enters with air and that the air cannot be exhaled, and therefore it is diffused throughout the body. Therefore, the dead body is toward the top. Is it not for this reason that a dead person seems to be heavier to carry than a living one? Therefore, it is not lighter, but heavier. But the fact that one seems heavier to carry results from the difficult nature of the carrying. If it can be carried because it is light, it still seems heavy, and conversely, etc.

Question 19: Why a certain wise man after being bled forgot all that he knew.

Again one inquires why it is that a certain wise man after being bled forgot all that he knew.

Response. When someone sleeps, the humors, impelled by the spirits, cross to a phlegmatic place. Thus, on a day when one is being bled one ought not to sleep, on account of the bubbling up of the humors traveling to the aforementioned place or ascending with an overpowering heat to the head. The bubbling of the humors turns itself toward the chamber devoted to memory and, after the nerve has been blocked through which the animal spirit passes, memory ceases.

Question 20: Why someone suffering a fever lost the knowledge of letters.

Again one inquires why someone suffering a fever lost the knowledge of letters.

I reply that the fever was in the brain or in some part having a connection to the brain, and the pores closed up on account of the accumulation of humors that occurred near the chamber devoted to memory, through which the animal spirit was conveyed from the rational chamber to the chamber devoted to memory, or perhaps a drying out of its nerve occurred.

Question 21: Why some people who eat more are thin, whereas others who eat only a little are fat.

Again one inquires why some people who eat more are thin, whereas others who eat only a little are fat.

I respond: Thin people are very hot and dry, and thus the food is consumed by the heat, but the dryness blocks the incorporation of the foods. Others are hot and moist, and the moisture resists the heat so that digestion and the incorporation of foods occur.

Question 22: Why were eyes and ears absent in a boy that was born?

Again, why were eyes and ears absent in a boy that was born?

Response: there can be a defect in the sperm's [complexional] balance or with the formative power either in the womb or at the beginning of generation.

Question 23: Why was it that someone who saw an enemy dropped dead?

Again one inquires why it was that someone who saw an enemy dropped dead.

Response: Melancholy, aroused by fear, squeezed the heart, and the vital spirit was suffocated. For just as the heart expands with joy, so too it contracts with fear.

Question 24: Why is it that must, although it is clearer, does not cause drunkenness?

<Again one asks> why it is that must, although it is clearer, does not cause drunkenness.

Response. For drunkenness to occur, sharp heat and subtlety have to occur simultaneously. Although must is warm, it does not, on account of its thickness, have a penetrating power, and thus it does not cause drunkenness. Water, however, licks away at the subtle smokiness and does not cause drunkenness because it lacks a penetrating power, etc.

BOOK FIVE

Question 1: Whether coition is necessary for the generation of animals.

"ABOVE WE DISCUSSED the disposition of all the members of animals," etc.[1] In this fifth book one asks about generation, and first one asks about coition and propagation.[2] First, whether coition is necessary for the generation of an animal.

1. It seems not. The generative power is bestowed upon every animal so that the species, which cannot be preserved in the animal itself, may be preserved in one like itself. Therefore, since this power is given to every animal, the animal does not require anything beyond this in order to generate.

2. Moreover, every animal is corruptible. Therefore, it needs something in the animal itself through which it can preserve its nature. Therefore, there exists a principle sufficient for generation in every animal *per se.*

The Philosopher says the opposite, and it is also apparent to the senses.

To this, one must reply that coition is necessary for the generation of some animals because the more perfect something is, the more things are required for its generation. Thus, for the generation of some imperfect animals a universal agent with properly disposed matter is sufficient, as is evident in the generation of those from putrefaction. But for the generation of perfect animals there is required an agent of the same species as well as a universal agent. But generation is of two types: one is the change of the whole into [another] whole, and in this way water is gen-

1. Ar., *HA* 5.1 (538b28f.). Cf. Avic., *DA* 5.1 (fol. 33vb).
2. A., *DA* 5.1.1.1–3 (*SZ* 1: 487–88).

erated from fire.[3] The other type of generation entails the separation of a part from the whole, and the generation of some animated things is like this. But in some animated things there exists only the active power of the one generating, and in others there exists only passive matter. And this is why, since every generation occurs from matter and from an agent, it is necessary that the two concur for the generation of ones like this. And this is why coition is necessary for the generation of ones like this.

1. On to the arguments. To the first, one must reply that the generative power is given to an animal for the sake of the preservation of the species. But this power can be understood in two ways: either as that which reproduces one like itself by itself (and this is not given to every animal) or as that which reproduces one like itself in another, and this can only occur by means of the contact of the male's penis with the woman's vulva, and this is why coition is necessary.

2. To the second argument one must reply that because every animal is corruptible, this is why it does not require something by which to preserve itself but instead requires something by which the species may be preserved in another, and this can occur by the propagation of the two [animals] and the action of one on the other through coition, by which a third animal is generated.

Question 2: Whether coition is necessary for every animal.

One asks further whether coition is necessary for every animal.

1. And it seems so. For the operation of a nobler power is itself nobler. But an animal is nobler than a plant; therefore, the

3. Although counterintuitive, this claim seems to refer to the possibility that one element will be entirely transmuted into another. In the background may also lie the belief that the fire of the Temple's altar was miraculously transformed into water and the water then back again into fire at 2 Mc 1.21–22. Indeed, Peter Damian remarks that "water is born of fire, and fire in turn is produced from water." See *Ep.* 153.52, trans. Blum and Resnick, Fathers of the Church, Mediaeval Continuation 7 (Washington, DC: The Catholic University of America Press, 2005), 44.

generative power is nobler in the animal than in the plant. But this would not be the case if generation did not occur differently in an animal than in a plant, and therefore, etc.

2. Moreover, a power does not lack an operation. But the operation of the vegetative power is to generate. Therefore, since the vegetative power is in every animal, each animal has a potential for generating. But this generation is present properly through coition with an animal, and therefore, etc.

The Philosopher implies the opposite in books four and five.[4] For he says that the differentiation into male and female is not present in every animal, which would, however, be the case if this type of generation were suited to every animal.

One must reply that coition is not present in every animal, because the more perfect a thing is the more its perfection extends itself to many. But the operation of a perfect animal is to produce one like itself. But, according to Aristotle, every thing has been perfected when it can produce one like itself. But many animals are imperfect, and thus it is not surprising that they cannot produce ones like themselves. This is why they do not have the members that are properly designated for generation, but rather are produced by a universal agent, that is, by the sun, from mud or from putrefied matter.

1. On to the arguments. To the first, one must reply that a plant is more imperfect than an animal, and this is why fewer things are necessary for its generation. Thus the parts of a plant come to be from nutriment, over which the plant has power, and this is why if some parts are cut off from the plant they can be regenerated. But an animal is more perfect, and more things are required for its generation, and this is why the power that is in one like itself is not sufficient for the generation of a given animal in the same way that the power which is in each part of the plant is sufficient for the plant's generation.

2. To the second argument one must reply that wherever there is a power there is also the operation proper to the power. But the proper operation of the vegetative power is not to gen-

4. Ar., *HA* 4.11 (537b22f.), 5.1 (539a24f.). Cf. Avic., *DA* 2.4 (415a26f.).

erate but to nourish and cause growth. Generation, however, is its proper operation only in the more perfect beings, and this is why, although every animal has a vegetative potency, nevertheless not every one can generate one like itself.

Question 3: Whether the greatest pleasure occurs in generation by means of coition.

One asks further whether the greatest pleasure occurs in that generation that takes place by means of coition.[5]

1. And it seems not. For pleasure results from the union of one thing suited to another, and sadness occurs from their separation. But in coition the semen, which is nearest to conversion, is separated. Therefore, there is more sadness in coition than pleasure.

2. In addition, each and every one takes more delight in preserving itself than in preserving another. Now, nutriment exists for the sake of preserving itself, whereas coition or generation exists for the sake of another. Therefore, there is more pleasure in nutrition than in coition, and therefore the greatest pleasure does not occur <in coition>.

3. And this is confirmed in this way. Pleasure follows sensation. But two senses operate in nutrition, namely, touch and taste, but in coition only one operates, namely, the touch from the contact of the penis with the vulva. Therefore, etc.

The Philosopher implies the opposite in the text. For he says that animals change their calls and are in motion at the time of coition and call out to one another, and many animals are solitary, and this is why, etc.[6]

One must reply that each one rejoices at its proper operation, as it is said in the *Ethics*.[7] For the just rejoices in just operations, and the temperate in temperate operations. But nature ordains nutriment for the sake of the preservation of the indi-

5. Cf. A, *DA* 5.1.1.8, 9.1.7.66, 9.2.3.104–5, 15.2.3.60.80–81 (*SZ* 1: 490–91, 798, 814–15; 2: 1120–21). See above *QDA* 1.13.

6. Ar., *HA* 5.14 (544b22f.).

7. Ar., *Eth. Nic.* 1.8 (1099a15f.).

vidual and the sexual act for the sake of the preservation of the species. And this is why nature joined the greatest pleasures to these operations, and the more it intends the preservation of the species than the preservation of the individual, the more it ordered the greater pleasure for the sex act than for the nutritive act [*in opere venereo quam in opere nutritivae*].

Moreover, pleasure results from the union of one suitable thing to another suitable thing. But semen is caused by the superfluity of aliment, which aliment is nevertheless suited for and close to conversion. Thus, when the semen courses through those members that are especially sensitive and nerve-filled, it causes the greatest pleasure. And in addition, the expulsion of what is superfluous is pleasurable, generation occurs through a superfluity, and semen arises from the superfluity of aliment. This is why its expulsion, which occurs in coition, is pleasurable.

1. On to the arguments. To the first, one must reply that if semen were suited for a part, and if it were on the way to conversion into a member, then its separation from that part into which it is convertible would not be pleasurable. But this is not the case in coition, when the semen courses through the nerve-filled members which are sufficiently restored and do not require its conversion. Thus its flow causes a certain titillation in the members, and it is not converted substantially into the members themselves, and this is why it can be expelled without pain.

2. To the second argument one must reply that although two senses concur in nutrition, nevertheless pleasure in coition is greater than that in nutrition. For although sensible pleasure may be greater in nutrition owing to the plurality of the senses [operative in it], nevertheless natural pleasure is greater in the sex act owing to the contact with that which is suited to it, or owing to the contact of those things which are more suited to it.

3. To the third, one must reply that nature is of two types: universal and particular. The preservation of the individual is the intention of the particular nature, and the preservation of the species is the intention of the universal nature. And this is why pleasure is greater in that act through which the species is preserved than in that act through which the individual is preserved.

Question 4: Whether pleasure is greater in men than in women during coition.

One asks, further, whether pleasure is greater in men than in women during coition.[8]

1. It seems not. Because the Philosopher says in the first book of the *Physics* that "matter seeks form just as a female does a man and as baseness does the good."[9] But each one desires most greatly that by which it is formed. Therefore, the greatest pleasure and the greatest desire are in the woman. This is confirmed by the fact that love is greater in women than in men.

2. In addition, a doubled pleasure is greater than a simple pleasure. But in men there is pleasure owing to the emission of the semen, but in women there is pleasure owing to the emission and reception [of the seed]. Therefore, etc.

To the contrary. Pleasure arises from the emission of the seed.[10] Therefore, where the seed exists more truly, there pleasure exists more truly. But the seed exists more truly in the male, and this is why, etc.

To this, one must reply that pleasure and desire are greater in men than in women. And the reason for this is that if pleasure arises from a certain activity, there will be greater pleasure where that operation exists more formally. But pleasure follows the sex act, that is, the act of generation. Since, therefore, this operation exists more formally in a man than in a woman, he will have greater pleasure.

Moreover, the cause of desiring coition is heat and a large quantity of semen, because heat moves from the center to the circumference and this is why heat is the motive [cause] in the emission of semen. But heat is more abundant in the man than in woman, and so too is the semen. For although there may be a great deal [of semen] from the superfluous and undigested

8. As an aid to the reader, *femina* is consistently translated "female," and *mulier* as "woman." *Vir* is translated as "man," and *mas* as "male." The resultant awkwardness helps to point out A.'s need for a consistent vocabulary.

9. Ar., *Phys.* 1.8 (192a22–23).

10. Cf. A., *DA* 9.2.3.104–6 (*SZ* 1: 814–15).

moisture, this is nevertheless evacuated in the menstrual flow, which does not occur in men, and pleasure is therefore more intense in men than in women, although perhaps its is more extensive in a woman than in a man, because she takes pleasure in several ways, namely, in both the emission and the reception of seed.

1. On to the arguments. To the first, one must reply that matter is said to desire form, and a female a man, not because the female desires intercourse with the man but because it is understood that every imperfect thing naturally desires to be perfected, and a woman is an imperfect human [*homo imperfectus*] in comparison to the man, and this is why every woman desires to exist under [the category of] manhood [*virilitas*]. For there is no woman who would not wish to put off the basis of her femininity [*femineitas*] and naturally to put on manhood. And in this same way matter desires to put on form, and it is the same for baseness.

2. To the second argument one must reply that pleasure is quantitatively greater in a woman, but is intensively greater in a man. As confirmation one must state that desire, like love, is of two types: namely, true and apparent. True love and desire are greater in the man, and this is why a truer pleasure exists in him. But apparent desire and love are greater in a woman owing to the weakness of her judgment. For just as matter desires to exist under every form, and thus that matter which exists under a noble form desires to exist under a vile one, so is it the same for a woman, who desires to exist under the one she does not have, since, owing to the weakness of her reason, she judges that the one she does not have is better than the one she has. And this is why she desires intercourse more often than a man, because when she is not joined in intercourse she desires to be joined in intercourse owing to the corruption of judgment, etc.

Question 5: Whether pleasure is greater in brute beasts than in the human at the time of intercourse.

One asks whether pleasure is greater in brute beasts than in the human at the time of intercourse.

1. It seems that it is greater in the brute beast. Because one very intense power draws another away from its operation, for intense imaginings perceive less of what is presented to the external senses. But a human inclines more to the animal power than does the brute beast. Therefore, he inclines less to the generative power, etc.

2. In addition, each takes more pleasure in its own proper operation and therefore takes less pleasure in what is contrary to its own proper operation. But the proper operation of a human is [to live] according to reason, and, according to the Philosopher in the third book of the *Ethics,*[11] the generative operation and the operation of taste work in opposition to this. This is why childish sins[12] consist in these things and why a human takes scant pleasure in them.

The Philosopher implies the opposite. For he says that "a human has intercourse at any time," whereas other animals do not.[13] This would only be so if the human found more pleasure in intercourse than the beast, and therefore, etc.

One must reply that the human has more pleasure than the brute beast. The reason for this is that the pleasure of intercourse consists in touch. Where there is a better and surer sense of touch, there is more pleasure. But touch is surer in the human than in the brute beast. This is why, etc. A further indication of this is the abundance of blood and sperm in the human in comparison to the brute beast.

1. On to the arguments. To the first, one must reply that one power is only an animal power and another is only a natural power and some are composed of both of these. The power that is only a natural power draws the animal power away from its operation, as is evident in people who are asleep and, contrariwise, in those who are awake. But the generative power is composed of both the natural and the animal power, and on account of this it is neither blocked by nor drawn away from its

11. Ar., *Eth. Nic.* 3.11 (1118a30f.), 3.12 (1119b5f.).

12. *Peccata*: "misdeeds" or "peccadilloes," not necessarily with a theological overtone.

13. Ar., *HA* 5.8 (542a26–27).

operation by either one of them. This is because it requires digestion of the aliment, whose superfluity is the semen.

2. To the second argument one must reply that the generative operation and the operation of taste do not work in opposition to an operation according to reason, unless it shall depart from the mean, because the moral power, which does not work in opposition to reason, can reside in these operations. Thus the sensitive appetite works in opposition to reason the most when it is not subordinated to the command of reason, but the generative power and the power of taste themselves do not, nevertheless, work in opposition to reason in this way.

Question 6: Whether desire in the operation of intercourse is greater in men or in women.

One asks further about desire itself in this operation, whether it is greater in men or in women.

And it seems to be greater in women, because in intercourse the male and female emit sperm and the menses. That is why there is no menstrual flow in virgins, and this is why they are sick more often. Since a woman naturally desires the menstrual flow—since this is a cause of health and not of trouble, which the retention of the menses causes—and such trouble does not occur in males from the retention of the sperm, then a woman naturally desires [coition] more than a man.

Moreover, the natural and animal powers in women are likewise weaker and more suppressed than in men, as will be said below.[14] The generative power is therefore more intense in them, and, as a result, so too is desire.

Moreover, love burns more ardently in a woman than in a man; therefore, she has more desire.

To the contrary. Desire occurs on the basis of heat. But the male is hotter than the female, and therefore, etc.

To this, one must reply that two things come into consideration with respect to desire: namely, pleasure and a judgment about what is desirable. If we speak of desire in terms of plea-

14. See *QDA* 18.1.

sure, this is naturally greater in the male than in the female.[15] If we speak in terms of the judgment, then one must distinguish, since one judgment is direct and another is indirect. A direct judgment is when something is judged to be pleasurable which is pleasurable. And judgment like this flourishes more in males than in females. An indirect judgment is when something is judged to be pleasurable which is not pleasurable, and this sort flourishes in women. And this is why, when she displays herself to one man, she still desires to display herself to another, for she judges this to be more pleasurable although in fact it is not. Thus, desire in terms of indirect judgment is greater in women, and the arguments show this, but in terms of direct judgment it is greater in males.

Question 7: Whether desire for coition is caused by the hot or the cold.

One asks further, whether desire for coition is caused by the hot or the cold.

1. It seems that it does not occur because of heat, because repletion aids in coition, and this is why gluttony [*gula*] and wantonness [*luxuria*] have a great deal in common. And a thing whose function it is to consume and to empty does not contribute to coition. But heat is like this, and therefore, etc.

2. Moreover, the causes of contraries are contrary, or the same thing existing in diverse modes. But attraction and expulsion are contraries, and the attractive power operates with a mediating heat; therefore, the expulsive power does not. But expulsion is present in generation or coition, and therefore, etc.

3. Moreover, the desire for food or drink occurs with a mediating cold. Therefore, so too, for the same reason, does the desire for coition.

To the contrary. Desire is greater in a woman than in a female brute beast. But this is only because a woman is hotter, and therefore, etc.

One must reply that desire in coition occurs because of heat.

15. Reading *si* for the nonsensical *sie.*

And the reason for this is that the motion of the hot and the motion of the cold are contrary to one another. For heat moves from the center to the circumference, but cold moves from the circumference to the center.[16] But in coition the motion of the semen occurs as if from the center to the circumference, because it moves from inside to outside. Therefore, it occurs with a mediating heat.

Moreover, the opening of the pathways is required for generation, so that the semen may exit or cross over more freely. But heat opens and cold constricts, and therefore, etc.[17]

An indication of this is that hot animals, like sparrows and doves, copulate and generate a great deal, and likewise, women who are hotter have greater desire than cold ones.[18] And this is why it is said that coition occurs because of heat, as if because of an instrument.

1. On to the arguments. To the first, one must reply that heat is of two types.[19] One is elemental heat, and its business is to consume and to empty. Another heat is in something composed of an element and it does not exist in extremity, the way it does in an element, but it is rather ruled by the power of a superior. Heat such as this is in animals, for it is ruled by the soul. Thus, whatever aliment is digested in the animal occurs through heat; but the fact that it is digested this way or that occurs through the soul's power. And heat like this can contribute to repletion and not only to emptying something.

Or one can state in another way that heat's operation is twofold. One operation is to open the pathways, and the other is to consume what is superfluous. It contributes to coition or is a medium for coition in the first way, but not in the second.

2. To the second argument one must reply that by virtue of its double operation heat can both expel and attract. Because it consumes and empties, it can attract; but because it opens the pathways, it can expel.

3. To the third argument one must respond that motion in

16. A., *DA* 20.1.7.36 (*SZ* 2: 1378).
17. Cf. A, *DA* 20.1.4.20 (*SZ* 2: 1368).
18. Ar., *HA* 5.13 (544a25f.); A., *DA* 5.1.6.40–43 (*SZ* 1: 507–9).
19. Cf. A., *DA* 12.1.3.32–42 (*SZ* 2: 906–9).

desire [*appetitus*] (that is, in the consumption of food) is from the circumference to the center, and this is why it occurs by virtue of the cold. But in the emission of the seed it is from the center to the circumference, and this is why it occurs by virtue of the hot. In this way a solution is apparent.

Question 8: Whether the operation of coition flourishes more in a cold or hot region.

Further, one asks in which region the operation of coition flourishes more, whether in a hot or a cold one.

1. It seems that it flourishes more in a cold one. For when the nutritive operation comes to an end, the generative operation begins. For at the point where the nutriment is converted into some member, as it were, sperm and semen arise from its superfluity. Therefore, where the nutritive operation flourishes more, there too will the generative operation flourish more. But this occurs in cold places, because in cold places the internal heat is fortified owing to the surrounding cold. Therefore, in a cold season an animal requires more nutriment. Since the nutritive power is fortified in cold places, so too will the generative power be fortified.

2. Moreover, an abundance of one superfluity diminishes another, for the abundance of urine diminishes the excrement, and contrariwise. But semen and sweat are superfluities of the third digestion. Since, then, sweat increases in hot places, in these same places a diminution of the semen will occur.

3. Moreover, human or animal semen is moist. But in cold places moisture is more abundant than in warm places; therefore, etc.

The opposite is apparent to the senses. For humans in warm lands are stimulated to coition more.

To this one must reply that two things are required for generation, which occurs through coition: namely, a superfluity of the aliment acting as the matter, and a heat opening the pores and pathways acting as an efficient cause. But the pores and pathways to the outside are closed in cold places owing to the surrounding cold, and in warm places they are opened. And this is why, with respect to the efficient cause, coition flourishes more in warm than in cold places. But in cold places the nutritive power is fortified, for a contrary is fortified in the presence of its contrary, as is said in the first book of *On Meterology*.[20] And this is why the water in some springs is colder in summer and warmer in winter, because during the summer the cold flees through the surrounding heat to the bubbling water of the spring, and this is why the water is cold then; and contrariwise, in winter, because the heat flees inward through the surrounding cold, and is incorporated into the source of the spring. Thus in cold places, if a lot of nutriment abounds, then there is a lot of superfluity there, and as a result a lot of the matter for the semen as well. And so too the generative power flourishes more in cold places than in warm places as far as the material cause is concerned. An indication of this is that, of all the animals, fish multiply most. If, however, nutriment should be lacking in cold places, then the whole nutriment is converted into the substance of the members, and little remains there of the superfluous nutriment, and as a result there is little for the semen's matter.

One can say, however, that some animals are naturally cold and others naturally warm. In cold animals the powers are not strengthened in cold places, as is evident in the case of flies and serpents, but are further weakened instead. In warm animals, however, the interior powers are strengthened in cold places. And this is why the generative power flourishes more in warm animals in cold places than it does in warm places, yet in warm places the animals are more stimulated for coition, while in cold places the generative power is intensified more.

1. On to the arguments. To the first, one must reply that in cold places the nutritive power is so intensified that the entire aliment is converted into the substance of the members, and there is little there that is superfluous, and therefore the nutritive power is intensified or flourishes there, whereas the generative power slacks off.

20. Ar., *Meteora* 1.12 (348b30f.); cf. A., *Meteora* 2.1.31.

Or one can put it another way and say that the generative power flourishes more in cold places with respect to the material cause, whereas with respect to the efficient cause it flourishes more in warm places.

(2, 3) And all the other arguments prove this.

Question 9: In which season coition flourishes more.

One asks further in which season coition flourishes more.[21]

1. And it seems that it flourishes more in summer. For the Philosopher says in the second book of *On Generation* that "the sun's approach to us is the cause of generation, and its retreat is the cause of corruption."[22] But during summer the sun is nearer to us. Therefore, the generative power particularly flourishes then.

2. Moreover, it seems that it flourishes more in spring. For it is then that each one is perfected, when it can reproduce one like itself, according to the Philosopher in the second book of *On the Soul.*[23] Thus, when the powers are more perfect, they are then more generative. But the powers in animated beings are more perfected in spring, for it is then that nature especially expels superfluities, for this is when rashes and scabs especially appear, and these attest to the power of nature as an expelling force. Therefore, the generative power especially flourishes in spring.

That it flourishes in winter can be proved through the arguments adduced to the previous question.

To this, one must reply that the generative power is twofold. One is superior, and this consists in the sun and the moon and the other planets, and the other is inferior, and this is different in different beings. The first generative power especially flourishes in spring, then in summer, third in autumn, but least of all in winter. But it exists in some animals in various ways, be-

21. Cf. Ar., *HA* 5.8 (542a20f.); Avic., *DA* 5.2 (fol. 34va); A., *DA* 5.1.2.25 (*SZ* 1: 499).

22. Ar., *DG* 2.10 (336b17–18).

23. Ar., *De anima* 2.4 (415a26f.).

cause in some the inferior power agrees with the superior, and in these the generative power especially flourishes in spring, for example, in plants and in some middling animals. But in others, such as flies and serpents, the inferior power does not agree with the superior, owing to the weakness of the heat, and in such as these it flourishes more in summer than in spring. And the same thing occurs in large animals that live on things born of the earth, since in winter, owing to the lack of nutriment, they become very thin, and they are not stimulated to generate until they have become a good bit fatter, which does not occur in them before summer or autumn. In ones like this the generative power especially flourishes in summer or autumn.

1. On to the arguments. It is apparent from this that the first argument proves that in summer the generative power does not flourish as much in animals that are not hot in themselves.

Or, it can be better said in another way that neither an excess of heat nor an excess of cold is required for the generative power, because if the heat were excessive it would consume the moisture, which is the semen's matter, and if it were cold the pores would close through which the semen has to exit. And this is why a temperate heat is required, which is particularly present in spring, because spring has a warm and moist complexion naturally, just as the semen does.

2. To the second argument one must reply that it particularly flourishes in spring except in those animals that are too cold. Ones like this wait for the summer before they have intercourse.

Question 10: Whether the male and the female are especially urged on at the same time to generate.

One asks further whether humans, male and female, have the same time for generation or are especially urged on or moved to generate at the same time.

1. It seems so. For a human naturally has a warm and moist complexion. Therefore, the powers in a human are determined for the same time. But each one does better in the season dis-

similar [in complexion] to him and does worse in the season that is like him. For the author of the *Six Principles* says that humans, who have a dry and arid complexion, do worse in autumn and better in spring.[24] Therefore, the human, male as well as female, is especially aroused to intercourse in autumn, during a dissimilar season.

2. The same thing is clear in other animals as well and similarly in plants of the same species.

The Philosopher says the opposite.[25] For he says that "a man has intercourse especially in the winter, and woman in summer."

To this one must reply that a human, unlike other animals, does not have a season determined for coition. And the reason for this is that a human lives according to reason, which is capable of opposites. Now, "the genus of humans lives both by reason and by art," as he said in the first book of the *Metaphysics*.[26] But other animals are ruled by the instinct of nature, which always acts in the same way unless it is impeded.

Another cause is that the human has a nutriment that is appropriate and adequate for him in every season, and the same is true for his other necessaries. But this does not occur in other animals. And this is why there can be material for semen and the power for intercourse in the human in every season. Nevertheless, intercourse flourishes in the male more in winter with respect to his own makeup than in the other seasons, and in a female this occurs in summer. And the reason for this is that generally a man has a hotter complexion than a woman,[27] and this is why in summer, on account of the surrounding heat, the interior heat is dulled, because the exterior, surrounding heat causes the natural interior heat to disperse and to escape. But in winter the heat is put to flight by the surrounding cold and is gathered together on the inside, and the united power is stron-

24. Gilbert Porreta, *De sex principiis* 4 (PL 188, 1263A); A., *De sex principiis* 4.6. Note that Gilbert seems here to contradict the widely held view that humans are naturally warm and moist.

25. Ar., *HA* 5.8 (542a32–b1).

26. Ar., *Metaph.* 1.1 (980b27–28).

27. Cf. A., *DA* 12.1.2.32, 12.1.7.85 (*SZ* 2: 906, 926).

ger than one dispersed on its own.[28] But the opposite is the case in a woman, because in winter, on account of the surrounding cold, the pores and pathways of the semen are constricted and blocked, and in summer they are opened by the accidental heat, and this is why a man is especially potent for intercourse in winter, whereas a woman is so in summer, and this is clear.

1. On to the arguments. To the first, one must reply that a man and a woman can have intercourse in the same season, but nevertheless a man of his own accord has intercourse more in winter than in other seasons, and a woman has intercourse more in summer.

2. To the second argument one must reply that animals and plants live and are ruled by nature, which acts in a uniform manner, but the human is ruled by reason.

Questions 11–14: Related questions about intercourse.

Related to this, one can ask why the woman can conceive at twelve years of age, whereas the man does not usually generate before the fourteenth year, and why at the other end the woman ceases to conceive before a man [*homo*] ceases to generate.[29]

Again, one may inquire why things more quickly deteriorate which are made by those having intercourse secretly, frequently, or with pleasure. For a cheese made by a woman who secretly has intercourse quickly putrefies, and a sword made by [such] a man rusts more quickly.

Again, one may ask why dogs follow such people so often.

Again, one may ask why men are more ashamed of this sort of act than of another.

To the first, one must reply that a woman conceives more quickly for two reasons: one is owing to the matter, like the abundance of menstrual blood, and the other is owing to the efficient cause. For a woman is colder and moister, and, owing to the weakness of her heat, she therefore more quickly overflows with menses or abounds with semen, which is a superfluity of

28. Cf. *Liber de causis* 16 (Bardenhewer, 1882; reprint 1970), 179.
29. Cf. A., *DA* 5.2.1.55, 9.1.1.7–15 (*SZ* 1: 514, 776–79).

the aliment. But in a man the heat is more potent and digests the aliment better by converting it into the substance of the members, so much so that little of the superfluity is left over, and as a result there is little semen, and this is why a woman is more quickly apt to conceive than a man is apt to generate.

Another cause is that more is required for acting than for undergoing. And this is why it is required that the power of the man be more fortified, since it is active, than the power of the woman, since it is passive, and this is why he needs more time [to generate]. But the less dense something is, the more quickly it is penetrated even by something corrupting, and a woman has a less dense complexion, and this is why her power is more quickly consumed by an exterior corruptive agent, and this is why she ceases to conceive more quickly than a man ceases to generate.

To the second argument one must say that just as the eye of a menstruating woman infects a mirror (because the humors then abundant in the eye infect the air near the eye, and this infects the next bit of air, and so on up to the mirror),[30] so too in people who have intercourse a great deal it is as if all their parts were infected, and therefore things that are touched by them receive an infection and as a result tend more quickly toward corruption.

To the third, one must reply that dogs take more pleasure in a strong odor and follow after dead bodies. Further, the body of a person who has a great deal of intercourse closely approaches in its disposition the nature of a dead body owing to an abundance of corrupt semen. This is why dogs, which have a very good sense of smell, follow them.

To the fourth, one must reply that a human is naturally ashamed by evildoing.[31] Thus a human is not ashamed of those things which are morally neutral [*indifferentia*] or which are not harmful to others. But because intercourse is turned to the detriment of another and frequently to oneself, and similarly because pollution occurs from intercourse, a human is especially ashamed of it.

30. Cf. A., *DA* 7.2.5.133 (*SZ* 1: 647).
31. Cf. A., *DA* 22.1.5.11 (*SZ* 2: 1446).

BOOK SIX

Question 1: Whether flyers naturally have to generate by laying eggs.

"IT IS NECESSARY that in addition to what we said, . . ." etc.[1] This sixth book is mainly concerned with egg-laying. And one asks first whether flyers naturally have to generate by laying eggs.

1. And it seems not. For subtle and light things are generated more quickly, and this is why males are formed in the uterus more quickly than are females. But flyers are subtler and lighter than walkers. Since, then, walkers generate without an egg acting as a medium, and since live young come forth from a womb, how much more so should this be the case in flyers?

2. Moreover, coitus is the path to generation. Now, where coitus occurs more quickly, generation occurs more quickly. But flyers have intercourse most quickly, and therefore, etc.

3. Moreover, if flyers were generated from eggs, this would occur on account of their complexion. The conclusion is false, and therefore the premise [is false]. The conclusion is evident, because the first qualities are the causes of all posterior acts; for odors and colors and tastes are caused on account of these. The falsehood of the conclusion is evident because, if complexion were the cause of egg-laying, all egg-layers would have the same complexion. But this is not the case. For fish are cold and moist, and flyers are warm and dry, but they lay eggs notwithstanding.

The opposite is evident to the senses, that flyers are generated by laying eggs.

1. Ar., *HA* 6.1 (558b10f.). Cf. Avic., *DA* 6.1 (fol. 34vb); A., *DA* 6.1.1.1–2 (*SZ* 1: 525–26).

One must say that flyers are generated by egg-laying, and the same is true for certain animals living in water, like fish, and for certain ones living on land, like serpents and lizards. There are many reasons for this. For these animals are swift of movement, and for that reason they have a narrow body and a narrow womb, or nature provides for such a narrowing of the body and the womb, so much so that the animal cannot be formed in the womb but the animal is formed with something serving to mediate the process, namely, an egg.

Another reason lies in the great number of fetuses, because these animals produce a lot of sperm, and all of the semen cannot be absorbed at the same time in the womb on account of its narrowness; rather one part of the semen is expelled before another is formed, and for that reason they lay eggs in succession and lay them down underneath themselves. Scantiness of nourishment also contributes to this. Whence it can be said briefly that such animals generate by laying eggs on account of the weakness of their power. For there is a great number of fetuses in them, but they do not possess potent power for retaining the semen wherever the fetus is formed. As a result, it does not wait to be formed in the uterus, but rather is first expelled into the egg and is formed externally, and this is why nature provides a web and a shell for the egg, so that the fetus will be preserved until its formation is complete.

1. On to the arguments. To the first, one should say that although flyers are subtle and light in comparison to walkers, their powers are nevertheless weaker, and because of the weakness of their powers they are generated through a medium, namely, through the egg and its shell.

2. To the second argument one should say that it is just this way with regard to generation and to coition, because the opening of the pathways is sufficient for coition. This is especially found to be the case in flyers since they are warm and dry, and this is why they have intercourse very quickly. But generation results immediately from the strength of the [generative] power, and this is deficient in fish and serpents or reptiles owing to their lack of heat and is deficient in flyers because of their

lack of moisture, because life demands proportionate heat and moisture.[2]

3. To the third argument one must respond that complexion is the cause of egg-laying in flyers. Yet this complexion is not the precise cause, but the complexion regulated by the power of the soul. And this is why it is not necessary that all egg-layers share in the same specific complexion. But something is found to be shared by all things that lay eggs, for example, swiftness of motion, narrowness of the womb, a great number of fetuses, a great quantity of sperm, and a weakness in the retentive and formative power.

Question 2: Whether the egg white is generated from some superfluity of the egg or from some substance of the one laying the egg.

Next one asks about the parts of the egg. And first one inquires into the white of the egg, whether it is generated from some superfluity of the egg or from some substance of the egg-layer.

1. It seems that it is from the substance of the egg-layer. Now the one generating generates something like itself. But a flyer is generated from the white [*albugo*] and the white [*albumen*] is generated from the flyer.[3] Therefore, the white is like the flyer in substance.

2. Moreover, the frequent emission of eggs weakens the egg-layer, for it is said in the text that "hens laying a great deal die more quickly."[4] But this would not happen if the white of the egg were not derived from the substance of the egg-layer; therefore, etc.

3. Moreover, spirits and heat are derived from the one generating, as will be explained in book sixteen of this work.[5] But spir-

2. A., *DA* 16.1.1.9; 20.1.3.13 (*SZ* 2: 1157, 1364–65).

3. Throughout this section A. uses *albugo* and *albumen* interchangeably.

4. Ar., *HA* 6.1 (558b20–21).

5. Ar., *GA* 2.3 (736b33f.); cf. A. *DA* 16.1.13.68–71 (*SZ* 2: 1192–94).

its would not be derived from the spirits of the one generating, nor heat from its heat nor power from its power unless the one generated from it were of the same substance as the one generating. Since, then, a fetus comes to be from the white, as the text says,[6] the white will be of the substance of the one generating.

To the contrary. The white is in egg-layers just as the sperm is in those generating and the seed is in plants. But the sperm is not of the same substance as the one generating, nor is the seed of the same substance as the plant. These come rather from a superfluity of their nutriment. Therefore, etc.

One should say that the white in the egg stems from the superfluity of the third digestion and is not from the substance of the egg-layer, because if it were from the egg-layer's substance, then it would be derived either from one part alone or from the whole. If from one part alone, then the animal could not be produced from the white, but only a part of an animal could be so produced. If it were actually derived from the whole, then the white would itself be a small animal that would have in itself the nature of every animal, or it would be composed from all the parts of the animal.

Moreover, separation from something continuous does not occur without pain; if, then, sperm were of the same substance as the one generating it, its separation would not occur without pain, and as a consequence there would be no pleasure in that act, which is false. But the white in egg-layers is analogous to sperm in viviparous animals.[7] Therefore, the white will not be from the substance of the one generating, but rather from some superfluity of the nutriment.

1. On to the arguments. To the first, one should reply that when one says, "the one generating generates something like itself," this should be understood to mean generating in a univocal sense, and principally, and not equivocally and instrumentally. But the white in the egg does not have the nature [*ratio*] of

6. Ar., *HA* 6.3 (561a24f.).

7. "Viviparous animals": *generantibus animalia.* Or, perhaps, "those that produce by generation."

a generating principle, but is rather an instrument of the principal agent.

2. To the second argument one should respond that frequent egg-laying weakens the egg-layer, just as the frequent emission of sperm or frequent coition weakens one generating, not because something belonging to the substance of the egg-layer is in the egg, but because the white of the egg comes from superfluous nutriment, and when there is frequent egg-laying, whatever should be converted into the substance of the members is converted into the white before it is converted [into the members]. This is just as occurs in those having frequent intercourse: what should be converted into the substance of the members is emitted through the seminal vessels [*vasa seminaria*].[8]

3. To the third argument one should reply that the spirit, heat, and other power of the egg-layer is derived along with the egg, because these are the means of reducing to act that which exists in the egg potentially. But this is not the case for the white itself, and for that reason, etc., as has been said.

Question 3: Whether the yolk of the egg is generated from menstrual blood or from some other moisture.

It is asked further concerning the yolk [*vitellum*] of the egg, whether it is generated from menstrual blood or from some other moisture.

1. It seems that it is not from blood, because menstrual blood is found in the human nature alone; therefore, the yolk of the egg is not generated from menstrual blood.

2. Moreover, the seed and the nutriment come from the same thing, because seed [*semen*] is a superfluity of nutriment. But the yolk is the nutriment for the fetus. Therefore, it comes forth from the same root source as the seed. But the seed is not from menstrual blood, and therefore, etc.

To the contrary, the thing generated attests to its principle. But the yolk is yellow [*citrinum*] or reddish, and therefore it is

8. A., *DA* 16.1.13.68–71 (*SZ* 2: 1192–94).

generated from a reddish moisture. But the blood is such as this, and therefore, etc.

One should reply that the yolk in egg-layers is analogous to the milk in those who lactate, since, just as the milk is the fetus's nutriment, so too the yolk is the nutriment for the fetus, that is for the chick, when it is in or enclosed in the egg, because the chick is generated from the white and the yolk serves as its nutriment. Nevertheless, milk is white and the yolk is red, because milk receives its whiteness from the breasts containing it, which are white and glandular. But the yolk is enclosed in the middle of the white, and it cannot assume a color external to it but may only retain the color nearest to the principle from which the yolk is generated, and the yolk is generated from menstrual blood, as is milk, and for that reason it retains the color of blood.

1. On to the arguments. To the first, one should reply that there are many parts of menstrual blood. One part is converted into the spermatic members, and another part is converted into the members filling up the empty spaces in the body—for example, converted into fat [*zirbum*] and flesh—and still another part is converted into the nutriment for the fetus, and another part is superfluous, and that, capable of providing nourishment, is retained in the uterus while the fetus is in the uterus; after its birth, it is cast out. The menstrual blood is not present in flyers with respect to this part, but it is with respect to these other parts. Therefore, since the yolk is the nutriment for the fetus, there is menstrual blood in flyers, from which the yolk is generated.

2. To the second argument one should say that semen and the nutriment of the one generating do proceed from the same root source, because semen is a superfluity of nourishment [*alimentum*]. Nevertheless, the seed and the nutriment [*nutrimentum*] for the one generated do not come forth from the same root source, except perhaps indirectly, and the yolk is the nutriment for the one generated and not of the one generating, and this is why it is not necessary that it come forth from the same root source as the seed does.

Question 4: Whether the white of the egg is warmer than the yolk, or contrariwise.

It is asked further whether the complexion of the white or that of the yolk is warmer.

1. It seems that the white has a warmer complexion. Because life depends on heat and moisture, then that which is nearer to the origin of life is warmer, and this is the white in comparison to the yolk, and therefore, etc.

2. Moreover, according to Aristotle's fourth book *On Meteorology*, digestion proceeds by a thickening process.[9] That which is more digested is warmer, and the white is of this sort; therefore, etc.

On the contrary, every astringent is colder. But the white is an astringent, for it is introduced to wounds in order to restrain the flow of blood, whereas the yolk dilates [the blood vessels]. Therefore, the yolk is warmer.

One should respond that among all the humors blood is warmer, with a tempered heat, and therefore whatever is nearer to blood is warmer. But such is the yolk, and therefore, etc.

Moreover, the liver is warmer than the breasts, or blood is warmer than the milk. But the yolk, in comparison with the white, is analogous to blood in the liver. It is therefore warmer.

Nevertheless, a twofold heat can be distinguished. One heat is accidental on account of the member that retains it, and this heat is greater in the white than in the yolk just as the blood that is in the heart is warmer than the blood that is in the liver because of the organ where each is contained. Another heat is a natural heat, and this one is greater in the yolk than in the white.

1. On to the arguments. To the first, one should say that the principle of life is double: near and remote. The principle of life that is nearer is warmer in proportion to its proximity. But both the white and the yolk are a principle of life, since from the white a life comes forth as the result of a power that is in the

9. Ar., *Meteora* 4.2 (380a3f.). Cf. A., *Meteora* 4.1.15.

white, and the yolk is converted into the substance of the one generated, and the yolk is better disposed for conversion into the substance of the one generated than the white is disposed for conversion into the one generated.

2. To the second argument one should say that what is more digested is warmer in comparison to the same thing when it is digesting. This is not the case for the yolk and the white, and therefore the solution is clear.

Question 5: Whether the animal is nourished from the yolk.[10]

One asks further whether the animal is nourished from the yolk.

1. It seems not, because "we are nourished by and come from the same things," according to Aristotle in book two of *On Generation*.[11] But an animal does not come to be from the yolk, and therefore it is not nourished by it.

2. Moreover, the relationship of the white and the yolk is like that of the flesh and the white. But flesh is more suitable as nourishment than the white, and therefore the white is more suitable than the yolk.

3. Moreover, just as the fetus is nourished in the womb of those generating live young, so too are young nourished in the egg of the ones that lay eggs. But in those generating live young the nutriment is not located in the middle of the womb, and in egg-layers, therefore, neither is it found in the middle of the egg.

4. Moreover, every thing that is nourished has an instrument through which it receives nourishment. But the chick in the egg does not have an instrument through which it might receive nourishment, because this only occurs by means of the mouth. Now, however, in the egg the animal's entire head is covered, and the mouth is closed. Therefore, it is not nourished.

10. Cf. A., *DA* 1.1.6.83; 6.1.4.31–32; 17.1.2.14–15 (*SZ* 1: 76, 539–40; 2: 1237–38, 1243–44).

11. Ar., *DG.* 2.8 (335a10f.).

The Philosopher asserts the opposite.

One must respond that the yolk is the nutriment for the chick in the egg. According to what the Philosopher says in book nineteen of that book,[12] motion and order depend on the end. But order is double: general and special. According to a general order all animals are ordered to the human, but according to a special order every animal whatsoever is ordered to the continuation of its own species. But in themselves animals are ordered to the human, but they are ordered to the conservation of their own species through their superfluity. For an ox has a power, through its own superfluity, to generate another like itself, but the ox itself passes over to become human nourishment. It is the opposite for plants, because they conserve the species using a part of themselves, but by means of their superfluity, their fruit for example, they are ordered to the human and its nourishment. In the same manner there are two [orders] in the egg. One is ordered to the conservation of the species, and this is the white, out of which the animal is generated, and the other is the one that serves as the nutriment for the animal that is already generated, and this is the yolk. An indication of this is that the more the chick grows in the shell, the more the yellow is diminished, so much so that the whole of it is consumed just as if in a crack in the shell.[13]

1. On to the arguments. To the first, one must say that "we come from and are nourished by the same things," etc. The Philosopher means by this that just as we do not come from just one element, or even from just two, but rather come from them all, thus neither are we nourished by simple elements (for example, from pure air or pure earth, etc.), but rather from a mixture of all the elements, so much so that, just as the four elements come together for our fashioning, so too the four come together for our nutrition or nourishment. Therefore, the Philosopher does not mean that "we come from and are nourished by things the same in number," but from things the same in their proportions.

12. Ar., *GA* 5.1 (778a29f.). Cf. A., *DA* 19.1.1.3–4 (*SZ* 2: 1333–34).

13. The Latin here is quite turgid but seems to suggest that the yellow begins to disappear, just as happens when there is a crack in the shell.

2. To the second, one must respond that each thing is ordered to a proper end, and because flesh is directed to nutriment and the white is not, flesh is therefore a more suitable nutriment. And in the same manner, because the yolk is ordered to serve as nourishment for the chick and the white is not, the yolk is therefore more suitable as nutriment than the white, since the latter is not ordered to serve as nourishment.

3. To the third, one must reply that the fetus in the womb and the chick in the egg or the egg's shell are quite different things. For the fetus is nourished from without, and the nutriment for the mother and for the fetus are one and the same. And this is why it is unnecessary for the fetus's nutriment to be located in the middle of the womb, whereas the chick in the egg cannot receive its nutriment from the outside because it is enclosed in a shell. And for this reason nature has ordered that the yolk be in the middle, so that nutriment can be sent to all the parts equally.

4. To the fourth argument one should say that just as the fetus in the womb is nourished not by means of its mouth but through the umbilical cord, so too even a chick is nourished in the egg through an umbilical cord. The umbilical cord in flyers, however, is not visible because of its small size. Nevertheless, it does have an umbilical cord near the anus. This is why the Philosopher says[14] that two pathways go forth from the umbilicus, one to the yolk, through which it receives nutriment, and the other to the web, through which respiration occurs.

Questions 6–13: On certain properties of eggs.

With respect to this, one asks:

1. Why the yolk sinks when placed in water but the white floats.

2. Moreover, why the yolk is not suitable nutriment until it is cooked although milk and fruit, to which the yolk is analogous, are suitable nutriment without cooking.

14. Ar., *HA* 6.3 (561b5f.).

3. Moreover, why is it that eggs among flyers are larger in size but fewer in number than are those of fish?

4. Moreover, why do the eggs of flyers have a hard shell, and the eggs of fish have a soft shell?

5. Moreover, why do the eggs of flyers have various colors whereas in fish they are but one color?

[6.] Moreover, why do eggs in flyers have an oblong shape whereas in fish they are round?[15]

[7.] Moreover, why is it that males are more often generated from eggs with an oblong shape, whereas females are more often generated from those that are round?

[8.] Moreover, why do some eggs placed in fire make noise, and some do not?

To the first, one must reply that egg white is very viscous and adheres to that to which it is joined; but the yolk is more permeable and, in addition to this, is denser.[16] Thus, a yolk placed in water is immediately penetrated by the water on account of its permeability, and it sinks on account of its density. But the white floats owing to its viscosity, just as oil does, and because it is strongly adherent, it has the power of clarifying that to which it is added. For it draws all the gross and unclean parts upward and firmly adheres to them, and this is why it clarifies the residue. If, however, vinegar is introduced first, this does not happen because the vinegar is very good at breaking things up. That is why when it is applied to something it breaks it up so effectively into very small pieces, so that the thicker and earthy parts cannot be separated out by the white.

To the second, one should reply that in order for something to nourish, it must be digested. Now, however, fruit is digested by a double heat: both one that is interior, which is natural, and another that is exterior, which is accidental. And this is why fruit can nourish without cooking, but the yolk is only digested by interior heat. Therefore, for it to nourish fittingly, it has to be digested further by fire. And besides this, moisture is repugnant

15. The numerals 6, 7, and 8 do not appear in the text but have been added by the translators to facilitate understanding.

16. A., *DA* 17.1.2.15 (*SZ* 2: 1243–44).

to digestion, and a great deal of moisture abounds in the yolk, since it is generated from the menstrual blood. But earthiness is more abundant in fruits, which is why they especially have a pontic taste owing to their dryness.[17] This is why fruit can be more easily digested than yolk, because the heat of the sun or of the air that contains the heat is adequate to digest fruit, but to digest a yolk a much more piercing heat is needed, like the heat of fire.

Moreover, milk is digested in the breasts, and for that reason it can nourish without further digestion. But the yolk is not digested in something else in this way; this is why, etc.

To the third argument one should reply that fish disperse their eggs in a liquid, and this is why they do not require hard shells. But flyers disperse their eggs in something hard like the earth and this is why they require hard shells, to block contraries and harmful things until the fetus achieves complete formation. Another cause will be revealed.

To the fourth, one should say that variation in color is caused by variation in the first qualities.[18] Now, however, fish have a weak heat, and this is why they do not have a power sufficient to vary the color. But in the flyers there is a stronger heat, powerful enough to separate out things of a different nature, and it can thus vary the colors.

Or, one may say in another way that three things must come together in flyers for the generation of the egg: namely, the male's seed; the white, which is the female's menses; and third, the yellow, which is the nutriment for the fetus. And this is why different colors occur in the eggs of flyers. Nevertheless, because moisture impedes heat and dryness sharpens it, whenever this occurs, the eggshell will be of a different color. For the shell of some birds is white, as in the hen, and in some it tends to black, and in some it is spotted, which occurs when more and then less heat causes separation and purification. But an interior heat is not as strong, and this is why those [eggs] that are interior appear the same in all instances. But it is different in fish, since the

17. "Pontic taste": i.e., a sharp, bitter taste. Cf. A., *DA* 1.2.21.400, 3.2.3.111, 8.4.1.132 (*SZ* 1: 196–97, 398–99, 726–27).

18. A., *DA* 17.1.2.12–13 (*SZ* 2: 1242–43).

male's seed is not necessary for the generation of the egg and the females have eggs from birth [*a principio*], and after they disperse them the male follows and spreads his semen over them, and this is perhaps the reason why fish eggs do not have hard shells. And this is why fish eggs do not have different colors.[19]

To the fifth,[20] one should reply that a thing's magnitude depends on the strength of the agent, but its multitude depends on an abundance of matter. But the egg's matter is more abundant in fish, since they are moist, but the active power is stronger in flyers since they are warm and dry. This is why fish have many eggs, but small ones. They are many on account of the abundance of the matter, but small owing to the weakness of the active power. In flyers, contrariwise, they are large but few in number.

To the sixth argument one must reply that heat moves from the center to the circumference and moves best in a pyramidal shape. This is evident in a fire's flame which, even though it may be broad and wide at the base, tends to form a cone. And for this reason, because heat is stronger in a flyer than in a fish, a flyer's eggs are oblong and a fish's eggs are round, since a weak heat spreads them out into a circle [*gyrum*].

To the seventh, one should answer that males are naturally warmer than females. A male could be produced from the same seed from which a female is produced if the heat were powerful enough. Therefore, because eggs are oblong due to the strength of the heat and are round owing to its weakness, and because in males there is stronger heat than in females, then males are generated from oblong eggs and females from round eggs.[21] And one must understand that since the pointed end of the egg is warmer and more subtle and the other end is colder and thicker, then the larger part is placed below in the womb and emerges first at the time of laying.

To the eighth, one must reply a great windiness is present in some eggs, and when the shell is violently split in eggs like these, the windiness exits with a great force or noise. And this occurs

19. A., *DA* 17.1.6.42 (*SZ* 2: 1258).
20. This is actually a response to point 3 (above).
21. Ar., *HA* 6.2 (559a28f.); A., *DA* 6.1.1.8–9 (*SZ* 1: 528–29).

when the exterior heat is strong. Thus, if a flyer's egg is placed in the middle of a fire, the shell quickly and violently splits, and the windiness exits with a loud noise. But this would not happen if the fire were smaller or the shell had already been split. But great windiness is enclosed in fish eggs, and this is why they make a great noise if they are placed in or spread over fire, as is clear from the herring [*allec*].[22]

Questions 14–18: On the properties indicating coition.

One inquires further of the properties indicating coition.[23] First, one asks whether in coition animals naturally have to be gentle or wrathful.

1. It seems that they have to be gentle. For one power draws another away from its operation. But in coition the concupiscent power is especially vigorous, and therefore the irascible power is drawn away from its operation. Therefore, then, they are hardly wrathful at all.

2. Moreover, contraries do not exist at the same time. But the concupiscent and the irascible are contraries and are divided by opposition; therefore, since the concupiscent is dominant in coition, the irascible will succumb.

The Philosopher says the opposite.[24] Now it is clear in many animals—for example, the camel, bull, boar, wolf, lion, and others of this sort—that during mating season they attack all those passing by and do not permit males of their species to approach the females.

Related to this, one asks why at the time of coition [*tempus coitus*] frenzy [*furor*] is more prevalent in males than in females.

Second, why humans are more liberal at this time.

Third, why they want less food at this time than at another time.

22. At *DA* 24.1.8(2) (*SZ* 2: 1660) A. pronounces the *allec* a "tasty" fish.

23. A., *DA* 6.3.1.96–99 (*SZ* 1: 571–72).

24. Ar., *HA* 6.18 (571b8f.). Cf. A., *DA* 6.3.1.97–102 (*SZ* 1: 571–74); Avic. *DA* 7.1 (fol. 36ra).

Fourth, why pigs move at this time to the north and not to the east or west.

To the first, one should respond that a well-ordered boldness arises out of strength of power, and this occurs in those confident in their own strength. But during mating season the spirits in the body are moved a great deal, and this is why they create a certain confidence in the animal with respect to their own strength. And this is why at this time they are bolder. But disordered boldness arises from a warm and dry complexion. This is why a choleric man sometimes attacks four men stronger than he. Thus one ought to respond briefly that the spirits and humors are in motion during mating season, and this is why animals then rely on their own powers and are bold.

1, 2. On to the arguments. To the first, one should say that a concupiscent power and an irascible one can be related to the same or to different things. If to the same, then they cannot exist at the same time. For no one desires something and at one and the same time wants to avoid that same thing. If they are related to different things, then they can exist at the same time, for one does not withdraw the other from its operation. For one who desires something has anger towards its contrary. And this is why, since at that time animals greatly desire coition, they have anger towards other animals who can prevent their coition. This response is clear for each argument.

To the second, one should respond that a male can impregnate more times than a female can conceive, because, among those generating live young, a female, after she has conceived, does not desire coitus again until she has given birth. But a male, on the same day that he has impregnated one female, can impregnate another. Therefore, during mating season, an uncontrollable frenzy [*furor*] rages more in the males than in the females.

Besides this, males naturally have to do combat for the females, and this is why they are naturally more wrathful. Nevertheless, it should be understood that in every species the female—except the human female—has a diminished desire after conception. And the reason for this is that the seed [*semen*] is less abundant in females other than a woman. For in a woman there is an abundance of menses after conception just as there was before. But after conception the opening of the womb is closed, and the menses cannot exit, and for this reason the woman especially desires intercourse after conception, so that, as a result of intercourse, the menses can exit, because otherwise these sting and gnaw at the womb before they go forth.

To the next, one should say that customs are attendant upon the natural complexion. The movement of the hot is, however, from the center to the circumference, and the movement of the cold is the opposite. But a liberal substance pours forth from itself, and as a result it follows the motion of the hot, but an illiberal substance is retentive and does not release anything, and so it follows the motion of the cold, which constrains. This is why the elderly are naturally illiberal. Since, then, heat abounds at the time of coitus, this is why at that time every sensuous person is liberal, because his substance gives [of itself].

To the next, one should say that the generative power is comprised of the animal and natural powers. An example is the power of the stomach, for the stomach's power not only desires food, as a natural power, but also senses and perceives the mode of desiring, as an animal power, and so the generative power acts in much the same way. And this is why, when the generative power is extended, the appetite for food is diminished, because the one causes the other to depart by virtue of its operation.

To the next, one should say that the planets are moved from the east to the west by the movement of the firmament; but brute animals naturally and especially desire a temperate place, and because a place that tends away from the planets toward the north or toward the south is more temperate than a place that is positioned directly under the planets, this is why pigs and horses are distemperate and cannot rest during mating season, on account of the movement of their spirits, and they move to the north or south as if they were seeking a temperate place and not to the east or west, because they always hold themselves equidistant from the planets or the location of the planets.[25]

25. A., *DA* 6.3.1.100 (*SZ* 1: 572–73).

Questions 19–22: On the generation of animals, in relation to wind.

One asks further about the generation of animals, in relation to the wind, whether the south wind is more suitable for generation than the north wind.

1. It seems that it is. Life depends on heat and moisture. Therefore, a hot and humid wind better results in generation. But the south wind is such as this in comparison to the north wind, and therefore, etc.

2. Moreover, according to Avicenna brute beasts flourish and grow more in a warmer region though it is just the opposite with the human.[26] But the same reasoning seems to obtain both for the region and the wind. Therefore, a warmer wind results better in the generation of beasts. But this is the south wind; therefore, etc.

3. Moreover, the south wind better results in the generation of plants, because the north wind consumes and burns the leaves of herbs and plants.[27] Therefore, the same reasoning holds concerning the generation of animals.

The Philosopher says the opposite and so do the medical writers.[28] These say that the only good the south wind brings is to loosen the bowels [*venter*].

Next, one asks why the north wind is calm and the south wind is rain-filled.

Second, why the north wind causes the generation of males, and the south wind the generation of females.

Third, why the north wind constrains the bowels and causes the urine to flow, but the south wind loosens both.

To the first, one must say that life depends on heat and moisture. But these occur in two ways. For something may be hot and moist naturally, and one such as this is conducive to generation and life. Or something may be warm and moist accidental-

26. Avic., *DA* 8.1 (fol. 36rba).

27. A., *DA* 6.1.3.19 (*SZ* 1: 534).

28. Ar., *HA* 6.19 (573b32f.); cf. Avic., *Can. med.* 3.16.1.2. Galen, *In Hippocr. de humoribus* 3.13 (Kühn 16: 412).

ly, such as when water is made warm or earth moist, and such a thing, occurring by accident, does not result in generation or in the conservation of life. Now then, the south wind is warm and dry naturally, because it issues forth from just such a place, but it becomes moist on account of the places through which it travels, and is therefore moist accidentally. Or perhaps every wind is cold and dry naturally, as Aristotle says in book one of *On Meteorology*.[29] For the wind's matter is a dry vapor, compressed and congealed by the cold, as the Philosopher says in book two of *On Meteorology*.[30] This is why the north wind has, or preserves, natural properties, because it is cold and dry, whereas the south wind is warm and moist accidentally, because it passes through places that are warm and moist. This is why it is less conducive to generation, because on its own it weakens the powers that are in the body, while the north wind fortifies them. Because the north wind constricts the external parts, it puts the heat and spirit to flight inwards and consolidates them there, and the unified power is stronger than when it is dispersed, whereas the opposite occurs from the south wind. And this is why the north wind is better suited for generation, as the Philosopher intimates.[31]

1. With this the response to the <first> argument is made clear.

2. To the second argument one should respond that the south wind is warm and moist *per accidens*, and life does not depend on this type of heat and moisture.

3. To the third, one should reply that plants are generated from cold and dry matter. Now, they take their origin from the earth, and this contributes some measure of warmth and moisture for their generation. Thus the cold and dryness of their matter are greatly increased by the north wind, and their matter is tempered rather by the south wind. But animal semen is hot and moist in its own right. As a result, it is distempered by an external warmth or moisture, and the south wind thus does

29. Cf. Ar., *Meteora* 1.13 (349a12f.).
30. Ar., *Meteora* 2.4 (360b30f.); cf. A., *Meteora* 3.1.8.3.
31. Ar., *HA* 6.19 (574a1f.).

not contribute to the generation of animals as it does to the generation of plants, and if it does contribute, this will be for the generation of cold animals like the serpent and the bloodless animals.

To the next inquiry it is already clear that because the north wind is dry, it is therefore calm, because it cannot raise the vapors and fumes up into the air but rather presses them downward; and because the south wind is moist, it is therefore rainfilled.

To the next, one should say that because a male is generated when the power is strong, and a female is generated only when the power is weak, and since the north wind fortifies the powers and the south wind weakens them, this is why the north wind causes the generation of males, and the south wind the generation of females.

To the next, one should say that the north wind constricts and the south wind softens, and this is why the north wind is conducive to constipation, and the south wind, as a result of its heat and moisture, causes diarrhea. Moreover, the north wind strengthens the powers, so that the digestive power is fortified and digests quite well, converts a great deal, and similarly exercises the retentive power, so that it will retain well. But the south wind brings about the opposite effects. For it softens and weakens the digestive, converting, and retentive powers, and thus it relaxes the stomach so that it cannot retain superfluity.

Moreover, the north wind dries out the nerves and, as a result, renders them excellent for motion, whereas the south wind softens, and this is why in the south all the animals are more ponderous and less well suited for movement. Nevertheless, because the bladder is fragile and situated near the external parts, when the north wind strikes it, it injures and constricts it and provokes urination. Now, both the north wind and the south wind provoke urination, although for different reasons: because, namely, the north wind constricts the belly and the south wind stimulates it.

Question 23: Whether urine is a simple or mixed body.

One asks further whether urine is a simple or mixed body.

1. It seems that it is mixed. This is because any part of a mixed body is itself mixed. But urine is derived from the nutriment, which is a mixed body, and therefore, etc.

2. Moreover, in book four of *On Meteorology* the Philosopher says that nothing simple putrefies or coagulates.[32] But urine putrefies and thickens when it is changed and corrupted. Therefore, it is a mixed body.

To the contrary, every mixed body can nourish. But according to the physicians, urine cannot nourish. Now if this is true, then it is not derived from the nutriment; therefore, etc.

To this, one should respond that one should consider two things with respect to urine. One is that it is a moisture and that it trickles out of the members, and the other is that this [moisture] is resolved from parts of the food and mixed together with urine. But I say that according to the first reason it can be a simple body, but according to the second reason it is a mixed body. By means of this distinction I have refuted a certain physician of Cologne, who was accustomed to say that urine is absolutely a simple body.

1. And from this, then, to the arguments. To the first, one must say that something can be derived from a mixed body in two ways: either by a division of quantity, and in this way any given part of a mixed body is mixed; or by resolution, as happens when a mixture is resolved into its elements, and in this way it should not be necessarily true that a part of a mixed body be mixed. But urine can be derived from the nutriment by resolution and separation, and then it should not be necessarily true that it is a mixed body.

2. To the second argument one must respond that if urine thickens, this does not happen by itself, but happens because of the admixture of food and of an incorporated heat and spirit,

32. Ar., *Meteora* 4.1 (379a2f.).

which thicken or render deadly and cannot render it as clear as it was before.

Questions 24–26: On the urine of brute beasts.

Next, one asks why the urine of brute beasts is thicker than that of humans.[33]

Second, why the wolf and the dog raise their leg or their hind foot when they are urinating.[34]

Third, why dogs cannot see before a set time, for example, before the ninth or the fourteenth day.[35]

To the first, one should respond that the more powerful the digestive power is, the more subtle and clear the urine is. Now, however, according to Avicenna the digestive power in brute beasts is not as tempered as it is in the human, because God gave humans the noblest complexion that could be found.[36] This is why, etc. And because brute beasts eat coarse things, like things born of the earth, and chew them by ruminating, but the human eats more subtle things and has an especially tempered heat, and his other natural operations [are also tempered], a human's urine is therefore thinner than that of another animal.

To the second, one should say that the dog and the wolf are melancholic animals and a great deal of bile [*cholera*], which is pungent and biting, is mixed into their urine, and for this reason it damages the nerve-filled places in the bladder through which it passes.[37] And because elevating the foot opens the pores and enlarges the passages through which the urine is emitted, for this reason, etc. And this is why they all urinate against something elevated, like a rock or something else, because in this way the earth receives the urine more quickly and quickly makes a rivulet visible to the dog's gaze, and this is why the dog believes it emits the urine more quickly and with less damage, and for this reason, etc.

33. Ar., *HA* 6.18 (573a17f.); A., *DA* 6.3.1.105 (*SZ* 1: 575).
34. Ar., *HA* 6.20 (574a17f.).
35. Cf. Ar., *HA* 6.20 (574a20f.); A., *DA*, 6.3.2.111 (*SZ* 1: 578).
36. Avic., *Can. med.*, 1.1.6.3.
37. A., *DA*, 7.1.5.48 (*SZ* 1: 606–7).

Moreover, it is quite remarkable that after dogs and wolves have passed excrement,[38] they scratch the ground with their feet and cover it up. And this is why dogs often suffer colic from constipation, because they are gluttonous animals and swallow food without chewing it, and this swells the intestines. Therefore, because they suffer cramps when defecating, they tear at the ground as if they are angry and cast forth their excrement as if it were something poisonous and harmful. An indication of this is that when they have had a mild, soft bowel movement they do not do this, and if they do this, it is out of habit or they want to clean the anus. Thus according to Galen a dog's excrement is useful as an enema for every type of colic.[39] And this is clearly shown in our book *On Animals*.[40]

To the third, one should respond that at birth a dog has a narrowing in its nostrils and an abundance of moisture in its head. Now, nostrils act just like an overflow drain for the brain, and for this reason, because the brain and the head cannot be purged adequately through the nostrils, owing to the abundance of the material, the moisture flows to the optic nerve through which the visual spirit flows to the eyes, and this superfluous moisture renders the nerves poorly suited for motion, as can be seen in paralytics.[41] And this is why, when the optic nerve is softened, the eye cannot see; but as time passes, the natural heat is fortified and consumes these humors and expands the nasal openings, and, as a consequence, the optic nerve begins to harden and is rendered better suited for motion or for the visual spirits, and then the whelp begins to see.

This is a summary of the opinions found in the book *On Animals*.

38. Lit., "have cleaned themselves," *mundaverunt*.

39. The editor of *QDA* was unable to identify the appropriate passage in Galen, but suggests instead Avic., *Can. med.* 2.2.614, 3.16.4.14.

40. For some medicinal uses for canine feces, see A., *DA*, 22.2.1.35, 90 (*SZ* 2: 1464, 1502).

41. For the medicinal, "drying" properties of beaver musk and leopard flesh, useful for paralytics suffering from this superfluous moisture, see A., *DA*, 22.2.1.40, 109 (*SZ* 2: 1468–69, 1514–15).

BOOK SEVEN

Question 1: Whether the non-living is naturally prior to the living.

"THE NATURE AND generation of animals," etc.[1] In this seventh book the Philosopher makes a determination concerning the disposition and behaviors of animals. One asks first in this seventh book why the Philosopher says in the text, "nature proceeds from the non-living to the living,"[2] and whether the non-living is naturally prior to the living.

1. It seems not. In the order of the universe there is one first [thing] which is prior by nature to that which is nearest to it. Thus the intelligences are naturally prior to the celestial bodies, and the celestial bodies are prior to those below them. But in the natural order the living, since it is analogous to intelligence, is nearer to the first [thing] than is the non-living; therefore, etc.

2. Moreover, that thing is prior by nature that exists more according to nature's intention. But nature intends to produce living things more than non-living. An indication of this is that it does not send forth an embryo before life is introduced into it. And this is also proved because by nature we seek to desire that which is better and we also seek to create something better out of things which are contingent, according to the Philosopher.[3]

To the contrary. Whatever enters into the composition of a thing is prior to it. But the non-living enters into the composition of the living and not the other way around, just as the ele-

1. Ar., *HA* 8.1 (588a16f.); cf. Avic., *DA* 8.3 (fol. 38ra); A., *DA* 7.1.1.1–3 (*SZ* 1: 586–87).

2. Cf. Ar., *HA* 8.1 (588b4–5).

3. Ar., *GA* 4.1 (766a5f.); A., *DA* 18.1.2.18 (*SZ* 2: 1289–90).

ments enter into the composition of something mixed; therefore, etc.

Moreover, whatever adds to another is posterior to it. But life adds something to the <non>-living. Therefore, the living is posterior to it.

One must reply that this is said in many ways, as is evident from the fifth book of the *Metaphysics.*[4] Thus, something can be prior by nature in two ways: either with respect to intention or with respect to execution. With respect to intention, the end is prior to those which exist for the end, and, generally speaking, the perfect is prior to the imperfect, because nature operates for the sake of an end and for the sake of perfection. But with respect to execution, the imperfect is prior to the perfect. For nature operates and proceeds by composing and by proceeding from the more common things to individual things, as can be understood from the first and third books of the *Physics.*[5] Thus in nature the living is prior to the non-living when considered in the first way, but it is posterior to the non-living when considered in the second way.

By this [the response] to the arguments is apparent, for they proceed along their own paths.

Question 2: Whether there is something intermediate between living and non-living.

Next one inquires whether there is something intermediate between living and non-living.

1. It seems not. There is no intermediate between contradictories. But living and non-living are just like contradictories, and therefore, etc.

2. Moreover, species of things are incapable of mixing. But living and non-living are distinct with respect to species. Therefore, they are not capable of mixture and, as a result, there is no intermediary between them because an intermediary enjoys the nature of its two extremes.

4. Ar., *Metaph.* 5.11 (1018b9f.).

5. Ar., *Phys.* 1.1 (184a16f.), 3.1 (200b21f.).

3. Moreover, the relationship between the living and non-living is the same as the relationship between the rational and the non-rational. But between the non-rational, which is a brute beast, and the rational, which is a human, there is no intermediary. Therefore, neither will there be an intermediary between the living and non-living.

The Philosopher says the opposite in the text.[6]

To the first argument one must reply that an intermediary is spoken of in many ways: in one way by the negation of both extremes. In this way, an intermediary is placed between health and sickness and good and evil, according to the Philosopher in the *Categories*.[7] In another way, it is spoken of according to the mixture of the extremes, and thus the dusky is midway between black and white. In a third way, an intermediary can be understood as something equidistant between extremes, and in this way virtue is midway between two vices or a point is between the two ends of a line. In a fourth way, an intermediary can occur through the participation of properties, so that if something participates in the nature of one thing in some way and in the nature of another in another way, it can be called an intermediary between the two.

And in this way the Philosopher posits intermediaries between mixed living things and mixed non-living things.[8] And, seen in this way, plants are a sort of intermediary between the living and the non-living. For they are immobile with respect to place just like the non-living, and they do not change in a material sense; nevertheless, they are nourished and grow just like living things. Thus the genus of plants, when compared to the non-living, is living, but, compared to animals, is non-living.

And still more to the point. Between the non-living and plants there is some intermediary, like the mushrooms and truffles [*tuberes et fungi*], which are very abundant in woods around Cologne.[9] And in the same way there is something that is like an intermediary between plants and animals, like the marine

6. Ar., *HA* 8.1 (588b4f.).

7. Ar., *Cat.* 10 (11b38f.). See Boethius's translation (PL 64: 266CD).

8. Ar., *HA* 8.1 (588b4–27). Cf. A., *DA* 7.1.1.5–7 (*SZ* 1: 587–88).

9. The English "truffle" is, in fact, derived from *tuber* through Old Provencal.

sponge, for it moves by expansion and contraction just like an animal and yet has leaves in the manner of a plant, which we have seen with our own eyes in the sea. And likewise a child is a sort of intermediary between a brute beast and a human, since he drinks and eats the whole day just like a beast. Thus children, in their manner of living, participate in the nature of the brute beast—this is why drunken and intemperate men are said in the third book of the *Ethics*[10] to have childish vices—and yet they differ in that they participate in the rational soul.

1. To the first argument one must reply that non-life can be spoken of in two ways. It can be as a negation lying outside the genus (and in this way there is no intermediary between life and non-life), or there can be a negation within the genus, such that it may indicate the negation of some particular life. In this way an intermediary does fall between life and non-life. Or it can be, as we propose, that matter is an intermediary between being and non-being and yet it is not an intermediary between contradictories, because it is established as an intermediary[11] between being in act and what is being in no way at all, that is, being in potency. So it can be said in the proposition that a mushroom is an intermediary between the living with respect to the perfected life of a plant and what is living in no way at all, like a rock. And in the same way a sponge can be said to be an intermediary between plant and animal, and a child can be said to be an intermediary between a human and a brute beast.

2. To the second argument one must reply that an intermediary is not established between the living and the non-living by means of a mixture of extremes, as the dusky is an intermediary between white and black, but rather according to a participation in properties.

3. To the third argument one must respond that a child can be said to be an intermediary between a human and a brute beast in a certain sense, because a human "does not understand without a phantasm," as is clear in *On the Soul*.[12] But there are

10. Ar., *Eth. Nic.* 3.15 (1119a33f.).
11. Or: "which is placed midway between . . ."
12. Ar., *De anima* 3.7 (431a16–17).

no ordered phantasms in a child, owing to its excessive moisture, a moisture that is in the first rank of the soul's confusion according to Galen in the first book of the *Book on Internal Illnesses*,[13] because the power of the phantasm [*virtus phantastica*] is receptive of the sensible species, and this is why the child does not understand, just as a brute beast does not.

Question 3: Whether a power regulating life exists in a human.

One inquires further whether a power regulating life exists in a human.[14]

1. It seems not. For a human is not ruled in only one operation, but in diverse operations. But there is no single power in the human which can rule all his operations; therefore, etc.

2. Moreover, a regimen corresponds to life. Therefore, whatever gives life, gives the regimen. But form is the principle of life, and therefore, it is the principle of the regimen. Therefore, it is not necessary to posit a regulative power beyond the form itself.

3. Moreover, if there is such a power, it will either be a vital, natural, or animal power. But none of these can be [the regulative power] because no one of them can have dominion over another. Therefore, etc.

To the contrary. That one seems to be a regulative power of life which all the others presuppose and which does not itself presuppose another. But the vital power is one like this; therefore, etc.

Yet it seems to be the natural power. Because that seems to be the regulative power which attracts what is suitable and which expels what is noxious. But the natural power is one like this, and therefore, etc.

Yet it seems to be the animal power. Because an animal is distinguished from a non-animal by sensation and motion. But

13. The citation seems to be to Galen's *De locis affectis* (Kühn 8: 1–452). In the ancient translation it is entitled *De interioribus passionibus membrorum.*

14. A., *DA* 7.1.2.8–9 (*SZ* 1: 589).

the animal power confers sensation and motion, and therefore, etc.

One must respond that in an animal (or in the human) there is some power that is regulative of life because certain parts exist in the animal that are opposed to one another. For, unless it is blocked, the natural heat has to consume the radical moisture in which it exists as if in a subject, but it does not consume the moisture itself. And this will only be the case if it is ruled by some power.

Moreover, each power in an animal follows its proper function, as it is born to do, which will only be the case if there is some power in the animal that is regulative of life itself.

But doubt exists as to what the regulative power is, and this quite commonly is doubted by a great number of physicians in various places. For some say that it is one power formed into a whole out of all the other powers, and others say that it is one common power that is distinct from the others.

Nevertheless, it seems to me that one should respond that it is not one power, but several, and these are the natural, the vital, and the animal. Life is not ruled sufficiently by any one of these apart from the others, but rather each of these is required for a complete regimen, and each in some way rules another. For it is said—and it sounds extraordinary, although it may be part of nature [*licet sit physicum*]—that an offspring often resembles what the woman imagines or what falls into her imagination at conception.[15] In this way, Avicenna says, a certain woman, owing to this, gave birth to a dwarf.[16] But this would only be the case if the imaginative power ruled the generative power in some way, and so too for the others. In truth, just as the heart is the principal part in the animal, so too can the vital power, which is immediately situated in the heart, be called principal and regulative with respect to the others because a more noble power and operation is due to the more noble member and because without heat, which proceeds from the heart, no other power can operate adequately.

15. A., *DA* 22.1.4.7, 22.1.5.9 (*SZ* 2: 1443, 1445).

16. Avic., *Can. med.* 3.21.1.2.14; *DA* 9.5 (fols. 43rb–44rb), 18 (fols. 62vb–63rb).

The arguments prove that there is not just one power in the animal that is sufficient, without the others, to regulate life.

To the second argument one can respond that in no [principle] beyond the first is the substance of something also its operation. Thus the soul is not the immediate principle of its operation except through an intermediary that is the power or potency, just as one must see in the book *On the Soul* and as we have explained there.[17] Thus if life is received on behalf of the first act, the soul is its immediate principle; if it is received on behalf of the second, then the soul is not the principle of life except as a mediating power.

Question 4: Whether food will be necessary to the animal.

One inquires now about the regimen of life with respect to the nutriment. And first one inquires whether food will be necessary to the animal.

1. And it seems not. That which is especially distant from opposition [*a contrarietate*] can be maintained longer on its own.[18] This is clear from a celestial body.[19] But the body of an animal is especially distant from opposition, and as a result it is rendered fit for life. It can therefore be maintained more on its own from moisture, and as a result it does not need nutriment.

2. Moreover, that which is corrupt cannot confer anything on another. But nutriment, before it nourishes, is corrupted. Therefore, after its corruption it cannot confer anything on the animal and therefore it cannot nourish.

17. Ar., *De anima* 1.5 (411a21f.); cf. A., *De anima* 1.2.15.

18. Although the phrase *a contrarietate* is cumbersome, Albert seems to mean that when the qualities or elements are properly balanced in a living thing's complexion, that thing will require little or no nutriment to restore that which otherwise is lost through the action of contraries (e.g., hot and cold, moist and dry). Indeed, as he mentions below in *QDA* 7.5, thirst results from an appetite for the cold and moist, and hunger is an appetite for the hot and dry, but these are in opposition to the foundational principles of living beings, i.e., heat and moisture. This implies that if a being is properly balanced in its complexion, its natural heat and moisture should be self-sustaining.

19. The editor indicates that something seems to be missing from the text here and in the lines immediately following.

3. Moreover, the nobler something is, the more solicitous nature is toward it. But the living is nobler than the non-living. Therefore, since the non-living does not require nutriment, how much less so will the living.

The contrary is readily apparent, and the Philosopher makes the same determination.[20]

One must respond that an animal requires nutriment. And the reason for this is that every finite quantity can be quickly consumed by the continuous removal of some finite quantity, unless it is restored. But in an animal the natural heat is situated in the radical moisture just as in a subject and in its nourishment.[21] Therefore, its own nourishment is properly moist just like the unctuous moisture of fire. Thus, the heat does not cease to feed off the moisture itself. But that moisture is finite, and therefore, unless it is restored, it will be quickly corrupted. This is why food is necessary for the animal's preservation; because once the food is taken in it resists and blocks the action of the heat, lest the heat act upon its own proper subject—namely, the radical moisture—corrupting it and, as a consequence, destroying itself as well.

1. On to the arguments. To the first, one must respond that there are four [things to consider] in an animal body. The first is its distance from opposition, and this is why an animal has the sense of touch. The second is its weak coagulation, since soft bodies are a result of weak coagulation. Third is the frequent action of heat, because heat does not cease to act on the proper object of its action if it should come near to it. Fourth is the continuous operation of the animal, by means of which the animal's powers are weakened. Thus, food is not necessary on account of the first of these, but it is necessary on account of the other three. For because bodies are weakly coagulated they are prone to injury and corruption, and because heat acts on them continuously there is continuously some loss there, and because the animal operates continuously there is a continuous weaken-

20. Ar., *HA* 8.2–10 (589a10f.). Cf. A., *De homine* q. 10–11.

21. See also *QDA* 13.2 (245, 13). On radical moisture as food and *pabulum* for the natural heat, see *DA* 20.9.511 (*SZ* 2: 1386–87).

ing there. Thus food is a necessary remedy against these three.

2. To the second argument one must reply that what has been corrupted, to the extent that it is corrupted, confers nothing. But food, although it is corrupted as far as its own form [*species*] is concerned, assumes the form [*forma*] of aliment itself, and, as a result, that which was lost earlier from the thing it is nourishing is now restored by the food. So, if it were not corrupted previously, it would be unable to restore what has been lost. Thus, to the extent that it is corrupted, it confers nothing at all, but insofar as it assumes a new form [*species*] through a transmutation of its nature, [it does confer something].

3. To the third argument one must reply that inanimate things are generated from a determinate matter, and this is why they acquire a perfected quantity in their first generation. But animated beings are generated from semen, which first has a small quantity. Thus in their first generation they do not have a perfected quantity and for this reason animals have their own proper powers: one, by means of which they are moved to a perfected quantity, is the augmentative power; and another, by which they are preserved in their being, is the nutritive power. Thus, the argument can lead to the opposite conclusion, that among these inferior things the more perfect a thing is, the more things are required for its being.[22] And this is why it does not follow that animated beings are <not> nourished, although inanimate things are not nourished. In this way a solution is evident.

Question 5: Whether drink is necessary for an animal.

One inquires next whether drink is necessary for an animal.

1. And it seems not. What is opposed to the first foundation of life is unnecessary for the animal. But life is grounded in the hot and the moist. Drink, however, is properly cold, for it is said in the second book of *On the Soul*[23] that "thirst is an appetite for the cold and moist"; therefore, drink is not necessary.

22. "Inferior things": that is, not celestial bodies, but those found here below.

23. Ar., *De anima* 2.3 (414b12–13).

2. Moreover, nature does not do with two whatever can be done by one. But with food, an animal can restore and preserve what has been lost. Therefore, an animal does not require drink.

On the contrary. There are two appetites in an animal: that for the hot and the dry, and this is hunger, and that for the cold and the moist, and this is thirst. Therefore, if the object of the one appetite is necessary, so too is the other. But food is the object of hunger, and drink is the object of thirst. Therefore, drink is as necessary as food.

To this, one must reply that drink is understood in two ways: one is as both food and drink, and another is as drink alone. Wine, milk, beer and all those that are compounds are both food and drink; simple water, however, is drink alone. That which is both food and drink nourishes, just as food does, and in addition to this it tempers the natural heat by means of its active coldness, and it irrigates the members with its moisture. And it similarly causes the food, which is thick, to penetrate the members more easily and this is why drink is necessary, both on account of irrigation and on account of the thickness of food particles which, without the moisture, cannot easily travel to the individual members.

Moreover, animal life is grounded in certain humors and in heat. That which conserves the natural moisture and by which the heat is tempered is necessary, then. But this is drink. And this is why drink is necessary.

1. On to the arguments. To the first, one must reply that an excessive and dominating cold corrupts the heat, whereas a temperate cold supports the heat, because, if it dominates the heat, it will convert it into its own nature. Thus, although drink may be cold, if nevertheless it is consumed in a temperate fashion, it does not corrupt the natural heat but rather will support it and will preserve the radical moisture from being quickly consumed. Thus the natural heat can be consumed in many ways: by being dominated by its contrary, by a defect in its subject, and by its own excess, because if it is excessive it will act on its own proper subject, feeding on it. And drink will be necessary

for these two reasons, because it conserves the heat with its subject and blocks its excess.

2. To the second argument one must reply that food alone can not accomplish what food and drink can accomplish together. Although the restoration of what has been lost can be accomplished by food alone, nevertheless the dryness and aridity of the members and the sharpness of the heat cannot be tempered by food, but they can be better tempered by drink.

Question 6: Whether food or drink is more necessary.

One inquires third whether food or drink is more necessary.[24]

It seems that drink is more necessary. What is nearer to some necessary thing is itself more necessary. But air is more necessary to an animal for the sake of respiration, and drink is nearer to it than is food. Therefore, it is more necessary.

Moreover, that is more necessary whose lack is more oppressive. But this is drink because the longer thirst lasts the more intense it becomes and, as a result, if it is oppressive earlier, at the end it will be more oppressive. But this is not the case for food, as is evident in one suffering from fever.

On the contrary. The conservation of heat and moisture is more necessary for an animal's preservation. But these are conserved by food and not drink, because whatever is simply drink neither nourishes nor restores what has been lost. Therefore, etc.

To this, one must reply that something can be said to be necessary to an animal in two ways: either *per se* or *per accidens*. Food is more necessary *per se* because food is converted into the substance of the one being nourished and restores what has been lost. Drink, however, is more necessary *per accidens*, because the more the animal is fed, the more its heat is sharpened, and if it lacks drink the heat will consume its own subject. And this is why drink is more necessary *per accidens*, to temper the heat. This is evident from this analogy. A house is composed of wood

24. "Third": appears in the text without explanation.

and not of water, and nevertheless if the house were burning the presence of water nearby would be more necessary than that of wood because wood situated nearby would only spread the fire, whereas water would extinguish it. So the conclusion is correct. Thus one must understand that whenever the heat in a body intensifies without the emptying out of the members, as occurs in those suffering from fever, then one especially desires what is drink alone, like water. Nevertheless, when the members have been emptied, then one desires both food and drink. This is evident in the example of a healthy person who is hungry in warm weather, for as a result he desires drink more than food.

The arguments proceed along these paths.

Question 7: Whether fish are nourished by water.

One inquires further about the nourishment of things that swim and, first, whether fish are nourished by water.

1. It seems so. For that which enters into fish and nourishes them is their nourishment. But they are nourished by the entry of water, as is said in the text, and therefore, etc.[25]

2. Moreover, "we are nourished by the same things from which we take our existence." But fish take their existence from water in terms of their domain. Therefore, they are nourished by water.

3. Moreover, they cannot live without water. This would not be the case if they were not nourished by water. Therefore, etc.

To the contrary. According to Aristotle in the second book of *On Generation*,[26] a simple element does not nourish. Therefore, fish are not nourished by water, since it is a simple element.

One must respond to this that things that swim are not nourished by water alone either as food or as an object. But some of them are nourished by the slimy matter [*muscillago*] that is in the water, others are nourished by the dust of the earth, still

25. Ar., *HA* 8.2 (589b16f.). Cf. *DG* 2.8 (335a10f.).

26. Ar., *DG* 2.8 (334b31f.).

others are nourished by plants, and others by flesh, and this occurs according to the diversity of their complexions.[27] Thus those in whom the heat becomes more intense live on flesh; those in whom it becomes less intense, live on slimy matter, and those that approach more to an elemental nature live on the dust of the earth. Nevertheless, although they are not nourished by water as if by an object, water still contributes to them in two ways. The first has to do with the distribution of the nutriment, because they take in their nutriment with water mediating it. Moreover, they use water to temper the natural heat, just as breathers do with air.

But one must understand that although fish take in seawater, they still do not draw in that which is earthy in the water but only that which is sweet.[28] For the sweet can be separated from that which is earthy, as is evident in a wax-like instrument that is very concave and almost completely closed. If this is left in the sea for a day and a night, it will be filled with fresh water, and the salty and earthy parts will cling to the outside, as we showed in the book *On Meteorology*.[29] And this is why the meat of marine fish is not salty but sweet, although they live in salt water.

1. On to the arguments. To the first, one must reply that water does not enter into fish in the same way as nutriment, but rather for the distribution of the nutriment.

2. To the second argument one must reply that "we are nourished on the same things from which we take our existence" applies to principles near to us but not to those that are remote from us. Now let us grant that animals arise from the elements as if from remote principles, but from compounds like the humors as if from proximate principles. Therefore, they are not nourished by the elements but rather from the humoral moistures which are their proximate principles. Or one can say that

27. *Muscillago*, CL *mucilago*, implies a moldy, slimy substance. It must refer to the suspended, murky matter that exists in all but the clearest waters.

28. "Sweet": Lat. *dulce*. The saltiness of water was attributed to its earthy elements, and without these elements it was considered fresh water [*aqua dulcis*].

29. Ar., *HA* 8.2 (590a24f.); *Meteora* 2.3 (358b34f.); A., *DA* 7.1.2.16 (*SZ* 1: 592); A., *Meteora* 2.3.16.

just as they exist on water from their domain, so those compounded are nourished by other [humoral moistures] from their domain.

Question 8: Whether fish eat their young.

One inquires further as to the nutriment of fish. And first one asks whether fish eat their young.[30]

1. And it seems not. Because every animal has a natural love for its offspring [*fetus*]. But eating the offspring does not stem from love, and therefore, etc.

2. Moreover, other animals do not eat their offspring; therefore, for the same reason, neither do fish.

The Philosopher says the opposite.[31]

One must reply that of all the animals fish do eat their young rather a lot. The reason for this is that fish are very gluttonous and have impaired senses. Gluttony arises from the cold of the stomach, for a hot stomach is good for digestion, but bad for the appetite, according to Galen. But a cold stomach is bad for digestion but sharpens the appetite. But fish have a cold stomach, and this is why they are so gluttonous.

Moreover, they have impaired senses, and this is why, when fish see their young, they are not able to discern whether they are their own or belong to another. And they are especially gluttonous. And this is why they eat their own young as well as those of another without distinction.

1. On to the arguments. To the first, one must reply that an animal's love for its own offspring is natural, nevertheless in an accidental way an animal can be blinded to or impeded in this in some measure, namely, owing to an inability to distinguish them because it does not recognize them. This even occurs in perfect animals like the human, who sometimes kills his brother thinking that he is someone else.

2. To the second argument one must reply that although

30. Cf. A., *DA* 7.1.3.17–25 (*SZ* 1: 592–95).

31. Ar., *HA* 8.2 (591a7f.).

this occurs in fish in particular, nevertheless it can also occur in others. For, if a wolf is wounded, it is said that all the others surround and devour it. Thus, since a wolf is a very gluttonous animal, when it perceives its prey and sees that another wolf is incapable of seizing it, it does not spare it more than it does any other animal. And the same holds true for the rapacious animals with hooked talons.[32] And, moreover, some pigs have been seen in Cologne to eat their young, and this was owing to their gluttony and to the absence of nourishment, etc.

Questions 9–11: On rainwater.

One inquires further about something said in the text, namely, that rainwater is sweeter than other water.[33]

And it seems not. Because rainwater is generated from vapors from the sea and from marshes that rise to the middle interstices of the air. These vapors are salty, and therefore so too is rainwater.

Second, one inquires whether rainwater causes constipation.

And it seems not, because every like increases like, and rainwater is fluid and moist, and therefore, etc.

Moreover, salt is astringent whereas rainwater is sweet; therefore, etc.

One inquires third why it injures birds more than fish even though birds receive less of it and for a briefer period of time.

To the first question one must reply that sweetness arises from moist air. Thus, those things that are initially sweet, through the exhalation of the moist air, later increase and grow bitter. But rainwater descends through the air and frequently swirls in the air while it falls. Thus it has a great deal of moist air mixed in with it and this is why it is sweeter than other water.

To the argument one must respond that when the sun raises and evaporates the vapors from the sea and the marshes, it raises the subtle portion and leaves the gross and earthy portion below in the sea. And this is one cause for the saltiness of the sea.

32. On the *gampsonicis* see *QDA* 2.30, note.

33. Cf. Ar., *HA* 8.19 (601b10f.); A., *DA* 7.2.1.94–97 (*SZ* 1: 628–30).

Thus the vapor is not salty, although it is raised up from the sea, because what is sweeter and subtler is raised up. Or, given that something bitter may be raised up, nevertheless when it passes through the sphere of the air, its bitterness is removed and it is sweetened.

To the second question one must reply that rainwater does cause constipation both because it is subtle and because it is sweet. For it is subtle on account of its frequent swirling through the air, and it is sweet for the reason already mentioned. Since it is subtle, it penetrates the individual members easily, and it leads the gross humors that are in the members along with itself to the bladder, and this is the reason for the increase in the urine and, as a result, for the reduction in the excrement, because an increase in one is a cause for the reduction in the other, and when the excrement is reduced, the constipation is relieved.

Moreover, only the sweet nourishes. Thus, the individual members attract this water owing to its sweetness, and all the powers are strengthened by it as well and, as a consequence, the retentive power. The expulsive power, however, is weakened because everything that is amicable to nature resists those that impede nature, promotes those promoting it, or aids those contributing to it. And since the increase in the expulsive power during diarrhea is inimical to nature, whereas the retentive power is well disposed and friendly to nature, rainwater therefore strengthens the retentive but weakens the expulsive power.

To the argument one must reply that although rainwater is fluid and moist, still it causes constipation for the reasons already mentioned. Or one can say that diarrhea, when it arises from a cold and moist cause, like slimy plants, is intensified by rainwater. But when the diarrhea arises from a warm cause, such as from an abundance of blood and bile, rainwater is good for this. Thus physicians specify that all the medications that should be taken with liquid for this illness must be taken with rainwater, if they can get it.

To the third question one must reply that fish live in water and that sweet water particularly benefits them because only the sweet nourishes. Thus some fish are nourished solely on water mixed with mud, like eels. But birds live in the air, and they seek nutriment by extending their wings and feathers, and these get stuck together by rainwater, and this is why water of this sort injures them more than it does fish. And another reason is that birds do not require drink as often, and because when they do they drink avidly. Thus, they incur a flux like lienteric flux, and this is why, etc.[34]

Or one can say otherwise that overabundant rains injure them both because they suffer a swelling of the nutritive members owing to the sweetness of the water, because sweet things inflate the nutritive power, according to the interpretation of the physicians and especially Galen.[35] And this is why, etc.

Questions 12–16: On the nutriment and the drink of flyers.

One inquires further into the nutriment of flyers. And first, whether flyers have a big appetite.[36]

It seems not. This is because a cold stomach has a great appetite, but birds have a very warm stomach, and therefore, etc.

Second, one inquires why, when they are nourished by flesh, salty flesh injures them.

And it seems the opposite is the case, since salt cleanses and consumes the feces. But these functions are useful to birds, especially to those with curved talons [*gampsonici*], as well as to walkers; therefore, etc.

Third, one inquires whether they only drink a little.

And it seems not, because an appetite for drink is an appetite for the cold and moist. Now, however, birds are particularly warm and dry. Therefore, drink is particularly necessary for them.

Fourth, one inquires why some birds do not raise their necks when they drink until they have finished drinking, and some do.

Fifth, one inquires why birds and fish do not ruminate, as walkers do.

To the first, one must reply that the appetite's intensity arises

34. The term *lientericus* refers to the spleen. On lienteric flux, see *DA* 15.2.7.114 (*SZ* 2: 1136).

35. Cf. Galen, *De methodo medendi* 13.14 (Kühn, 10: 908).

36. Cf. Ar., *HA* 8.3 (592a28–594a3); A., *DA* 7.1.4.30–41 (*SZ* 1: 599–604).

from the emptying of the stomach. But the stomach's emptying can arise from two causes: in one way, from the consumption of the stomach's contents, as occurs from heat, and in another way from the constriction of those things that are in the stomach, just as occurs from cold. Thus, those who have a warm stomach have a large appetite and can eat a lot, whereas those who have a cold stomach have a large appetite but can only eat a little. Thus, just as fish are gluttonous owing to the coldness of their stomach, so birds have a big appetite owing to their heat.

By means of this, a response to the argument is clear.

To the second argument one must reply that salt is warm and dry and birds are warm and dry. Thus salty flesh intensifies their warmth and dryness, and, owing to this, it is a source of death to them since the heat sharpened by it quickly consumes their moisture. Thus, although salt cleanses and consumes feces, nevertheless it also sharpens the natural heat and introduces dryness. And this is why salty flesh is noxious to birds. It is not noxious to walkers, however, because they have a stronger complexion, etc.

To the third argument one must reply that birds of prey live on raw flesh in which there is a great deal of moisture, and this is why birds like these require less to drink, since the rawness of the flesh tempers their heat.[37] Nevertheless a sparrow hawk [*nisus*] has been seen which has drunk deeply, but perhaps this occurs in such as these owing to some imbalance resulting from fever, and not naturally.[38]

To the fourth argument one must reply that some birds have a great appetite for drink and have a short neck, and these do not raise their necks before they have drunk, like the doves, sparrows, jackdaws and the like, owing to their gluttonous nature. There are, however, other birds that do not have a great appetite for drink, and ones like this may have a long neck, like the goose and the crane, or a short one, like the birds of prey, and they do raise their necks while drinking, interrupting their

37. Cf. Ar., *HA* 9.3 (593b28f); A., *DA* 7.1.4.41 (*SZ* 1: 604).

38. A. clearly had spent significant time observing falcons and hawks firsthand and had consulted experts such as a colorful Alpine hermit who trapped birds of prey for a living (*DA* 23.8.57–58 [*SZ* 2: 1582–83]).

drink. Others, like Pliny, say that this is because it is more pleasurable for them to drink this way, and I believe that this is owing to the narrow esophagus in such animals. This is why they raise their necks, in order to gulp their drink more easily.

To the fifth argument one must reply that animals that walk ruminate, but birds or fish do not.[39] There are many reasons for this: because walkers, more so than fish or birds, live off things born of the earth, which are difficult to digest, and this is why they require several digestions, and this is why too they recall the food to the mouth and ruminate. Another reason is that in order for an animal to ruminate it is necessary to retain the food for a long time and chew it well. But birds lack teeth because their weight would impede their flight, which is why they cannot chew. Moreover, should fish retain their aliment in their mouths for a long time, too much water would enter the heart and suffocate them, and this is why wise nature [*natura docta*] has ordered them to obey the Creator, according to Avicenna in the third book, and not retain the food, which would happen if they ruminated, and this is why, etc.

Questions 17–19: Whether reptiles can live long without food. Why serpents have a special appetite for milk and wine. Why the horse has a special appetite for drinking muddy water and the cow has a special appetite for clear water.

One inquires further about reptiles, whether they can live long without food.

And it seems not. Whatever has a big appetite cannot go for long without food. But serpents are like this, because they are cold. Therefore, etc.

The Philosopher says the opposite about the serpents sold by apothecaries, who sometimes keep them for two months without food.[40]

39. Note that at *DA* 13.2.1.81 (2: 1023) A. parallels Ar., *Part. An.* 675a4, to state that the *astaros* (Gr. *skaros*, a parrot-wrasse) ruminates. On the rumination of the parrot-wrasse, cf. *GF*, 239.

40. Ar., *HA* 8.4 (594a21–24).

Second, one inquires why serpents have a special appetite for milk and delight in it, and likewise concerning wine.

Last, one can inquire why the horse has a special appetite for drinking muddy water and the cow has a special appetite for clear water.

To the first, one must reply that reptiles are cold and have a cold stomach and this is why they are particularly gluttonous [*avida*]. But the heat in them is weak nonetheless and consumes only a little of the radical moisture, and this is why they can survive a long time without nutriment, a thing that does not occur in warm animals. For if a human were not to eat for a day, or at least to fast on bread and water, as we and other religious sometimes do, on the next day his urine will be very reddish and tinted, as if burned by a natural heat.[41] Thus, the warmer an animal is, the less it is able to endure without food. But serpents and animals like these are cold, and this is why, etc.

And by this it is clear for the argument that although they have a large appetite, owing to their modest consumption of the radical moisture and the natural heat, they can nevertheless endure and be preserved in life for a long time.

To the second question one must reply that the sweet is very nourishing, and milk is sweet and makes a good nutriment for a clean stomach, whereas it is easily converted into corrupt humors in an unclean stomach. And this is bad for serpents. Reptiles therefore have a special appetite for milk because of its sweetness, so that they may be nourished by it more easily and so that their slithering [*lapsus*] will be tempered by it.[42] In the same way, reptiles are very cold, and wine is warm. Thus their heat is intensified by the wine and their coldness is diminished, and they rejoice in this as if they know what they are doing. I

41. "Other religious": that is, members of religious orders other than the Dominican order.

42. *Lapsus* is difficult here. It may be related to a verb meaning to "slide" or may somehow refer to a falling away from some state or other. The text seems to contradict itself at this point, with the suggestion that milk is bad for serpents whereas it easily nourishes reptiles. Albert usually treats serpents as reptiles, although at times he does distinguish between the two. See *DA* 21.1.7.42–44 (*SZ* 2: 1433–35).

have experienced this myself with a certain serpent I had in Cologne: I got him intoxicated on wine, and he swayed hither and thither through the cloister as if half alive, because "drunk"—*ebrius*—is said from *e*, that is, "beyond" (*extra*), and *bria*, that is, "measure," as if placed beyond the measure of nature.[43] And this is why they have an appetite for wine.[44] Thus, animals like this very often take in so much milk or wine that they immediately vomit it up, because they receive more than they can retain, which is one of the causes for vomiting.

To the last question one must reply that an ox has thick blood and a good stomach, and this is why it does not need drink that remains a long time in its stomach, but it does need thin drink by which its turbid and thick blood may be purified. And this is why it has an appetite for clear and thin water, following the sagacity of nature and a natural instinct, because according to Avicenna God has armed each animal with natural powers by which it desires whatever is suitable for it, and avoids what is naturally unsuitable, etc.[45] The horse actually has a more subtle blood and a weak stomach, and this is why it needs something that will remain a long time in its stomach and digests well, before it is delegated to the other members. For if its drink were thin, it would subtly penetrate from the stomach to the other members before digestion is completed, and since its blood is particularly thin, it would be made thinner by a drink of thin water. It is therefore useful that its drink be turbid in order to temper the blood. Thus an illness often affects horses in the feet or the other members from clear water, owing to its rapid penetration of the members, and this happens especially when they drink after a lot of movement, because then their blood is rendered particularly subtle. This is why smart people, after they have been riding, send the horses to rest, so that the blood may settle and the pathways constrict, and only after this do

43. See A., *Summa* 2, q.121, m.1 a.4 part. 1 ad 2, where A. specifies that *bria* is from a Gr. word, though the etymology seems fanciful. Cf. also *De bono* 3.2.10 (Kühle, 1951).

44. Cf. A., *DA* 7.1.4.42 (*SZ* 1: 604).

45. Avic., *Can. med.* 1.1.4.2.

they give the horses water, because then the water does not penetrate so quickly, etc.[46]

Questions 20–21: Whether local motion particularly agrees with birds.

Whether this local motion is for the sake of the nutriment or the disposition.

"And some people," etc.[47] First one inquires with which animals local motion from place to place agrees.

1. And it seems first that it agrees with fish. For everything is received in a recipient according to the mode of the recipient. Since, then, the sea is especially fluid and movable from place to place, such motion must especially suit fish; therefore, etc.

2. It seems to agree with the walkers. Because motion such as this agrees with an animal that has very good estimative power, because they feel [*aestimant*] it is better to be in one place than in another. But a good estimative power only suits walkers, and therefore, etc.

The Philosopher says the opposite.[48] For he says that it especially suits birds.

After this one inquires into the cause of this motion, whether it is for the sake of nutriment or for the sake of the disposition in one that possesses it.

3. It seems that it is not for the sake of the nutriment. Because this would agree equally with the large and the small, with the birds of prey and those that are not birds of prey, and this is not true. For the large birds migrate and not the small, and similarly the birds of prey do not migrate but others do.

Nor does it seem to be for the sake of the disposition in one possessing it, for the same reason. And in addition to this, all those that are in the same region migrate this way.

46. This disease and cure are reminiscent of the disease A. calls *infundatura* at *DA* 22.2.1.91 (*SZ* 2: 1502–3).

47. Ar., *HA* 8.12 (596b25f.); A., *DA* 7.1.6.61–72 (*SZ* 1: 612–17).

48. Ar., *HA* 8.12 (597a4f.).

One must reply first to the second argument that there is a twofold reason why animals migrate. One stems from the nutriment, and the other from the one possessing it or from that one's complexion. It stems from the nutriment, because animals with a good estimative power change to another location when the aliment appropriate for them is lacking in one place, moving to where it is more abundant. And this is why many fish move from the bottom of the sea to the shore, because the earthborn things on which they are nourished are more abundant there.

Another reason stems from complexion. For those, like the crane, that greatly fear the cold, following a natural instinct, move to warm places before winter, and those that fear the heat move to cold places, and this is why fish that fear cold move from the bottom of the sea to the shore, whereas those that fear heat move in the opposite way.

And for the same reason some animals remain on land during the day but spend the night in the water, because the land is warmer during the day owing to the surrounding warm air and the appearance of the sun, and the water is warmer at night for a period of time because just as the air quickly receives heat in the presence of the sun, so it loses it in its absence. But water receives heat with more difficulty, and for this reason it retains it longer.

Moreover, at night in the absence of the sun, the cold of the middle interstice flees the heat that was first received in the air at the surface of the water.[49] And this is one reason why water is warmer at night than it is during the day, in comparison to the land. Thus both the nutriment and, similarly, the animals' individual complexion cause migration among them. This change, however, principally suits birds and fish next, because birds are better disposed to movement than are walkers and have better estimative power than swimmers. And similarly birds exist in an element, namely, air, that is especially suited for motion. Thus,

49. On the middle interstice or interval of the air, see *QDA* 7.9–11. Recall that in this Aristotelian cosmos the earth, the heaviest of the elements, is surrounded by water, which is by nature cold and somewhat lighter than earth, and then by air, and finally by the lightest of all the elements, fire.

on account of these many reasons, this is more appropriate to birds than to other animals.

1. On to the arguments. To the first, one must reply that although the sea or the water is movable, air is nevertheless more movable and more divisible, and although water sometimes contributes more to a swirling motion, air nevertheless contributes more to straight movement.

2. To the second argument one must reply that although the estimative power is better in walking animals, still they possess more weight. Because of this the argument lacks force.

We have already spoken to the other question.

3. To the first argument one must reply that in order for animals to move from region to region it is necessary for them to be good-natured and strong, because they must pass through many dangers. Moreover, it is necessary for them to have good estimative power, so that they can better take precautions. These things are not found in small birds as much as they are in large birds. For large birds—like cranes, geese, swans, and the like—migrate more than do small birds. Still, small birds like sparrows, which are confident in their wings, do this.

Moreover, birds of prey are not as good-natured, nor do they remain in a flock, because they fear the loss or theft of their prey, which they acquire only with great effort. Similarly, they know how to live in places known to them better than in ones unknown.

And moreover the change of seasons has a greater impact on things born of the earth than on animals because things born of the earth grow for half the year and for the other half they wither.[50] But birds of prey do not live on things born of the earth and this is why it is not as necessary for them to migrate, as do birds like cranes and ones like them that live off things born of the earth.[51]

50. "Things born of the earth," *terranascentia,* would seem mostly to be plant-like, but may also include things thought to emerge from the ground, often by spontaneous generation. The crane was much admired for its migration.

51. See *DA* 23.1.24.113(49) (*SZ* 2: 1626).

Question 22: Whether animals lie dormant in winter on account of fear or weakness.

One inquires further into the concealment of animals, and whether they lie dormant in winter.[52]

It seems not, because they would lie dormant either on account of fear or on account of the weakness of the complexion. But it is not on account of fear, because serpents lie dormant and yet they are quite bold; nor on account of weakness, because the bear lies dormant for a large part of winter and nevertheless it is strong and rapacious.

Moreover, animals like this do not take nourishment during the time of concealment. But this would not be the case if they were weak; therefore, etc.

One must reply that animals lie dormant both on account of the character of their surroundings and the weakness of their complexion.[53] Those that lie dormant have a cold complexion and porous composition. And because a cold and weak heat is mortified quickly by excessive cold, animals like this seek out caverns in the ground when cold weather is imminent and rest in them during winter. And because both serpents and animals with a ringed body have a cold complexion, this is why this suits them both. In the same way, the bear has a porous composition[54] and weak motion and a cold complexion. Thus when winter is imminent, its spirits and motive nerves are mortified, and this is why it does not make an appearance either in winter or at the time of parturition.

One must reply then to the argument that serpents do not lie

52. Cf. Ar., *HA* 8.13–17 (599a4f.); A., *DA* 7.1.8.82–89 (*SZ* 1: 621–26). A note on terminology is in order. A. uses verbs like *latito* and nouns such as *latitatio* to indicate what we call hibernation. His vocabulary thus stresses the fact that the animal drops out of view. Thus, while it is tempting (and less awkward) to use the modern "hibernate," we will instead translate more literally to preserve the flavor of the original.

53. "Surroundings:" See *QDA* 12.2 where *continens* represents the water in which a fish lives.

54. One suspects that this should this be "a loose composition," that is, one that is not dense, or should be omitted as it is, in fact, in the T manuscript. For other views on the bear's complexional makeup see *DA* 7.1.8.89 (*SZ* 1: 626), 7.2.3.154–59 (*SZ* 1: 658–59), and 22.2.1.144(107) (*SZ* 2: 1540).

dormant on account of fear but on account of an external cold that dulls their weak heat. Neither does the bear lie dormant on account of weakness *per se,* but rather on account of the mortification of its spirits by the external cold.

Question 23: Whether animals are alive at the time of concealment.

One inquires further whether such animals are alive at the time of concealment.

1. It seems not. To live is to move oneself. But animals such as these do not move themselves; therefore, etc.

2. Moreover, life is based on a radical—that is, natural—heat. But heat is extinguished in them at the time of concealment. Therefore, etc.

On the contrary. "To live, for living things, is to be."[55] But the existence of one and the same thing does not occur at intervals. Since, then, such animals are alive before their concealment and after it, then, they are alive during it as well.

To this, one must reply that animals like this live because the soul is the principle of life and life does not depart unless the soul departs. But these animals do not lose the soul at the time of concealment, because if they did they would immediately experience corruption just like other dead animals. Therefore, they have life and exist as if they were dead, but are not dead.

1. On to the arguments. To the first, one must reply that motion is of two types. One is external, by which an animal is moved from place to place, and another is an internal motion, by which the natural heat, spirit, and life flow from the heart to the other members. Thus, the first motion is not present in them at the time of concealment, but the second is and it is sufficient for them to remain alive.

2. To the second argument one must reply that heat can be in something in three ways: either *per se,* as in fire, or *per accidens,* as in water that has been heated, or, as in cold water, po-

55. Ar., *De anima* 2.4 (415b12–14).

tentially—otherwise, fire would not be generated from it[56]— or as in wine. Thus, heat is in these animals *per se,* although the operations of the heat do not appear in their exterior parts because the heat flees to the interior parts on account of the external cold. Thus, although the exterior members appear dead, nevertheless the character [*habitus*] of heat exists in them according to the first act, and not according to the second act.

Question 24: Whether these animals have the first act of life without the second.

One inquires further whether these animals have the first act of life without the second.

1. And it seems so. In every genus, the first [species] can exist without the later ones, and therefore the first act can exist without the second.

2. Moreover, the Commentator and Avicenna say, concerning the second book of *On the Soul,*[57] that some animals have the first act of the soul without the second, like shellfish [*conchilia*]. But this seems to agree especially with these animals at the time of concealment, and therefore, etc.

On the contrary. Every thing that has life has the regimen of life. But the regimen of life depends on some operation, and operation is the second act [of life], and therefore, etc.

To this, one must reply that animals like this have both the first and the second act. But the second act can be understood as existing with respect to the sensitive power or with respect to the natural or vital power. They do not have the second act in terms of the sensitive—that is, animal—power, because they do not have sensation or act according to the senses, but have only the first act. But they do have the second act (but not the first act) in terms of the vital or natural power: in terms of the vital power, because the heart has some motion and there is a flow

56. It is difficult to comprehend how fire could be generated from water, although one can easily see how heat is generated from water that has been warmed. Cf. *QDA* 5.1 and note.

57. Averroes, *De anima* 2.1; Avic., *De anima* 2.3.

from the heart to the other members. Similarly, the second act exists there naturally because some digestions occur in them at the time of concealment. An indication of this is that when they lie dormant or begin to lie dormant they are lean and when they come out they are fat, and some are just the opposite, as will be evident.

1. On to the arguments. To the first, one must reply that existing without another can occur in two ways: either essentially or in effect. The first act can exist without the second essentially, because sensation is not from the essence of the first; but in effect, that is, in act, there is no first act without the second.

2. To the second argument one must reply that the soul can exist in first act without the second. This occurs in this way, such that when the soul has many operations it can exist for a period of time without some of them, but it cannot exist without all of them. Thus one having sensation is able not to see or to hear, and so too for the other senses. But it is impossible for it not to be moved by the motion of the pulse, and it is not possible for it to be deprived of each and every second act. Nevertheless, the fact that an animal can be deprived for a period of time of some second act of the senses is clear in those that are sleeping, and that it cannot be deprived of some second act of the vital and natural power is clear because the vital and natural powers are more active than the sensitive power and they are the foundations for the other powers. This is why, etc. In this way a solution is apparent.

Question 25: Whether animals take nourishment when they lie hidden.

One asks next whether such animals take nourishment when they lie hidden.

It seems so. Because some thin ones grow fat when they lie hidden, and they nourish the offspring [*fetus*], which would not happen if they were not nourished; therefore, etc.

Moreover, digestion occurs in them, as has been said. But this only occurs if there is some nutriment. Therefore, etc.

To the contrary. To take nourishment is to receive something external. But animals like these receive nothing from outside; therefore, etc.

To this one must reply that in animals like this there exist viscous and superfluous moistures, as is especially apparent in the bear.[58] Now just as motion, by sharpening and exciting the heat, and by opening the pores and evaporating it, causes the consumption of that which is superfluous, so too does rest cause the generation and accumulation of superfluities. The bear and serpents and other such animals love rest. Thus, during the time of lying hidden, nature returns to their superfluities, and they are digested by the natural heat and converted into the substance of these animals, and so they are nourished during the time of lying hidden, not by something taken in from outside then, but by something taken in earlier. Thus there are as many superfluities in the bear as are required for the nutrition of the mother and the offspring [*fetus*]. It is said also during the time of lying hidden the bear licks the pads of its feet,[59] because what is superfluous from their aliment reposes in the muscles of their feet. And at this time the bear returns to what has been laid down there, and this is why it licks its feet, in order to extract the moistures and the superfluities that have been deposited there, and so it sucks them and is nourished as a result.

It is uncertain whether a human can survive for such a long time without food. Nevertheless, should this happen in a human, it is more fitting that it occur in a woman than in a man, because the moist and viscous superfluities abound more in a woman, and since the heat is weaker in a woman than in a man, it cannot, in brief, consume them during this time.

By this a solution to the arguments is evident.

Question 26: Whether animals breathe during the time of concealment.

One inquires further whether animals breathe during the time of concealment.

58. Cf. A., *DA* 7.1.8.89 (*SZ* 1: 626).
59. Cf. A., *DA* 22.2.1.144–45 (*SZ* 2: 1539–41).

And it seems so. The natural heat resembles the heat of a lamp. But the heat of a lamp is extinguished in two ways: in one way, by blowing it out; in another, by suffocating it. And the natural heat is likewise extinguished in two ways. If there were no ventilation in an animal through respiration when the animal lies in hiding, then the natural heat would be extinguished, and, as a result, animals like these would not live.

Moreover, an animal lying in hiding participates more in the operations of life than does a dying animal. But there is respiration in a dying animal, and there is, therefore, all the more reason for it to exist in animals lying hidden.

To the contrary, respiration is only necessary on account of the vehemence of the internal heat. But in animals like these the heat is dulled and is weak. Therefore, they do not have to breathe.

One must reply that, for the most part, animals like these—that is, serpents, swallows, the cuckoo, and the like—do not always breathe during the period of lying hidden, just as bees do not. One must understand, however, that there are three kinds [*genus*] of animals. Some are hot and have blood, and ones like this always breathe. Other animals are cold and lack blood, and ones like this never breathe because they do not have lungs. And some occupy a middle position: for they are warm, but weakly so, and this is why they lie as if stupefied in the presence of a penetrating cold and are aroused to movement and for external operations when warmth arrives. Thus in the period of lying hidden they do not have to breathe because the heat in them is not abundant. Nevertheless, during the period when they are seen, they do breathe, because then the heat in them is fortified.

And if one should object that a contrary is fortified in the presence of its contrary, and that therefore the natural heat of such animals is strengthened in the presence of an external cold, and that, as a result, they will need to breathe then even more, one must reply that some have a strong heat that is not overpowered by the presence of the cold, and such heat is fortified in the presence of cold.[60] But some have a weak heat, and

60. Cf. A., *De morte et vita* 2.8; *De nat. loc.* 1.11.

in ones like this (such as serpents and the like) the heat is not strengthened but is rather weakened by the cold.

And in this way a solution is apparent.

Question 27: Whether life is preserved longer in places that are warm and moist, cold and dry, or cold and moist.

One inquires further, since some places are warm and moist, and some are cold and dry, and some cold and moist, where life is preserved longer.

1. And it seems that life is preserved longer in moist places, whether they are warm or cold, than in dry places. Old age is a path to death. Therefore, the prolongation of old age delays death. But in warm places, people grow grey more quickly, for Avicenna says that in the lands of the black Ethiopians people grow old at age thirty. Therefore, they live for a shorter period.

2. Moreover, life is based on warmth and moisture, and it is shortened when these are dissipated. But in a warm region the external heat of the region penetrates to the interior heat, dissipates it, and causes the natural heat and moisture to evaporate. Therefore, etc.

3. The Philosopher says the opposite in the book *On the Reason for Shortness or Length of Life.*[61] For he says that animals live longer in warm and moist places.

4. And this is clear from the argument that philosophers choose to reside and to study in places near the sea, and this is only because these places are warm and moist.[62]

5. And it is also clear from another argument, that a thing is preserved by the same things on which it is based. But life is based on the hot and moist, and therefore, etc.

6. Moreover, life is preserved by nutriment, but nutriment is more abundant in warm places than in cold, as the physicians say, and this is clear to one inspecting warm lands in comparison to cold ones. Therefore, etc.

61. Ar., *De long. et brev. vitae* 1 (465a9–10).

62. Ar., *De long. et brev. vitae* 5 (466a18f., 466b2f.).

There is a disagreement between Aristotle and the natural philosophers [*physicos*] on this question. For the Philosopher proposes that life is extended in warm and moist places,[63] whereas the others propose that this occurs in cold places. As a result, in order to resolve this disagreement, one must introduce the distinction that some animals are warm and some are cold. Cold animals live longer in warm places than in cold places, because their own cold is intensified and their own heat dulled by the cold of the place. Thus in winter serpents lie hidden in warm burrows in the ground. And in the same way Avicenna says that asses do not live in Scythia, which is a cold region, because they are cold animals.[64] Warm animals exist in a contrary fashion, because they have a loose composition in a warm region. Thus the external heat easily penetrates through the open pores to the interior and causes the natural heat to escape and consumes and dissipates the natural moisture. This is why warm animals—and humans belong to this group—do not live as long in such warm places.

Nevertheless a distinction can be made about the heat and the moisture of a place. Now, one is a natural and tempered heat and moisture, and another is accidental and distempered. The first one contributes to life, whereas the second does not because every natural corruption arises from such a heat, as is said in the fourth book of *On Meteorology*.[65]

1, 2. A solution to the arguments is apparent in this way, because the first ones speak the truth about warm animals.

3. A solution to the first argument to the contrary is apparent, since the Philosopher is thinking about cold animals and about a tempered heat and moisture, and the others are thinking about the contrary.

4. To the second argument one must reply that places near the sea do not cause the extension of life. In these places two things come into consideration. Specifically, one is the proper tempering of the heat and moisture and how much this contrib-

63. Ar., *De long. et brev. vitae* 1 (465a9f.).

64. Rather, Ar., *HA* 8.25 (605a21f.); *GA* 2.8 (748a22f.).

65. Ar., *Meteora* 4.1 (378b12f.; 379a16f.).

utes to the longevity of life, but the other is the infection of the air from corrupt vapors released by the sea. Thus, such vapors or such infected air when it is inhaled often causes a corruption in the lungs because it infects the principle of life, namely, the heart and its spirits, and as a result it contributes to the brevity of life.

5. One must reply to the third argument that just as life is based on warmth and natural moisture that is tempered, so is it preserved by them, but these shorten life when they are unnatural and intemperate.

6. To the fourth argument one must reply that the conservation of life is twofold. The first way is by an internal principle, which is the warm and the moist, and this sort is greater in cold places. In another way, it is by an external principle, for example, by the nutriment, and this sort of conservation of life is more frequent in warm places. But there is no question about this one. Truly, although the nutriment is more abundant in warm and moist places, nevertheless sufficient nutriment is found in cold places. Indeed, people eat incomparably more at a sitting in cold places than in warm ones. Thus, one Pole or German eats more in a single day than a Lombard or a Frenchman does in four, etc. In this way a solution is evident.

Question 28: Why people in warm places have a smaller stature and less strength and boldness than those in cold places.

One inquires further why people in warm places have a smaller stature and less strength and boldness than those in cold places, when, however, the opposite thing occurs in other animals and in plants.

And it seems that naturally this ought not to be the case. Because heat, together with moisture, is the principle of growth. Therefore, there will be more growth in what is generated in warm and moist places than in cold ones.

Moreover, strength and audacity arise from heat. Thus choleric people are bold due to the heat. Therefore, for the same reason, people in warm places will have more strength and boldness.

Moreover, this is the case in other animals, like the lion. Therefore, it will also be the case in the human.

The opposite is evident, thanks to Avicenna. For he says that in Germany and Flanders and Poland, which are cold places, the people have a larger stature and are bolder. It is even commonly said that people of warm regions are naturally timid and incapable of war.

One must reply that things are different for the human, plants, and other animals. For human seed is naturally warm and moist, and this is why it liquefies and is consumed by the heat in a warm region, and as a result people of smaller stature are generated there. But the plant's seed is cold and hard, and [its life] cannot be extended except by an abundant heat, and this is why plants grow more, are nourished more, and age more in warm places than in cold, and they also have a larger size. And it is the same for the seed of brute animals in comparison to the human.

Moreover, strength and properly ordered audacity arise from a natural and tempered heat. But such heat abounds more in people in a cold region, because the heat is strengthened and focused by the region's cold and this is why they are stronger and better ordered for audacity and attack with discretion and deliberation, because the cold remits the force of the heat to some extent. But in people in a warm region the heat is dispersed and, if it is intensified, this occurs *per accidens,* by means of an accidental and disordered heat. Thus, if they are bold, this is inordinate, and they withdraw in a rush and quickly. Such are the French, who want to do wondrous things at the beginning and in the end accomplish nothing, and people like this are called *hardi* [bold] in French.

Through this a solution to the arguments is apparent.

Question 29: Whether youth can be restored.[66]

One inquires further into the renewal of age, and first, whether youth can be restored.

66. Cf. A., *DA* 7.1.9.90–92 (*SZ* 1: 626–27).

And it seems not. Age follows upon complexion. Thus, to the extent that the natural heat consumes more of the radical moisture, age approaches more quickly. But when the natural moisture is consumed, it cannot be restored. Therefore, neither can youth be renewed.

Moreover, if age could be renewed, then it would either be returned to a prior state, and if this occurs, then life could be preserved in perpetuity—which is false, and therefore so is its premise. Or, it could turn into a better state, and then the same will follow as before, or into a worse one, and then the renewal of age would not follow but more deterioration.

Moreover, age follows upon the complexion, because in the first age the complexion is warm and moist, as it is in children, and in the second age it is warm and dry, as it is in youths, and in the third it is cold and moist, and in the fourth age cold and dry. Therefore, the same situation exists among members generated from the warm and moist, or the cold and dry, as exists for age following on the complexion and, as a result, for life. But members generated from spermatic moisture cannot be regenerated. Therefore, neither can the age following such a complexion be renewed.

The Philosopher says the opposite.[67] For he says that some animals alter their old age [*senectus*],[68] as when serpents change their skins and birds change their feathers, etc.

One must reply that age is of two kinds: one is natural and one is accidental. Natural age arises from the consumption of the natural or radical heat and moisture, and this age cannot be renewed just as the natural heat and moisture cannot be regenerated or restored. The other age is accidental, and it arises from accidental causes and from the consumption of the nutrimental heat and moisture, and this is an apparent age. And in this age two things come to be considered. For either the consumption of the radical moisture accompanies the consumption of the nutrimental moisture, and in this respect age [*ae-*

67. Ar., *HA* 8.17 (600b19f.).

68. In what follows, *senectus* is, for the sake of clearer English, sometimes translated simply as "age," but see the one instance in which A. does indeed use *aetas*.

tas] cannot be renewed. Or it does not accompany it clearly and manifestly, but only in an unseen manner, almost imperceptibly, and in this respect it can be renewed. And this very often occurs in serpents and birds. For serpents are cold. Thus many superfluous, viscous humors are generated in them. And this is why when they lie hidden, in winter, nature falls back upon these humors, digesting them, and as a result a certain empty spot, as it were, is left between the flesh and the skin where the humors had been. Thus nature was previously dulled and blunted by humors like this, and this is why, once they have been digested, it is strengthened and then they come out of their entire old skin and put on a new one, which nature generates in them.

And it is the same for the birds. For birds are hot and dry. Thus in old age there abounds in them a great deal of melancholy matter generated by the burning from which their feathers are generated. And this is why their feathers grow longer in them and, due to their old age, are less solidly attached to the skin. And this is why the feathers fall out in old birds, as is clear in cranes, which are sometimes captured plucked, almost nude. But because matter still remains in them, this is why new feathers are regenerated and birds like these appear to grow younger. Nevertheless, because a continual voiding of natural heat and the consumption of the radical moisture occur in them, this is why they are in reality growing older even though they appear to be growing younger.

Thus one must reply to the question that in reality youth cannot be renewed, and the arguments reveal this. Nevertheless it can be renewed in an apparent way, and the Philosopher understands this.

Question 30: Whether old age can be slowed down.

One inquires whether old age can be slowed down.

1. It seems not. Because when the path to an end is slowed, then the end is also slowed. But old age is the path to death. Therefore, if old age can be slowed, then death can be slowed.

2. Moreover, according to the Philosopher in the second

book of *On Generation,*[69] the lifespan of each individual is measured according to a certain period, and therefore old age cannot be slowed.

3. To the contrary. Old age arises from the consumption of heat and moisture, and this consumption is radical. But this consumption can be hindered. For it happens that one receives nutriment in a quantity proportionate to what has been lost, and it therefore works out that it slows old age, because it never reaches old age.

To this last argument, one must reply that old age is of two types: accidental and natural. The accidental is the one that occurs from certain accidental and fortuitous causes, and this type of aging can be slowed and can therefore be blocked by the removal of these causes. The other is natural aging, which arises from the voiding of natural heat and the consumption of the radical moisture, and this can be slowed but it cannot be avoided because the lifespan of each individual is measured in a certain period, as has been argued, and life is divided among the ages. Thus, the ages cannot be avoided.

1. On to the arguments. To the first, one must reply that just as aging can be slowed, so too can death. But just as aging cannot be avoided, neither can death.

2. To the second argument one must reply that although the lifespan of each individual is measured with respect to superior causes, just as the Philosopher and the Commentator intend in the second book of *On Generation,* nevertheless the period can vary with respect to inferior causes.[70] Thus, if this period is compared to superior causes, it is fixed [*certa*]; but when it is compared to inferior causes, it is variable.

3. To the argument to the contrary, one must reply that although one happens to absorb nutriment in a quantity proportionate to what has been lost, still one does not avoid aging because the natural power does not always have a power equal to converting the nutriment, since at the time of youth it converts more than has been lost and at the time of *stasis* it converts an

69. Ar., *DG* 2.10 (336b10f.); cf. A., *DG* 2.3.4.

70. Averroes, *DG* 2.56.

equal amount, but at the time of old age it converts less than has been lost, and this is why, etc. In this way a solution is evident.

Question 31: Whether age can be renewed by taking the poison of a viper or some other poison.

One inquires further whether age can be renewed by taking the poison of a viper or some other poison.

1. It seems not. For nothing of the sort that is contrary to the natural heat and dissolves it and dissipates its harmony contributes to its renewal. But poison is a thing like this, and therefore, etc.

2. Moreover, poison that has been taken in naturally seeks the heart. But once the heart is injured, the animal is injured, because it does not endure grave suffering, according to the Philosopher in *On the Parts* [*of Animals*];[71] therefore, etc.

To the contrary. If age is to be renewed, it is necessary for the earlier parts of the flesh to be removed; but this cannot be done without a contrary. Since, then, a poison is particularly contrary to these parts, they will have to be removed by means of poison. Therefore, etc.

To this, one must reply that age can apparently be renewed by taking a poison. For example, a leper is cured in this way. For the animal's corruption can occur in three ways: to wit, in quantity, quality, and substance. But changes occur through opposites. Therefore, corruption in a substance will be removed through a contrary in the substance. But poison is something like this. And this is why the ingestion of poison rejuvenates an animal, as is said of the stag, which eats the viper, and this is why the leper is cured and then apparently is rejuvenated.

The reason for this is that if someone takes up a serpent and then removes its head and removes the liver from the area of the tail in the amount of two fingers, and strikes the remaining part well with a rod and cooks it well and gives it to a sick animal, then, if the power of the sick animal is weak, it will die im-

71. Ar., *Part. An.* 3.4 (667a33f.); cf. A., *DA* 13.1.4.37 (*SZ* 2: 1000–1001).

mediately, but if the power is still potent then it will repel the poison.[72] And every like naturally draws its like to itself. Thus, during the expulsion of the poison the poison draws with it all the corruption that is in the flesh, as far as the exterior parts, and as a result a change occurs in the exterior parts and even a change in the interior parts.[73] Nevertheless if the power is weak, then the animal dies, and this is why it is very dangerous to use medicine like this.

1. On to the arguments. To the first, one must reply that although poison is contrary to the natural heat, nevertheless the poison has a power for drawing to itself the infection which is in the flesh, and for expelling the old flesh.

2. To the second argument one must reply that although poison naturally seeks the heart, since the heart attracts a great deal by virtue of its heat, if the heart's power is potent it still prohibits the poison's penetration and expels it with other infections.

Question 32: Whether pestilential disease [morbus pestilentialis] *arises from an infection of the air.*[74]

One inquires further into the sickness of animals. And because the Philosopher says that fish do not suffer pestilential disease, this is why one asks whether pestilential disease arises from an infection of the air.[75]

And it seems not, because if it arose from an infection of the air, then it would appear equally in flyers and in both wild and domestic walkers, since all of these live in air. Aristotle says the opposite.

Moreover, just as the air can be infected, so too can water, as

72. On this poison antidote concocted from the flesh of serpents, cf. A., *DA* 1.1.8.102; 2.2.4.119; 7.2.5.133 (*SZ* 1: 84–85, 342–43, 646–47).

73. Averroes, *Colliget* 4.28.

74. Although one is tempted to translate *morbus pestilentialis* as "plague," the term would likely be misleading, coming before the Black Death or plague of the fourteenth century. On some of the difficulties with terminology, see Jon Arrizabalaga (1994).

75. Ar., *HA* 8.19 (602b12f.); cf. Avic., *DA* 8.2 (fol. 37vaC); A., *DA* 7.2.1.101–4 (*SZ* 1: 632–34).

is clear from standing water. If, then, pestilential disease should arise from an infection of the air, for the same reason it should appear in fish from an infection of the water.

The Philosopher says the opposite.

To this one must reply that pestilential disease arises especially from an infection of the air, because when infected air is breathed it infects the lungs, which have a loose composition, and, once the lungs have been infected, then the heart is infected. And this is why a sickness of this sort occurs in those that breathe in and out. But the water is not as easily infected, both on account of its saltiness, as with seawater, and on account of its flowing movement, as with flowing fresh water. Now, salt water is capable of cleansing, and so too is running water, and this is why they do not easily endure such infection. But standing water undergoes infection quickly, and this is why fish in standing water die more quickly than others. Thus this disease can occur in fish, although nevertheless less often than in flyers and walkers for the reason already cited.

Moreover, air is infected by the cadavers of dead animals and by other putrefying things, which are more abundant on the ground's surface than in the upper part of the air, and this is why this disease occurs less frequently in birds than in walkers.

Moreover, domestic animals are in a smaller space in large numbers than are wild animals, and this is why the air is infected more by the multiplication of their exhalation. In the same way domestic animals are moister than wild animals, because they are fat, and moisture is the material for corruption and putrefaction, according to Galen.[76] And this is why domestic animals die more quickly from this disease than do wild animals, and this is especially so for sheep and cows because the sheep is a dull animal, as is said in the text,[77] and lives on earthborn things and it does not discriminate well between things that cause or do not cause corruption. The cow, however, has lungs with a very loose composition, and this is why its heart and spirits are more easily infected once it has breathed corrupted air, etc.

In this way a solution is apparent.

76. Galen, *In Hippocr. Epidem* 3.1 (Kühn 17: 651).
77. Ar., *HA* 9.3 (610b20f.).

Questions 33–39: On the sicknesses of pigs, dogs, and horses.

One inquires further why scrofula particularly occurs in pigs.

Second, why a flux in them is only cured with difficulty.

Third, why gout of the feet occurs particularly in dogs.

And fourth, why strangury particularly occurs in horses.

And fifth, why a miscarriage occurs in mares when a candle is extinguished and why the same thing occurs in some women.

And sixth, why human saliva infects a toad, as is said in the text.

And seventh, why an arrow, tinged with human saliva, kills the one struck by it more quickly.

To the first question one must reply that pigs have superfluous moisture and weak heat, and this is why they grow fat so quickly and why their superfluous digested moistures are turned back and cannot be well incorporated. Thus these moistures are directed back into the parts that hang down, and, as a result, to the jaws. This is why the flesh or anterior parts of a pig are less healthful than the others, because they are moister as a result of the superfluous humors running through them, because, according to the physicians, flux more commonly occurs in a part that slopes downward. And in a similar way, the backbone is more healthful than the stomach because the moistures flow to the lower parts. Moreover, the front feet are more healthful than the back feet, because they are more engaged in motion and exertion, and for this reason they are more purged of superfluities. Thus a disease like this occurs in pigs owing to the abundance of the moistures because they are gluttonous and cold and have a bad digestion, like phlegmatics.

To the second question one must reply that flux is of three types. One arises from the moistures, and this is called *diarrhoea*. Another arises from superfluous nutriment, as when food is excreted in the same form as it was received, and this is called *lienteria*. And a third arises from the blood, and this one is called *dysenteria*. Thus the saying among the physicians: "Raw [stool means] lienteric; simple [means] diar(rhoea); bloody [means] dysentery." But humors abound in the pig, and this is why, just as it is difficult to remove a pig's proper complexion, so it is difficult to restrain a proper flux if it should occur in the pig, and the Philosopher is referring to this. Nevertheless, it can be cured if wine is poured in through its anus, because then its intestines are quickly restrained as if by a clyster, etc. If, however, it is poured into its mouth, it is converted into chyle before it reaches the intestines, because it may be changed and it does not come under its actuality. And this is why, etc.

To the third question one must reply that the dog is a dry and melancholic animal and especially given to intercourse, and this is why an illness arising from dryness in the extremities afflicts them particularly,[78] and gout is of this type because coition greatly weakens the nerves, according to Avicenna. And, moreover, they are dried out beyond measure because they are very much involved in motion. For excessive dryness corrupts the nerves, according to Galen in *On Illness and Accident.*[79]

To the fourth question one must reply that the horse very often wants to urinate, although nevertheless it is not allowed to do so by the one mounted on it or riding it, nor does he perceive its desire to urinate.[80] This is why its bladder is bloated, from the retention of the urine, and thus strangury occurs just as it does in humans, and this is why, etc.

To the fifth question one must reply that the womb is a very sensitive member. Therefore, when it senses something unsuitable in the body's lower part, it rises and flees upward, as is apparent in prolapse of the womb [*praecipitio matricis*], and if it should sense this in the [body's] upper part, it is pressed downward so that it suffers suffocation. And this is why, when it perceives the bad smell of a candle that has been extinguished, the womb is often pressed downward so much that the womb's cotilideons (those things which bind the fetus to the womb just as an apple is tied to a tree) are ruptured, and then a miscarriage occurs.[81] This also occurs in weak women.

78. Cf. A. *DA* 7.2.2.109; 22.2.1.31–35 (*SZ* 1: 636; 2: 1460–64).

79. Galen, *De nervorum morbis.* See Galen, *De sympt. causis* 1.8 (Kühn 7: 633ff.).

80. See A., *DA* 22.2.1.80–81 (*SZ* 2: 1495–97).

81. On the cotyledons, see *DA* 10.1.1.7 (*SZ* 1: 829–30), 9.1.5.56 (*SZ* 1: 794), and 16.2.7.127 (*SZ* 2: 1224–25).

To the sixth question one must reply that animals with a ringed body that lack blood and do not breathe have a very loose composition, especially if they do not have lungs, because animals such as this need to have the interior heat tempered by the exterior air, and this is why their bodies are porous, as is evident in bees and wasps. But the human especially abounds with blood, and the toad and animals like it lack blood. Therefore, their natures are contrary, and this is why if a person's saliva—especially that of a person who is fasting, because it is better digested—is poured on a toad, it very quickly penetrates the porous parts of the toad to reach the interior and kills it by virtue of its contrariety.[82] Thus, too, smeared-on spit removes *morfea*.[83] And the same thing happens with salt, because salt is pungent and penetrating. And this is why people sprinkle meat with it, so that it will not putrefy. Therefore, salt, when it is sprinkled on a toad, penetrates to its internal parts in the same way and kills it, because salt cleanses and expels the filth—and all the parts of the toad are filthy—and to purge the toad of its filth is to kill it, etc.

To the seventh question one must reply that when something is properly ordered it is amicable, but when it is beyond this order it is inimical, as is evident in many examples. For good blood is amicable in the body, but when it is outside the body, it is inimical to it. Thus, if an ox sees its blood outside its body, it incites it to madness, which I once proved before my brethren at Cologne.[84] But saliva is very amicable in a person, because one cannot speak nor can one take in food without saliva to mediate. But if it is outside the body, then it is very hostile toward it, that is, toward the human body. And this is why an arrow that has been tinged with human saliva kills more quickly than does an arrow that has not been tinged with saliva; etc.

82. See *DA* 22.1.5.12 (*SZ* 2: 1446–47).

83. At *DA* 1.3.5.608 (*SZ* 1: 278) we hear of *morfea nigra*, a skin disease. Cf. 9.1.6.62 (*SZ* 1: 796–97) with notes.

84. At *DA* 2.1.3.31 (*SZ* 1: 299), Albert notes that the color red causes an ox to go into a rage.

BOOK EIGHT

Question 1: Whether brute animals know friendship or enmity.

"NOW THE TYPES of animal vary," etc.[1] In this eighth book the Philosopher makes a determination regarding friendship or enmity among animals. This is why it is asked whether friendship and enmity are present in brute beasts.

1. It seems not. Those passions that surpass the sensible powers and operations are only present in those animals possessing a power that rises above sensation. But friendship and enmity surpass the sensible operations, and they are therefore not present in beasts.

2. Besides, friendship and enmity are not present in those whose operations are directed toward a single thing, because these exist with respect to multiple things.[2] But the operations of brute beasts are directed toward a single thing, because they lack reason, which is capable of opposites; for this reason, etc.

The Philosopher says the opposite.[3]

It should be said that friendship and enmity are found in brute animals. The reason for this is that friendship consists in the perception of what is agreeable and enmity in the perception of what is harmful. But these are found in brute animals, and therefore, etc. For birds perceive that a seed is agreeable to them, and a sheep or lamb perceives that a wolf is harmful to it but that a human or a shepherd is a friend.

1. Ar., *HA* 9.1 (608a11f.). Cf. A., *DA*, 8.1.1–3, 1–32 (*SZ* 1: 667–80).
2. The translation preferred reflects the ambiguity of the original Latin.
3. Ar., *HA* 9.1 (610a33f.).

1. On to the arguments. To the first argument one should respond that the sensitive power is double—internal and external. Sight, hearing, and the like are external; common sense and the imaginative, estimative, and recollective senses are internal. Thus, although friendship and enmity are not perceived by means of the external senses, nevertheless they are perceived by means of the estimative sense, because the estimative sense is properly perceptive of intentions undivided [from the percept], which do not fall under the purview of sense or which do not move sense *per se.*

2. To the second argument one must respond that although the power of brute beasts does not relate to contraries as contraries, as does reason or the will, nevertheless the power of brute animals does relate to diverse options: to one through choosing something, to another through avoidance of it.

Question 2: Whether friendship and enmity are present in all brute animals.

Next one asks whether these are present in all brute animals.

1. It seems not. Friendship and enmity are not found among those in which the estimative sense is absent. But many brute beasts lack the estimative sense. For the first book of the *Metaphysics* says that the life of animals is ruled by sensation alone and a human's life is ruled by art and reason.[4] Therefore, etc.

2. Besides, those possessing friendship and enmity perceive objects from a distance. But many animals, such as the immobile ones, lack sight, hearing, and smell; therefore, etc.

On the contrary: Every animal takes delight in that which is agreeable and is saddened by that which is harmful. But where there are grief, joy, and sadness there are also friendship and enmity. Therefore, etc.

It must be said that a certain natural agreement exists, and this is found among inanimate things, for example, be-

4. Ar., *Metaph.*, 1.1 (980b25f.).

tween one fire and another, and one rock and another. Another agreement follows upon sensation, or is perceived through sense. Friendship consists in this type of agreement. Thus, even though one rock "agrees" with another, the rock does not, however, perceive that agreement. But there is no animal that does not perceive that which is agreeable to it and equally that which is harmful to it, because even immobile animals expand and flow out over agreeable things, but if they are touched by something harmful they contract, and thus, with natural instinct they are adjudging the former things as friendly and the latter things as hostile, etc.

Question 3: Whether the lamb will flee the wolf without perceiving that there is something harmful in it.

Further one asks whether the lamb will flee the wolf without perceiving that there is something harmful in it.[5]

It seems not, because, just as there is an intellective power in the human, so too is there a sensitive power in a brute beast. But the human intellect apprehends nothing that was not sensed first. Therefore, all the more is it the case that the beast apprehends nothing that is not sensed first, and, as a consequence, this is also true for something harmful. As a result, the lamb flees the wolf because it was sensed first.

To the contrary: A lamb perceives nothing in the wolf other than color and common sensibles. But it frequently perceives things that are poorly colored or shaped in an unbecoming fashion, but does not flee from them. Therefore, it perceives nothing harmful before it flees.

One must respond that the estimative power in a beast proceeds by operating in three ways: by way of apprehension, by way of experience, and by way of analogy. It operates by way of apprehension, just as a child seeks and suckles at the breast, which it has not seen before. Similarly, a lamb flees the wolf, which it has never seen before. While the color and shape of the wolf move the exterior sense and after that the interior sense, the

5. A., *DA*, 22.2.1.128 (*SZ* 2: 1528–29).

interior sense naturally apprehends its hostile intent. Thus, an apprehension of the harmful or injurious precedes its flight. Sometimes it proceeds along the path of experience, as when a bird flees someone who is threatening to throw a rock or is holding or moving a stick. This is because it often happens that it was deceived or it suffered some trouble through movements of this sort. An indication of this is that a young bird does not fear this until it has experienced for itself some evil like this, or because it was taught by its mother and father to flee, etc. [The estimative power is proceeding along] the path of analogy when a bird flees scarecrows, a painted picture of a person, or a figure made in the likeness of one holding a bow.

Through these things the answer to the arguments is clear.

Question 4: Whether art or craft is present in beasts.

Further one asks about art or craft in animals.[6]

First it is asked whether art or craft is present in beasts.[7]

And it seems not. Art is a representative gathering together of certain things to create a resemblance to certain preceding things. But such gathering does not occur without comparison. Since comparing pertains to reason and not to sensation, art will not be present in brute beasts.

Besides, according to the Philosopher in the sixth book of the *Ethics,* there are five intellectual virtues, of which art is one.[8] Since intellect is not present in beasts, neither will art be present.

In the text the Philosopher indicates the contrary.

For it must be said that art is a proper understanding of things capable of being made, as is clear from the sixth book of the *Ethics.*[9] Making [*factio*] is, however, an operation that carries over into something exterior, and this is different from action

6. *Cautela,* normally "caution," can take on the sense of "cunning" (see q. 10 below), but would seem to indicate "craft" in this context.

7. Ar., *HA* 9.3–43 (610b20f.). Avic., *DA,* 8.4 (fol. 38rbA–39rbD); A., *DA,* 8.2.1.33–41 (*SZ* 1: 681–84).

8. Ar., *Eth. Nic.* 6.3 (1139b14f.).

9. Ibid., 6.4 (1140a20f.).

[*actio*], which is an operation that does not carry over into something exterior. Therefore, art is not present, properly speaking, in those in whom reason is not present. But reason is not present in beasts. Therefore, art is not in them. Now truly, although they do not have art, they nevertheless do have something that resembles art. For a weaver cannot produce a weave without art any more than a tailor can make a tunic or a carpenter a house, but a spider weaves a web and an ape makes a tunic and a beaver makes a house as if possessing an art by which they are guided. Likewise, the swallow builds a nest.

Thus, one may say in brief that brute animals make works of art, but nevertheless do not do so by art but rather by nature. And this is why they always act in a uniform way, as nature does, unless they are impeded by something external. Thus, every spider makes a web just like every other spider, and swallows similarly make their nests alike. But this is not the way that every weaver works, or every builder of houses, because these use art, which allows for variation. Those which always act in a uniform way are working by nature, as the Philosopher says in the second book of the *Physics*.[10] But not every builder makes the same house as another builder, nor does the weaver produce the same weave. Still, some doubt remains with respect to some animals, like bees, and more about this will be revealed below.[11]

From this a solution to the arguments is clear. For these proceed from art as it is properly understood, and this is true. But the Philosopher intends that art be understood here metaphorically and considered by means of analogy. And in the same way for the other.

Questions 5–9: On the sheep and its properties.

One makes further inquiry into the Philosopher's progress, specifically and first as to whether the sheep is the stupidest animal.[12]

It seems not. Customs follow complexions. The Philosopher

10. Ar., *Phys.*, 2.8 (198b34; 199b18f.).
11. *QDA* 8.15. See A., *DA*, 8.4.3–5.148–83 (*SZ* 1: 733–47).
12. Ar., *HA* 9.3 (610b20f.). A., *DA*, 8.2.1.33 (*SZ* 1: 681).

provides proof in the second book of *On the Soul:* "Those with soft flesh have a sharp mind."[13] But the sheep has the best complexion, because it is warm and moist, and it has a soft flesh, since its flesh is easily digested. Therefore, it is the most compliant animal and has good and wise customs.

The Philosopher says the opposite.[14]

Second, one asks whether sheep naturally desire to consume salt.

And it seems not, for nothing naturally desires something contrary to it. But salt, since it is dry, is contrary to the sheep's complexion. Therefore, etc.

The Philosopher says the opposite.

Third, one asks whether consuming salt results in an increase in milk.

It seems not, because milk is generated from the menstrual blood digested in the breasts. But salt absorbs menstrual blood and also other moist things. Therefore, it results in a reduction of milk.

The Philosopher says the opposite.

Fourth, one asks whether its pelt is the best and most suitable for a person.

It seems not, because the animal pelts are necessary for a person as a protection against the cold. But the coat of a fox works better than a sheep's for this purpose; therefore, etc.

Last, one asks why clothes made from sheep's wool are especially conducive to the generation of lice, and especially in the case of a sheep killed by a wolf, because the wool especially does this in such a case.

To the first question one must respond that a sheep, like a goat, is a very stupid animal. If someone picks up one sheep, all the others gaze upon it as if awestruck and stupefied. There is a threefold cause for their stupidity: excessive moisture; excessive fleshiness in its head; and the bending of its head toward the ground. For excessive moisture impedes the natural heat. But sensation does not occur without heat, and as a result too much

13. Ar., *De anima,* 2.9 (421a25–26).

14. Cf. Ar., *HA* 9.3 (610b20f.); 8.10 (596a13f.). A., *DA,* 7.1.5.58–60; 8.2.1.33–35; 22.2.1.116, 128–30 (*SZ* 1: 587–88; 681–82; 2: 1519, 1528–29).

moisture blunts the senses, as is evident in drunks. In the same way, a great fleshiness in the head results in excessive moisture in the same place, and, as a consequence, a shortage of heat. Third, because its head bends to the ground, moistures flow to its head, and this is why a sheep, more than any other animal, holds its head down to the ground as if it were weighted. For these reasons, it has obtuse senses and is a stupid animal.

Or one can put it another way and say that complexion is spoken of in two ways: temperate and elevated. The sheep, however, is an animal with a temperate complexion, and this is why it is a very gentle animal and sticks close to a person and does the same to all things agreeable to it. Other animals have a more elevated complexion, and this is why they are wrathful or in some other way depart from the mean. Thus, because the sheep is not elevated in its complexion, the Philosopher says that the sheep is a stupid animal, for it is not as clever as other animals, and for this reason the goodness of its complexion is not precluded.[15] In this way a solution to the argument is clear.

To the second, one must respond that the sheep lives on things growing in the earth, which are insipid, and this is why it desires salt as a condiment by a natural instinct and does so more than other animals because it has so much moisture. This is also why they do not drink much.

Salt, however, absorbs moisture, and of all the animals they therefore have a greater natural desire for salt. If they do not have salt, they chew in clay-filled areas and do so especially in places where they urinate, owing to the reasons already mentioned. Other animals do not have so much moisture, and for that reason, etc.

To the third argument one should say that milk can be produced in two ways: either from having tempered the abundance of menstrual blood or owing to the consumption of those things that impede the generation of milk. Salt causes the generation of milk in the second way, but not the first. In this way a solution to the argument is clear.

To the fourth, one must say that a sheep's pelt is best because

15. Ar., *HA* 9.3 (610b22); *DA* 8.2.1.33–34 (*SZ* 1: 681–82).

it naturally conserves and conforms to a person's heat. Now a fox's pelt is more abundant in heat, but a sheep's pelt is more temperately warm, and this is why a sheep's pelt is more temperate and better *per se*. Whereas right now a fox's pelt may be more necessary or work better because the present season is colder, and it is warmer. This is why, etc.

To the fifth argument one should reply that heat is the agent or father of corruption and moisture is like a mother of corruption. But sheep wool is warm and moist, and this is why it is very close to corruption, and why also vermin are so easily generated in wool. Examples include lice or nits, which are produced out of putrefaction, and this is more the case, or especially so, if a wolf has killed the sheep, because the wolf's bite infects the wool by means of its breath—corrupt, melancholic, or gluttonous—and especially prepares it for putrefaction and corruption. And this is why the Philosopher says that a garment made from this wool is most prone to the generation of vermin and lice.[16]

Question 10: Whether it is by nature, from customary activity,[17] *or by instruction that cunning is present in animals that walk.*

One asks now about the cunning of walking animals.[18] And first one inquires whether this is in them by nature, from customary activity, or by instruction.

1. It seems that it is in them from customary activity. That which is perfected by habitual behavior is acquired by customary activity. But the cunning of animals is perfected in repetition, for the Philosopher says that young bees do not make honey well because they are not accustomed to doing so, and it is the same for a swallow's nest and the spider's web. Therefore, etc.

2. Moreover, a nobler effect follows a nobler cause. But the

16. Ar., *HA* 8.10 (596b7f.).

17. "Customary activity" = *consuetudo*. Although awkward, the phrase attempts to preserve the author's contrast between characteristics or properties that are instinctive to nature, learned through instruction, or acquired from experience.

18. Cf. Avic., *DA*, 8.4 (fol. 38rbA–39rbD); A., *DA*, 8.2.1–2.33–51 (*SZ* 1: 681–88).

intellective power is more noble than the sensitive one and more noble than the other parts or powers of the soul. Since, then, the cunning acts of a human possessing intellect are not acquired without customary activity, it seems that the cunning of brute beasts will all the more be acquired by customary activity and not from nature.

To the contrary. According to the Philosopher in the third book of the *Physics,* nature does not habituate one to a contrary.[19] But customary activity can be habituated to a contrary. Since it is natural for fire to rise upward, if there is no impediment above it, it therefore immediately rises upward naturally. But brute beasts are directed to one thing to such an extent that they are not habituated to the contrary. Therefore, they act naturally and not from custom.

Besides, whatever a person acquires as something medicinal and useful to him, he acquires from counsel. But brute beasts acquire medicines without counsel, as is clear in the deer and in many others. For the deer eats *dracontea* in order to give birth more easily, and the roebuck and the goat when wounded eat fleabane, so that the weapon piercing it may fall out more easily.[20] And a wounded bird seeks wild oregano for the same purpose, and a blinded serpent or one with weakened sight seeks fennel in order to be restored. And animals that desire to fight with serpents, like the weasel, either eat rue or roll around in it beforehand.[21] And the leopard, if it should eat something poisonous, eats human feces and will then be healed. Since these animals were not instructed by others, it seems that these acts of cunning are present in them by nature.

To this one must reply that some acts of cunning are present in beasts naturally, and some by customary activity. For certain animals are teachable, like the elephant, the horse, the dog, the falcon, the bear, and others like these, and certain acts of cun-

19. Ar., *Phys.*, 3.5 (205b21f.); *Eth. Nic.* 2.1 (1103a19f.).

20. *Dracontea:* dragon-wort or cuckoo pint, *Arum maculatum.* Cf. *De veg.* 6.290 and *DA* 23.21.98 with notes (*SZ* 2: 1614–15). For the claim that the wolf eats *dracontea* to sharpen its teeth, see *QDA* 8.14.

21. Cf. *DA* 22.2.1.122(79) (*SZ* 2: 1524–25).

ning are present in such as these as a result of teaching and customary activity. But some animals are unteachable, like all those that lack hearing, and acts of cunning are naturally present in such as these. In truth the perfection of an operation is present in them from customary activity. As a result, in these it is necessary to consider two things: the first is industry for working, and this is present in them from birth; and the second is a perfection for working, and this is present in them by customary activity. This is why mature bees make honey better than young bees, owing to repetition. Thus, all animals with hearing are teachable, and for those lacking hearing the opposite is true, as is said in book one of the *Metaphysics*.[22]

1. From this a solution for the first argument is evident.

2. To the second, one should reply that the sensitive and vegetative souls are assimilated to nature owing to their materiality, and this is why, just as nature does not habituate one to contraries or opposites, thus neither will the sensitive and vegetative souls, since they are naturally ordered to one thing. But reason is capable of opposites, and this is why those things that are agreeable to some animals by nature are not established thus for the human, but he can habituate himself to these and their opposites. And that which the brute beast seeks by an instinct of nature the human can investigate through reason.

The arguments to the contrary ought to be conceded, because they reach the same conclusion.

Nevertheless, the causes of the things taken up in the second argument are these. *Dracontea* is warm and dry,[23] and, owing to its warmth, it opens, and, owing to its dryness, it cleanses, and there is a great deal of menstrual moisture in the deer after conception, and this is why before giving birth it eats *dracontea,* so that through its warmth the pathways of the womb will be opened, and through its dryness these will be cleansed, and the superfluities, which would suffocate the fetus at the time of its birth, are consumed. And in this way it can give birth more easily.[24]

22. Ar., *Metaph.*, 1.1 (980a27–b28). 23. A., *DA*, 8.2.1.36 (*SZ* 1: 682).
24. Ar., *HA* 9.5 (611a15f.).

Moreover, fleabane is effectively warm and moist.[25] And this is why if an arrow pierces a goat, it eats fleabane, so that by means of its heat the wound will be opened and by means of its moisture the flesh will be softened once more. Once this has been accomplished, the arrow will fall out more quickly. And it does this by a natural instinct.

Moreover, oregano is warm and dry. And this is why, if it is eaten by a wounded animal, it cleanses the wounds, and things that are cleansed are more easily cured, and this is why wounded birds seek it.

Moreover, fennel clarifies vision. When, however, serpents have lain sheltered in caves for a long time, they have weak and impeded sight owing to the long period of inactivity during which superfluities gather in their eyes. And this is why, when they go out at the beginning of spring, they especially seek out fennel.

Besides this, the serpent is cold and moist, and as a result is very sensitive [*passibilis*]. Rue, however is warm and dry, and as a result it is opposed to the serpent's complexion and frequently kills it. And this is why those animals about to fight with serpents especially seek out rue and arm themselves with it.

Moreover, the odor of rue is especially overpowering, and the serpent's sense of smell is very sensitive because the nexus is weak,[26] and this is why, just as something visibly overpowering corrupts a person's sight, and a harsh or overpowering sound corrupts the hearing, so in this way does the odor of rue corrupt the serpent's sense of smell and stupefy it. And this is why rue is planted around beneficial herbs, because toads, serpents, and other venomous animals flee from it.

Moreover, the pard and the panther eat human feces after having eaten something poisonous, because human feces are especially choleric and the abundance of choler provokes diar-

25. A., *DA* 8.2.2.42.46–49 (*SZ* 1: 685, 686–89).

26. This phrase, *quia debilis nexus est*, is found only in the MS designated T by the editor. *Nexus* normally indicates a connection and may refer poetically to a serpent's coils, but that does little to suggest a relevant meaning for the phrase in this context. It may refer to the interconnection of the animal's parts as at *QDA* 8.11–14 below, or it may be best to bracket the phrase as an intrusion.

rhea and, as a result, the purging of the poison. The wolf, however, to the contrary dies from eating this excrement. The reason for this is as follows. A person may take some things for use as medicine and others as food, and if the medicines are consumed as food they will be harmful, and vice versa. Now, then, the pard takes up this human superfluity as a medicine, and this is why, once the diarrhea has been provoked and the poison cast out, it ceases to eat it. But the wolf eats these superfluities as food, and therefore provokes excessive diarrhea, so much so that very often death befalls it. This is because it has very weak intestines, and thus the food is more quickly dissolved, and as a result death befalls it.

Questions 11–14: Whether the deer is an animal possessing judgment.

Whether the deer is a long-lived animal.

Why the wolf eats dracontea *at the time when it gives birth.*

And why the horse especially loves its own kind.

Further one asks whether the deer is an animal possessing judgment.

And it seems not. Customs follow complexion. But the deer has a bad complexion. Therefore, it has less judgment.

Moreover, the spider and ant are more capable of judgment, since they provide for the future by carrying off food, which the deer does not do. Therefore, etc.

The Philosopher says the opposite.[27]

Second, one asks whether a deer is a long-lived animal.

And it seems not, because a brief lifespan results from a deficiency of warmth and moisture. But this deficiency is present in the deer, and therefore, etc.

Moreover, a melancholic complexion in particular results in a brief lifespan, according to the opinion of the physicians. But the deer is a very melancholy animal, and therefore, etc.

27. Ar., *HA* 9.5 (611a15f.). Cf. Avic., *DA*, 8.4 (fol. 38rbA); A., *DA*, 8.2.1.36–41 (*SZ* 1: 682–84).

One asks in relation to this why the wolf eats *dracontea* when it conceals itself or rests at the time of giving birth.

Likewise, why the horse especially loves its own kind.

To the first, one must reply that judgment, that is, industry, can apply to something in two ways: either *per se* or *per accidens*. *Per se* "those with soft flesh have a sharp mind,"[28] as is said in book two of *On the Soul*, and as a result an animal with more blood has more judgment. But it happens *per accidens* that melancholic animals have better judgment, because melancholic ones have more fear, and, owing to this fear, they guard themselves against more harmful things, and this is part of judgment. Now the deer and the hare, because these are especially timid animals, especially guard themselves from harmful things by taking flight, because this is their defense, since, as is said in the text, when a stag has cast off its horns then it seeks out a very well wooded place. It hides there owing to its fear of predatory animals, since it is unarmed then, and it does this so that it will not be discovered to have lost its defenses. And in like manner when it sheds its horns, it hides them, because if they are found by others it will be apparent to them that the stag is without its defenses, and they may then attack it. And this is why Aristotle says in the text, in the form of a proverb: "Go where the stag has cast off its horns."[29] Thus the stag is an animal with some judgment only *per accidens*. And in this way a solution is evident to the argument. Likewise, the hare—before it sits down someplace—covers about a mile wandering hither and yon and confusing or overlaying (that is, crossing over) its steps so that it may not easily be found by means of its tracks.

But, to speak briefly, human reason is stunned and unable to investigate—nay, what is more, we cannot even relate the many cunning acts, hidden from us, of brute animals who act according to a natural wisdom bestowed upon them by the Creator. For I have found out in a certain secret book so many and such great cunning acts of animals, of which most of the philosophers have been ignorant, which is an amazing thing to say. For

28. See *QDA* 8.5–9.

29. Ar., *HA* 9.5 (611a26–27). Cf. A., *DA*, 8.2.1.37 (*SZ* 1: 682–83).

now we know "in darkness as if in a mirror" and not completely, as is said at Cor. 13.[30]

To the second argument one should reply that the cause of long life is fourfold. One is the balance of the complexion, and this is why those animals with a more balanced and temperate complexion live longer, and those that stray from this balance have a shorter lifespan. And this is why those with a sanguineous complexion live longer than others, because they approach nearer to a temperate complexion. Another cause is the abundance of heat and natural moisture, and this is why the Philosopher says in his book *On the Reason for Shortness or Length of Life* [*De causa longitudinis et brevitatis vitae*] that large animals have a longer lifespan than small ones, because heat and moisture—which are the principles of life—are more abundant in them.[31] The third cause is the hardness of the parts and the strength of their connection, since ones like these are not easily damaged or weakened. For this reason, both plants and trees have a longer lifespan than animals, and among the plants those that are harder have a longer lifespan than those that are soft. And this is why an oak tree has a longer lifespan than a willow, because owing to its hardness it is less damaged or weakened by something external. The fourth cause is the renewal of the complexion. And this is why serpents can live a long time, since they renew themselves. These latter two causes are present in the stag, but not the first two causes. Whereas the arguments are based on the first [two] causes, the Philosopher nevertheless based his understanding on the second [two] causes. This is why the stag, when it has grown old, eats a viper, attracting it with its nostrils, and swallows it, as is said, and then afterward drinks a large quantity of water; and then its fur and its horns fall off, and in this way it is renewed.[32]

30. Although Albert's text here recalls the passage at 1 Cor 13.12–13, "We see now in a glass darkly . . ." his Latin text—*in aenigmate et quasi in speculo*—departs from the Vulg. *per speculum in aenigmate.*

31. Ar., *De long. et brev. vitae,* 4 (466a13f.).

32. Cf. Pliny, *HN* 8.50.118; Isid., *Orig.* 12.1.18. For an illustration of a stag sniffing a snake, see George and Yapp (1991, 80). The same statement is found in several bestiaries (White, 1954, 38). Note the great age often attributed to

To the third argument one must respond that the wolf is a prized animal and very voracious and great-spirited, and this is why, when it goes into hiding at the time of giving birth and ceases to eat or feed, its teeth are rendered dull and almost without feeling, and this is why it eats *dracontea,* so that its teeth will be cured of their stupor and in this way they will be sharpened so that it may feed or carry things off with them.[33] Now because this stupor results from a moist cold, and *dracontea* is warm and dry, they are cured by chewing it, and for that reason, etc.

To the fourth argument one should reply that something is loved more if it is acquired with greater difficulty, as the Philosopher would have it in the *Topics.*[34] But a horse is infinitely more distressed than other animals when it is pregnant and when giving birth, because it is pregnant for a longer period of time—for an entire year—not like other animals. This is why a horse naturally loves its own kind more than others, because it acquired it with so much labor.[35]

And besides this, on the forehead of the young when it is born, there is an outgrowth of flesh which is called the *hypomanes,* which the Philosopher treated above,[36] with which wizards do marvelous things since, if one gives three ounces of it, dried, with sugar, to a woman, she will immediately follow you and love you, but if it is given while still fresh it kills. Likewise, when mixed with cheese it kills mice, flies, and ones like these. In the same way, if given to a dog, it will make it rabid. Likewise, if put in an apple, it causes sterility and causes many other things which there is no need to mention here. For there is a "wolf in the story."[37] Now that fleshy outgrowth, I say, produc-

---

stags by the ancients, but denied by Ar., *HA* 578b23. The stag's breath lured the serpent out of its den.

33. A., *DA,* 22.2.1.114–15 (*SZ* 2: 1518–19).

34. Ar., *Topica* 3.2 (117b29–30); see Boethius, *Interpretatio Topicorum Aristotelis* 3.2 (PL 64: 937A).

35. Cf. A., *DA,* 8.2.1.35; 22.2.1.53 (*SZ* 1: 682; 2: 1478).

36. Ar., *HA* 6.18 (572a21); 6.22 (577a9), 8.24 (605a2). Cf. A., *DA,* 7.2.2.115; 22.2.1.53 (*SZ* 1: 638–39; 2: 1478). *Hippomanes* was a classic charm sought as an aphrodisiac. Mares were so lascivious they could be impregnated by the wind (Zirkle, 1936, 97–104).

37. This saying is as old as the comedian Terence and is used by Cicero. Its

es great love in the mare. As a result, when the young is born, the mare immediately licks that fleshy outgrowth and is immediately seized by love for her young. Thus the mare is especially like the human being in its love for its own kind, and this is why women give this fleshy outgrowth to their friends, in order to be loved by them, and this is true. The fact, however, that the horse is greatly injured in coition is evident because after it has had intercourse it cannot work much, as is clear in stallions. This is not the case for other animals.

Questions 15–20: Whether flyers are more cautious than walkers.

Next one asks whether flyers are more cautious than walkers.

It seems not, because among the animals the human is the most cautious animal. Therefore, one that is more like a human will be more cautious. But this is the walker, in comparison to the flyer; therefore, etc.

The Philosopher says the opposite concerning bees, which nevertheless are flyers.[38]

Second, one inquires why the cuckoo places its eggs in another's nest.

And this seems contrary [to nature], because the generation of birds occurs through egg-laying. Egg-laying requires, however, the building of a nest. Therefore, since the cuckoo naturally generates one like itself, it naturally has a capacity for building a nest.

Third, one inquires why the eagle kills or casts its chicks out if they shed tears when exposed to the sun.[39]

And this seems contrary [to nature], because it is natural for an animal to love its newborn and not, therefore, to kill it.

Fourth, one inquires why a raven does not nourish its chicks in their infancy until they have become black. At that point it

---

usual meaning is akin to our "speak of the devil and he will appear." A. is apparently trying to avoid bad luck.

38. Ar., *HA* 9.40–41 (623b4f.). A., *DA*, 8.4.2.141–47 (*SZ* 1: 730–33).

39. For this tale, see also A., *DA* 23.1.7 (*SZ* 2: 1547), and note.

thinks that they belong to it, but previously it did not nourish them owing to their whiteness, and they live on dew and air.

And this seems contrary [to nature], because to nourish and to generate pertain to the same part of the soul, that is, the vegetative soul. Therefore, that animal [power] that is naturally the principle of generation will also be the principle of nourishment; therefore, etc.

Fifth, why does the crane stand on one foot when sleeping?[40]

And this seems contrary [to nature], because during sleep the senses are restrained. Therefore, its body, when it sleeps, needs greater support, and it therefore ought to stand upon both feet.

Sixth, why does a bird of prey, like the falcon, the kite, and others of this sort especially desire the heart of the animal and of its prey?

And this seems contrary [to nature], because the heart is difficult to digest and especially resistant to the digestive power; for this reason, etc.

To the first question one should reply that some flyers are more cautious than walkers, especially the bees. Three things are required for their caution: a fineness of their parts, a subtle complexion, and an aptitude among their members, and because these things are more present in flyers, for that reason, etc.

To the argument one must say that although walkers are more nearly like the human in terms of body size, they are less so with regard to other factors, and therefore, etc.

To the second question one must reply that the cuckoo is a timid bird, because it is melancholic.[41] This is why all the other birds put it to flight and pluck its feathers if they can capture it, just as they put the *bubo* and the *noctua* to flight by day.[42] And

40. A., *DA*, 23.1.24, 113.

41. Ar., *HA* 9.29 (618a8f.). A., *DA*, 6.1.6.51–52; 8.2.5.91; 23.38 (*SZ* 1: 547–48; 707; 2: 1567–68).

42. No certainty seems possible among the many names for "owl" in the medieval panoply, and they themselves seemed confused. At *DA* 23.28(13) (*SZ* 2: 1561), A. himself says that the *bubo* is a member of the *noctua* genus. Cf. George and Yapp (1991, 148–50) for a discussion of the *nicticorax/noctua* confusion.

on account of this the cuckoo would fear for its chicks if it had its own nest, and this is why it places its eggs in another bird's nest, so that the young produced will be guarded and kept safe.

To the third question one should respond that the eagle has good vision and flies high, and for this reason the glacial humor is especially abundant in its vision.[43] This is why, in order to determine whether the chicks are its own and have sound vision, it turns them and places their eyes against the sun, and if one of them sheds tears it determines that it can not be its chick or is merely weak or helpless and this is why it casts it out, as the Philosopher says in the text. Sometimes it kills it because it thinks it does not belong to it.[44]

To the fourth question one must respond that raven chicks in their infancy have white feathers, like other young birds.[45] This is the reason that during this period they feed on the dew of heaven, owing to the abundance of spermatic blood, and this is why the adult birds do not think that they are their own and do not feed them. But when the feathers begin to turn black, owing to the consumption of the sperm, then they feed them and think that they are their own offspring.

To the fifth question one should say that the crane is an awkward flyer, and this is why it has long legs, so that when settled on the ground it may see its nourishment, and harmful things, from a distance.[46] But the legs are slender and very sensitive to cold. During sleep heat withdraws to the inner parts, and this is why it elevates one foot to its belly, in order to warm it there, and then later it releases that one to the ground and lifts the other foot. Others say that this is because when it stands on one foot and whenever it begins to fall, it shakes itself and calls out, and in this way it wakes itself and the others, and as a result it better safeguards itself from harmful things, etc.

43. On the *humor glacialis*, see also *QDA* 12.20 and *DA* 1.2.7.197–204 (*SZ* 1: 120–23).

44. Ar., *HA* 9.34 (620a2f.). Cf. Avic., *DA*, 8.5 (fol. 39rbA).

45. A., *DA*, 8.1.2.14; 23.28 (*SZ* 1: 672; 2: 1561).

46. Ar., *HA* 8.12 (597a30f.); 9.10 (614a34f.). A., *DA* 8.2.4.66; 23.24.113 (*SZ* 1: 694–95; 2: 1626).

To the sixth question one must respond that every animal that is well disposed is best nourished by things like itself. When, however, it is badly disposed, it desires something contrary to itself. But of all the members the heart is the hottest, and birds with curved talons are very hot, and this is why they particularly desire the heart, while it is still hot.[47] And another reason is that birds such as these are particularly voracious and the heart effectively resists their digestive power; because soft flesh is more quickly converted in them into fumes, for this reason they desire the heart, etc.

Question 21: Why the owl [bubo] *does not see in the daytime.*

One inquires why the owl [*bubo*] does not see in the daytime, although it has good and clear eyes.[48]

One must reply that the owl has large, but not firm, eyes. As a result, its vision is diffused or dispersed during daytime, whereas the contrary occurs at night because then its vision is unified by the darkness, and this is why it does not see during the day as it does at night. For this reason, etc.

Question 22: Whether art and craft are more suited to ringed flyers than to other animals.

Further one asks about the industry of ringed flyers. And first, whether art and craft are more suited to ringed flyers than to other animals.

It seems not. Art and craft are present in brute beasts in the same way that they have the senses. Now in those in which the sensitive powers are more perfect, art and craft will be found to be more perfect. But the senses of other animals are more perfect than [than those of ringed flyers], and therefore, etc.

Moreover, art and cleverness [*ingenium*] follow upon the subtlety of the spirits, but the spirit's subtlety follows upon the

47. A., *DA*, 8.2.4.69; 23.18 (*SZ* 1: 695–96; 2: 1554).
48. A., *DA*, 8.1.2.14; 23.28 (*SZ* 1: 672–73; 2: 1561).

goodness of the blood and natural heat. Since, therefore, animals such as these lack blood and, as a result, are deficient in heat, it seems that they are less inclined to craft and are less clever than others.

On the contrary. Animals whose activity extends to things that lie beyond their own conditions are the more clever ones. The human, therefore, is the cleverest animal, because his activity extends to things that are above him. But the activity of this sort of [ringed] animal does extend itself this way, because the spider makes a web, and an ant gathers grain into a storehouse, and the bee builds a dwelling and makes honey and lives in society [*congregat*] as does a human. Therefore, they extend themselves to activities concerning things that are above them.

One must respond that art and cleverness can be found together in an animal for many reasons. In one way, it is due to an excellence in making distinctions and comparisons, and these traits are found especially this way in a human. In another way, it is due to the good disposition of the flesh, and thus among humans these may be present more in one than another, because, as is said in book two of *On the Soul,* "Those soft in flesh are sharp in mind."[49] In a third way, it is due to subtlety of the spirits and agility of the members, and in this way they are found more in flyers than in walkers. In a fourth way, it is due to a certain melancholy solicitude, with the result that, since such an animal as this fears everything, then for that reason it takes precautions or safeguards itself from all things, as we already said concerning the deer, but this is *per accidens.* In a fifth way, it is due to the elevation of the estimative sense and its goodness, and in these last two ways art and cleverness especially belong to ringed animals, because these are melancholy, venomous, bloodless animals and have a very elevated estimative sense. And other causes are found more in others; etc.

49. For other instances of this saying, see *QDA* 13.3; 15.9.

Question 23: Whether the ant gathers seeds.

Further one asks whether the ant gathers seeds.[50]

1. And it seems not, for if it gathered seeds it would gather them at a time especially appropriate for their collection, just as the bee does its gathering. But the ant rests during the day, although the daytime is more suitable for this activity. Therefore, etc.

2. Moreover, if it is directed to gather seeds, this is either owing to its complexion or to its sensitive power. But it is not owing to its complexion, for this is comparable to that of the spider and to others of the same complexion. Nor is it owing to its sensitive power, for the same reason. It therefore remains to us to ask what directs the ant to gather seeds.

The Philosopher says the opposite.[51]

One must reply that the ant gathers seeds. And it is directed to do so for the benefit of its own species, because the operation makes known the form just as transformation does the matter. Thus form is the cause of the operation, and this is why those of different species have different operations. Nevertheless, the ant does not gather seeds because it anticipates the future, but its present actions are linked to future ones, and this is why from the imaging [*imaginatio*] of the present they do some things that are linked to future things.

1. To the arguments. To the first, one should reply that the ant is an animal with little power, and this is why, owing to the fear it has of encountering obstacles or opponents during the daytime, it labors especially at night and particularly when the moon is full, because, owing to the full moon, it is able to see what it is doing at that time.

2. To the second argument one should respond that it is directed to this work by its own species.

50. Ar., *HA* 9.40 (623b4f.). A., *DA*, 8.4.1.131–32; 26.16 (*SZ* 1: 726–27; 2: 1749–50).

51. Ar., *HA* 9.38 (622b24f.). Cf. A., *DA*, 8.4.1.131–32; 8.6.2.238 (*SZ* 1: 726–27; 772).

Question 24: Whether a spider naturally produces a web.

Further one inquires about the spider, and first one asks whether a spider naturally produces a web.[52]

1. It seems not, because an effect attests to its cause. But the web's material is sticky [*tenax*]. The spider, however, since it has a dry complexion, does not have a sticky nature. Therefore, etc.

2. Moreover, the spider is a venomous animal. But the web is useful for caring for wounds. Therefore, a web does not belong to a spider naturally.

The opposite is clear.

One must reply that the spider makes a web because, just as the Philosopher says, some animals live by hunting and others live by storing food.[53] The spider seeks its nourishment by hunting, and this is why it prepares for itself an instrument suitable for hunting. The web is something of this sort, because the spider remains in the middle or just below the middle of the web, and when a fly enters the web and is held there on account of the web's strength and stickiness, the spider drags to itself the animal that has been caught and sucks out its moisture.

1. To the arguments. To the first, one should reply that it is not necessary that the effect and its cause be similar in every respect, for in animals generated from putrefaction the effects and their cause differ by species and sometimes even by genus. Thus, the [cause of the] spider's production of a web is not univocal but equivocal.

2. To the second argument one should say that two things are required to care for wounds, namely, a drawing together of the sides of the wound and a knitting together of the parts. But the cold and the dry draw together, and the viscous knits together, and the spider's web has a cold and dry nature, and similarly it is sticky, and this is why it draws together wounds by means of its

52. Ar., *HA* 9.39 (622b27f.). Cf. Avic., *DA*, 8.5 (fol. 39vb). A., *DA*, 8.4.1.133–40; 26.7–8 (*SZ* 1: 727–30; 2: 1742–44).

53. Ar., *HA* 9.40 (623b13f.).

coldness and dryness, and knits together the parts by its stickiness.

Question 25: Whether the web is produced from the spider's substance or from some superfluity it has.

In addition, one asks whether the web is produced from the spider's substance or from some superfluity it has.

1. It seems that it is not produced from some superfluity. This is because every superfluity in the body either remains in the body, like sperm or milk, or does not remain but must be cast off, and one such as this is not needed. But the spider's web comes from neither of these two types. It does not remain in the body, because if it did it would contribute to the conservation of the species, like milk and sperm, nor is it from the second type of superfluity, because if it were the web would be unnecessary, which nevertheless is not true.

2. Moreover, nature is an internal principle of operation, whereas art is actually an external principle. But the spider works a web by a natural instinct. Therefore, the web's principle is internal. But this would only be so if it came forth from some part of the spider itself, and therefore, etc.

To the contrary, the Philosopher says that the web is related to the spider just as the bark is to a tree or a scale to a fish. But these things are produced from a superfluity, and for that reason, etc.[54]

One must reply that the web comes from a moisture remaining after the third digestion, because part of the spider's nutriment is converted into the substance of the spider, and the superfluous and residual part is retained in its body for the composition of a web. Then, from this material the spider, although it is a multi-pedal animal, first using an extracting motion forms a thread, one end of which it affixes to a given place, and then it draws the thread to another place, and thus, in the manner of a weaver, it produces a web from a superfluous moisture. Thus it

54. Ar., *HA* 9.39 (623a30f.).

is not produced from the spider's substance, as some have proposed,[55] because no animal separates off some of its substance without pain. But the spider produces a web without distress. This is why it cannot be from its substance.

1. To the first argument one must reply that the question differs for the spider and for other animals, because other animals strive to expel wholly the moisture remaining from the third digestion. But the spider divides that moisture into two, since it reserves the more subtle part for the web, whereas it casts out or expels the more digested part, just as other animals do.

2. To the second argument one ought to reply that although nature is a principle within, it is not necessary that the principle of a natural action come from the substance of the one generating, but only that it come from a superfluity of the last food. This is clear concerning a man's semen.

Question 26: Whether the bee can produce honey.

Further one asks about bees, whether the bee can produce honey.[56]

1. And it seems not. For something whose complexion resists sweetness does not produce honey. The bee is such an animal, since it is cold and dry; therefore, etc.

2. Moreover, nothing produces something that corrupts it. But every oily thing corrupts the bee, since it dies in oil. Thus, since honey is somewhat oily, it is not produced by the bee naturally.

The opposite is clear.

One must reply that the bee produces honey, since everything is naturally nourished by what is suitable to it. But the bee is an animal with a cold and dry complexion, and this is why it naturally has a short life span. Now life is conserved by warmth and moisture, and these [two] are present in sweet things. For

55. Cf. Avic., *DA*, 8.5 (fol. 39vb).

56. Ar., *HA* 9.40 (623b28f.). Avic., *DA*, 8.6 (fol. 39vb–40rb); A., *DA*, 8.4.3.148–62; 8.4.4.163–79; 26.4–6 (*SZ* 1: 733–39; 739–45; 2: 1518–19).

this reason the bee naturally desires the sweet. But although sweetness is in flowers and fruits, nevertheless these are not a suitable nutriment for bees until they are digested better and converted into honey. This is the reason why the bee naturally makes honey as its nourishment from such redolent things.

1. To the first argument one must reply that precisely because the bee has a cold and dry complexion, it needs something that can conserve its life better, and honey is such a thing.

2. To the second argument one must reply that oil and honey are different, since honey is not oily the way oil is, which is clear because oil floats on water and honey sinks to the bottom. And therefore, etc.

Question 27: Whether honey is made in the bee's body or outside it.

Next one asks whether honey is made in the bee's body or outside it.

1. And it seems that it is inside. For that which occurs from a natural heat is to be found near it. But honey is produced from the natural heat of the bee itself, and the natural heat does not exist outside that bee. Therefore, neither is honey made outside the bee.

2. Moreover, honey is made for the sake of the immediate nourishment of the bee. But nothing nourishes until it has entered the body. Therefore, etc.

To the contrary. Nourishment is something that first has the power of nourishing externally, but honey is the bee's nourishment. Therefore, it is made outside of it.

It must be said that just as a physician grinds up various things so that from them and through them he will make potions, so too the bee extracts something from flowers and leaves and then grinds and mixes that outside itself and converts it into honey, and later it is nourished by the honey made outside itself, because honey will be its nourishment only if it is made outside the bee.

1. To the first argument one must reply that although honey is digested by the power of the natural heat, nevertheless it is unnecessary for that digestion to occur in the body. For just as the content of a vase can be cooked[57] by a fire outside of it even though it is not immediately and physically close to the fire, so too in this case.

2. To the second argument one should say that in the end nourishment is internal, but in the beginning it is external.

Question 28: Whether wax and honey are produced from the same material.

Further one asks whether wax and honey are produced from the same material.[58]

1. And it seems not. Because where the matter and the efficient cause are the same, the effect will be the same, according to Aristotle's eighth book of the *Metaphysics*.[59] But honey and wax have different natures, because honey sinks to the bottom of water and wax floats on the top, if they are poured onto water. Therefore, although their efficient cause is the same, it is necessary that they will have different matter.

2. Moreover, nature does nothing in vain or without purpose. But the honey suffices for the bee's nourishment. It therefore does not need the wax, and consequently it does not form the wax from honey's matter.

The opposite is shown to be the case.

One must say that some animals make a little dwelling for themselves, much as the bird makes a nest. And this is necessary for animals that conceal themselves. The bee is an animal that conceals itself in winter, and for this reason it needs a home in which it may hide. This is the reason it makes wax and honey:

57. A. uses the verb *digero* here, which commonly means simply "digest" as it does in lines just before and after this one. The alimentary sense is surely not appropriate here. The verb can mean "decoct" in an alchemical sense and probably simply implies "cook."

58. A., *DA*, 8.4.3.152; 26.4–6 (*SZ* 1: 735; 2: 1741–43).

59. Ar., *Metaph.*, 8.4 (1044a31f.). A., *Metaph.*, 8.2.1.

wax for the house and its chambers, and honey as nourishment. Thus just as a nest serves a bird, thus does the wax serve the bees, and this is why the bees that generate, which are the smaller ones, have larger houses than the kings, because the kings do not make young as the others do, and this is why they have smaller homes.

1. On to the arguments. To the first, one must say that just as four humors are produced in the body from the same nutriment, in which diverse elements are dominant (because air dominates in blood, fire in choler, and so on for the others), so too can the bee produce either honey or wax from the same matter. Nevertheless, that which is more earthy is converted into honey, because, although honey is effectively warm, as its color attests, nevertheless earth is predominant in honey, and this is why it sinks to the bottom in water. But air predominates in wax, and this is why it floats on the top, just as air does.

2. To the second argument one must reply (rather, it has already been said) that honey and wax are not produced for the same reason, but honey is produced for nourishment and wax for a house, and this is why, etc.

Questions 29–34: On the customs of the wolf.

One inquires further into the customs of certain other animals.

And first we make inquiry of the wolf, why it is more wrathful before a meal than after, since the opposite is so for a human.[60]

Second, one inquires why, when it has grown old, it preys upon a human more frequently than it did in its youth.

Third, why, if it seizes a human, it does not eat his face but only the other members.

Fourth, why a wolf is more timid in the forest than in a field.

Fifth, why they are so few even though several are produced from one act of coition, whereas among sheep only one young is produced from a single act of coition and yet they are more numerous.

60. Ar., *HA* 9.44 (629b3f.).

Sixth, why a wound caused by a wolf is difficult to heal.

To the first question one should reply that the wolf is an especially gluttonous animal, and this is why, when its hunger is not satisfied, it often lies in wait for its prey, and thus it attacks whatever is offered to it.[61] But when it is satisfied, it does not need prey, and this is why at this time it is gentler. But the opposite is the case for the human, because, when one power is chiefly occupied in a human, the operation of any other power recedes. An example is clear to see in a person preoccupied with imagining, who does not then perceive what is presented to his sight. But after a meal a person's natural power is very much intent upon digestion, and for this reason the animal powers are dulled at that time, and he is more easily provoked to anger, because anger is a bubbling-up of the blood around the heart, and there especially occurs a bubbling of the blood at that time; for this reason, etc.

To the second question one must reply that the wolf, when it is young, is better suited for motion, and at that time it lives off wild prey and seeks to seize wild prey. But when the wolf has grown old, it cannot seize wild prey, and this is why it then seeks out villages and ambushes humans and domestic animals; for this reason, etc.

To the third question one should reply that the wolf preys on a human and yet fears him a great deal, and because a person is known by his face, the wolf especially fears the face and does not dare to attack his face.

To the fourth question one must reply that because the wolf is a timid animal, it frequently thinks that trees are humans, but when it is in the field it can look around in all directions better and take precautions for itself, and this is why it is bolder outside the forest than in the forest.

To the fifth question one must reply that the wolf does not engage in coition often. For the young ones naturally do not have coition while their parents are alive, but this is not so for sheep or for many other animals, and this is why wolves are so

61. Ar., *HA* 9.44 (629b3f.). Avic., *DA*, 8.7 (fol. 40rb); A., *DA*, 8.5.1.198–204; 22.2.1.114–17 (*SZ* 1: 753–56; 2: 1518).

few. Moreover, a she-wolf rarely conceives from coition, because she is an animal that is frequently in motion, and this is why, even though she gives birth to many at one time, nevertheless this happens but rarely, and this is why wolves are few.

To the last question one should reply that just as a wolf's eye is poisonous (because its eye infects the air near to it and this infected air infects another, and so on, as far as the human's sight, and this is why the sight of a wolf often renders a person hoarse,[62] because the infected air that the person then breathes in infects the chest, and this is the cause of the hoarseness), so too its teeth are poisonous and sources of infection. This is why they cause wounds that are cured only with great difficulty.

Questions 35–38: On animals that ruminate and have long extremities.

Further one asks whether every ruminant animal lacks teeth on the upper part [of its jaw], that is, on its superior mandible.[63]

And it seems so. Rumination is due to a defect of digestion. But the lack of teeth causes a defect in digestion. Therefore, etc.

Moreover, every animal that lacks teeth on the upper part [of the jaw] ruminates. Therefore, contrariwise, every ruminant lacks [these] teeth. This is the reason why, etc.

But the Philosopher says the opposite.[64]

Next one asks why ruminants abound in waste, as it is said in the text.[65]

Third, why these animals are social animals.

Fourth, why animals that have long extremities, like the crane, abound in waste, as do the stork and goose and others of this type.[66]

62. The common belief was that if the wolf saw you first, you lost your voice. Your only hope was to spot a wolf before it saw you. The belief is quite old. See Plato, *Republic* 1 (336d–e).

63. See A., *DA* 8.5.4.220–21 (*SZ* 1: 763–64).

64. Ar., *HA* 9.50 (632b1f.).

65. Ar., *HA* 9.50 (632b9).

66. "Goose": *auca*. See note at *QDA* 1.8–9.

To the first question one should reply that not every ruminant animal lacks teeth, although the opposite may be said. And the reason is this: Rumination belongs to an animal owing to a defect in digestion. Now in these animals, what is received into the stomach cannot be well digested, so nature provides that the same food will be returned to the mouth to be more completely chewed. But a defect in digestion can occur for two reasons: in one way, naturally, because of a lack of teeth, and this is why the cow and the camel ruminate. In another way, this can occur from excessive gluttony, because some animals snatch up their food so greedily that they swallow it unchewed, and in order for it to be properly digested it is necessary for it to be returned to the teeth. It is for this reason that mice ruminate, as is said in the text.[67]

And thus a response to the arguments is clear.

To the second question one must reply that an abundance of waste occurs from an abundance of superfluous moistures. But this abundance is present in ruminant animals, since for the most part ruminants are domestic animals and are for the most part at rest and at leisure, and this is why moist superfluities multiply in them. But non-ruminants are for the most part wild, or at least they are in motion more, and as a result their nutriment is more completely digested, and less of it is converted into waste.

To the third question one should reply that ruminant animals have common food and communicate with one another, and for that reason they pasture in social groups [*socialiter*], but non-ruminants live for the most part by theft, and they are rapacious and fearful that what has been obtained by one will be stolen by another; and this is the reason that, etc.

To the fourth question one should reply that animals that have long extremities have a short body and, as a result, they have short intestines, and this is why their food and the excrement quickly pass through the intestines, and this is why they abound in waste.

67. Ar., *HA* 9.50 (632b9).

Question 39: Whether the camel and the horse distinguish their mothers from others.

One asks whether the camel and the horse distinguish their mothers from others.

1. It seems not. To distinguish is an operation of reason. But these animals lack reason. Therefore, they also lack the power of making a distinction.

2. Moreover, among such animals the father does not distinguish his daughter from others, nor does the brother his sister. Therefore, for the same reason, neither does the son distinguish his mother from others.

The Philosopher says the opposite. For he says that a certain camel had coitus with his mother, and after she knew that she was his mother, she killed the one who had covered her head.[68] And in the same way he tells of a certain noble horse in the northern region that had coitus with his mother and after he discovered it he fled and cast himself off into a ditch and died.

It must be said that, just as in the natural order something caused receives influence and power from its cause, so in the course of generation, the one generated receives a power from the one generating, but not contrariwise, and this is why the one generated owes more respect to the one generating than does the one generating to the one generated. But the act of coitus induces shame, and the horse and the camel strive to observe the order of nature by a natural instinct. And this is why the son naturally hates to have coitus with his mother. Nevertheless, this is not true of all horses, but only of noble horses. Therefore, just as a man hates to have intercourse with his mother, because it is prohibited, so does the horse, because it is repugnant to the order of nature.

68. The keeper who had contrived to have the son mate with his mother had covered her head to keep her from recognizing her son. When the mother realized what she had done, she killed the keeper. Ar., *HA* 9.47 (630b31f.). Cf. Avic., *DA* 8.7 (fol. 40va); A., *DA*, 8.5.2.209–10 (*SZ* 1: 758–59).

1. To the first argument one should reply that it is not an operation of reason to make universal distinctions among individual things. Rather, differentiation among individual intentions can be made by estimation, just as a bird differentiates between grain and a stone, and a sheep between a wolf and a lamb or a man. Thus such differentiation can occur without reason.

2. To the second argument one should reply that the father receives nothing from a daughter, nor does a brother from a sister, in the way that the son does from the mother, and this is why the same argument does not apply to each.

Questions 40–42: On the cock's song.

One asks further about the cock's song.[69] First, whether its voice has to be higher pitched than the hen's voice.

And it seems not. Because according to the Philosopher women have higher-pitched voices than men, and similarly for the hen.[70]

Second, one asks why before his song he strikes himself with his wings.

And this seems a contrary thing, because such striking is more a cause of pain and not a cause for song and joy. Therefore, etc.

Third, one asks why he sings at certain hours.

To the first, one must reply that the properties and parts of the male, or the disposition of the male's parts, very frequently differ from the female's in the harshness or gentleness of the voice. But now it is the case that the hen has a good complexion, and the smaller it is the better it is, and this is why it is given to sick people, because it is more temperate [in complexion]. But the cock departs a little bit from this temperament and inclines toward dryness; likewise, it is hotter than the hen. It is owing to the heat that its path for giving voice is opened, and on account of its dryness its voice is harsher. This is why the cock's voice is harsher than the hen's, and most often this is due to a greater abundance of the heat and spirit.

69. A., *DA* 8.5.2.214; 23.1.24.115–19 (*SZ* 1: 760–61; 2: 1627–30).
70. Ar., *HA* 7.1 (581b7f.).

To the second argument one must respond that the reason why the cock beats itself with its wings is the same as why people stretch out their arms when they yawn, especially if they are suffering from a quartan fever, because whenever the fumes are collected in humans under the arms they struggle to get out. They cause people to extend their arms, and, when the arms are extended many times, the fumes are purged, and as a result the mouth opens and the emission of such fumes follows. So too in the cock the fumes are collected under its wings, and they irritate and tickle it, and this is why it extends its wings and strikes itself with them. And because these fumes exit when it extends its wings, its nature is relieved, and the cock sings for joy after beating itself. Now the hen also lifts up its wings, its feet, and its tail just as other animals do for the same reason, etc.

To the third argument one should reply that those things below follow the motion of the celestial bodies.[71] Thus, although there are four humors in the body, their dominion varies according to the variation or change of the weather, since in hot and humid weather blood dominates, whereas in hot and dry weather bile dominates, and in cold and dry weather melancholy dominates, and in cold and humid weather phlegm dominates. But the cock is a choleric animal, to which the dryness of its brain and the small size of its head in proportion to the rest of its body attest, because it has a small head and a dry brain. But according to the physicians the movement of bile is by thirds: either every third year, or every third month, or every third day, or every third hour. And this is evident in plants, because certain plants in which bile dominates bear fruit only every third year, and some herbs bear fruit every third month, and one suffering a fever from an abundance of bile struggles every third day. And the cock, since it is a choleric animal, is moved every third hour. Thus, especially at those hours when bile dominates, it sings, because it is then that the bile bubbles forth in the gall bladder, and the heart draws much of this bile to itself, because the heart is nourished by bile, and this is why it sings at those hours. Moreover, the song often anticipates the motion of the bile, and this is evident because it sings especially when the weather is calm and it is not rainy, etc.

71. A., *DA* 3.2.3–4.100–31 (*SZ* 1: 393–408).

Still, one must understand that bile and indeed the humors are in motion at all hours, yet they are more in motion at one hour than at another, and it is then that choleric parts like the heart draw bile to themselves owing to their likeness, because every like rejoices in its like. And this is the reason why, if a person who has recently been killed is cast into the water and a red shield drifts down in the water, when it approaches the body lying a little beyond it in the water, it stops there or is moved in a circle and no further. This is not so for other shields, because of all the colors it is the red that especially multiplies itself, which is evident from a linen cloth suspended in a window in a ray of sun. Thus if one who has recently been killed is placed in the water, and if the blood either bubbles forth from him or is bubbling in him still, that blood multiplies its appearance [*species*] through the parts of the water, and likewise the red shield multiplies its appearance through the water. Thus when a shield floating through the water approaches reddened water, it does not withdraw from that place because it finds something like itself and like rejoices in something like itself. Thus Avicenna says that a certain person clothed in red crossed through a stream whose beaches were red, and since the appearance of his cloth was multiplied in the water, the water and its shore then rose up to the man and enveloped him. And the only explanation for this is that like draws its like to itself. Thus the water was reddened by the beaches, and the clothes were made more intensely red by this, and this is why both the water and the beaches moved at the same time to the clothes, by reason of their likeness. Thus too, when the bile is moved, the choleric members draw more bile to themselves, and bile is of a penetrating and bitter nature. This is why, in order to allay and shake off the penetrating bile, the cock extends its wings at that time and moves them to the front and to the rear and especially toward its chest, so that the choleric fumes may exit, and then, after their exit and after they have been shaken out, it sings. And because this movement of the bile occurs at certain times, as has been determined, it sings at these certain times and hours just like a clock, and especially after midnight, because then the sun begins to approach our hemisphere and to rise, and the warm humors begin to move then. For this reason, etc.

BOOK NINE

Question 1: At what time especially should human generation take place?

"A DISCUSSION, THEN, concerning the principle of human [generation],"[1] etc. In this ninth book one asks first, at what time human generation especially should take place.

1. And it seems that it should take place during adolescence, because generation ought to occur more or especially during a time when the material for generation is more abundant. But this material is moisture, and moisture is more abundant during adolescence. Therefore, etc.

2. In addition, heat is the effective principle of generation. But heat is more abundant during adolescence, which is clear from a human's growth pattern. And this is why, etc.

3. Moreover, pleasure is more abundant at this time, and the greatest pleasure occurs during a time of generation. Therefore, etc.

The Philosopher says the opposite. For he says that the period most suitable for generation occurs at the end of the third seven-year period.[2]

One must reply that generation is of two types. One is of a complete fetus, and the other is of an incomplete fetus. The first type in particular has to occur at the end of the third seven-year period, because when the nutritive power and the power of growth are bounded, then the generative power begins. But at this time growth is bounded, and this is why generation be-

1. Ar., *HA* 7.1 (581a9–11). Cf. A., *DA* 9.1.1–2.1–23 (*SZ* 1: 774–82).

2. Ar., *HA* 7.1 (582a27–29). Cf. Avic., *DA* 9.1 (fol. 40vb); A., *DA* 9.1.2.19 (*SZ* 1: 780).

gins, because the seed, which is the principle of generation, is a superfluity of the aliment. But during the period of growth there is little of this superfluity because the whole of the aliment is converted into the substance of the members. Generation in the second sense, however, begins at the end of the second seven-year period, because it is then that the seed begins to move and likewise the menses. Yet in truth these are not well ordered or made firm then, and this is why in this period the fetus is weak and poorly disposed.

1. To the first argument one must reply that too much moisture dulls the heat and a great deal of heat is necessary for generation. Therefore, although moisture abounds during adolescence, nevertheless it dulls the heat, and generation is therefore weakened. And in the same way one can say that moisture is of two types. One type is the material for the nutriment, and this is abundant in adolescents. The other is the material for generation, and this one is abundant after the third seven-year period.

2. To the second argument one must reply that heat can be said to be greater in two ways: either qualitatively or quantitatively. It is quantitatively greater in adolescents, but it is qualitatively less because it is curbed by moisture, which is its contrary.

3. To the third argument one must reply that pleasure is of two types: ordered, which occurs in men, and disordered, which occurs in adolescents.

Question 2: Whether a difference in ages ought to be calculated according to seven-year periods.

Further one inquires whether a difference in ages ought to be calculated according to seven-year periods.

And it seems not, for the moon has dominion over everything moist, because it is the mother of the waters.[3] And the moon varies according to its four phases. Therefore, ages and other dispositions containing moisture have to be calculated according to a number containing four.

In addition, illness is indicated through the number seven or

3. Cf. A., *DA* 9.1.2.21 (*SZ* 1: 781).

by a seven-day period, and therefore for the same reason so too is age.

The opposite is apparent according to the Philosopher.[4]

One must reply that a given disposition, such as illness, has its cause in matter just as in a subject and this is why it lasts for a shorter period and then ends. But there is another disposition that is caused by the consumption of moisture and natural heat, and a disposition like this requires a longer period of time for its onset, because the first one can be introduced from an indisposition of the moisture or of accidental heat. Illness is introduced in the first way, and age in the second way, and this is why illness is introduced according to days, and age according to years. But the moon has dominion over everything moist through its four phases, but its four phases are measured or counted in periods of seven days.

And in this way a solution is apparent for the arguments.

Question 3: Whether the period for generation among men lasts longer than does the period for conception in women.

Further one inquires whether the period for generation among men lasts longer than does the period for conception in women.

And it seems not, because the power of the man is weakened while generating since he emits something outside himself. But during conception the woman does not emit anything outside herself. Therefore, since whatever is more weakened fails more quickly, a man will cease to generate more quickly than a woman.

The Philosopher says the opposite.[5]

One must respond that the nobler something is, the more slowly does it acquire its perfection. For it is clear that a human fetus moves on its own later than does the fetus of another, and the more perfect something is, the more it preserves its own perfection. That is why a perfect animal has a longer lifespan

4. Ar., *HA* 5.14 (544b25–27), 7.1 (581a13–14, 582a16–17 and 27–29).
5. Ar., *HA* 7.5–6 (585b2–8). Cf. A., *DA* 9.1.6.59 (*SZ* 1: 795–96).

than an imperfect animal. Since, then, a man is more perfect than a woman, a man acquires the complete power for generating more slowly and conserves it longer. Therefore, in a woman it begins after the second seven-year period, and ends about the fiftieth year, but in a man it begins at the end of the third seven-year period and ends about the seventieth year.

There even may be other causes, because a man emits semen alone whereas a woman emits semen and the menses. And the woman is likewise more burdened with giving birth to and nourishing the fetus. Therefore, it often happens that she loses an appetite for intercourse, as a result of the pain of childbirth, and ceases to subject herself to it.

One must reply to the argument that although a woman does not emit seed during intercourse, nevertheless she admits something that is outside the womb, which is the matter for the fetus, into the womb.

Question 4: Whether a male grows more slowly outside the uterus than a woman does.

Further one inquires whether a male grows more slowly outside the uterus than a woman does.

1. And it seems not, because heat is the principle of growth. But heat is greater in a man than in a woman. Therefore, etc.
2. In addition, a male grows more quickly in the uterus than a female. Therefore, the same is true outside the uterus.

The Philosopher says the opposite.[6]

One must reply that the male grows more quickly within the uterus, but more slowly outside it. And the reason for this is that temperate heat is the active principle of growth, and temperate moisture is the passive principle. But the heat is greater in the male than in the female, as reason proves. Since, then, there is sufficient moisture in the uterus and sufficient heat in the male, he therefore grows more quickly in the uterus. But outside the uterus the heat is increased, but the moisture is diminished be-

6. Ar., *HA* 7.3 (583b23–28). Cf. A., *DA* 9.1.3.35–36 (*SZ* 1: 786–87).

cause the heat consumes the moisture, and this is why he grows more slowly, owing to the dryness of the members, a dryness that opposes the moisture's growth. But in the female the heat is not powerful enough to consume the moisture, and this is why the woman grows more quickly outside the uterus, whereas she grows more slowly within the uterus owing to the weakness of her heat, while outside her moisture may grow and is obedient and her heat is strong, etc.

1. To the arguments one must reply that the heat is greater in the man, but in the uterus he has proportionate moisture, and outside of it he consumes this moisture, and this is why he grows more quickly inside the uterus but more slowly outside of it.

2. In the same way a solution is evident for the second argument.

Question 5: Whether the menses are necessary for generation after impregnation.

Further one inquires whether the menses are necessary for generation after impregnation.[7]

1. It seems not. For the agent introduces form and disposes the matter, but the thing that is made conserves itself. For although "nothing generates itself, each thing preserves itself," as is said in the second book of *On the Soul*.[8] But the menses are neither form nor matter for the fetus. Therefore, they are not necessary.

2. In addition, some woman are masculine-like [*viragines*] who conceive and yet do not suffer menstruation, as is said in the text.[9] Therefore, etc.

To the contrary. Just as flowers exist for the production of the fruit, so the menses relate to the production of the fetus. But flowers are necessary for the sake of the generation of the fruits, and therefore, etc.

7. Cf. Ar., *HA* 7.3 (583a31–32); A., *DA* 9.1.3.31 (*SZ* 1: 784).

8. Ar., *De anima* 2.4 (416b16–17).

9. Cf. Ar., *HA* 7.4 (584b6–9).

One must reply that the menses are necessary in women for several reasons. Because one part of them becomes the radical spermatic members, like the heart, the brain, and ones such as these. And it is from another part of them that the empty spaces [*vacuitates*] in the embryo's members are filled in, namely, the spaces between the nerves and the veins and things of this sort, and these parts are necessary for the production of the fetus. A third part is delegated for the breasts and is converted into milk. A fourth part is enclosed in the womb and in the stomach and makes the stomach swell and softens the membranes [*pelliculas*] and the opening of the womb so that it may stretch out better at the time of birth, and it then exits after the birth. And a fifth part flows from the individual members in a given month, although in a modest amount, and sometimes it exits imperceptibly with the urine in pregnant women,[10] and this is not necessary for the sake of the fetus but only for the sake of cleansing the woman.

1. To the first argument one must reply that things are different for animate and inanimate things. For animate things are not preserved without nourishment, and this is why it is necessary for the fetus to be nourished in the uterus. But its nourishment while it is in the uterus is the menses, by which the fetus is nourished through the umbilical cord. Therefore, form, matter, and nutriment are required for the fetus's production, and it is preserved by this.

2. To the second argument one must reply that in a warm region the moistures that become the menses in other regions are consumed, and this is why the women there do not menstruate. This also makes clear why poor women or women who work hard and do not live a delicate life do not undergo menstruation, because whatever they assume as nutriment is barely adequate to sustain nature and this is why they rarely suffer menstruation, or hardly at all. And it is the same for masculine-like women, because the menses are consumed in them by the strength of their heat. This is why, etc.

10. Ar., *HA* 7.2 (582b19–20); *GA* 1.20 (727b26–29); A., *DA* 15.2.6.111 (*SZ* 2: 1534–35).

Question 6: Whether only women suffer menstruation.

Further one inquires whether only women suffer menstruation.[11]

1. And it seems not, because the menses are a superfluity of the blood. But this can appear in men just as in women, and therefore, etc.

2. In addition, every viviparous animal [*animal productivum sui similis*] has some part of itself on which the fetus is nourished. This is the menses, and therefore, etc.

Moreover, that which is converted into menses in a woman is turned into claws, horns, and other such superfluities in other animals, such as brute beasts. But it can happen that it is not converted into things like this. Therefore, it is necessary that their matter be converted into menses.

The Philosopher says the opposite.[12]

One must reply that the menses are a superfluity of the blood, and their color demonstrates this. This is caused by an undigested moisture flowing from individual members to the vessels for the blood. Thus the menses are caused by individual members, just as semen is. But moistures of this sort in the man are consumed owing to the excellence of his heat, and this is why there are no menses in the man. In the same way that part of the menses which is converted into parts of the fetus, and that part on which the fetus is nourished in the uterus, and that part which is converted into milk in the breasts, exist in individual animals. But that part of the menses that flows monthly does not exist in the females of other animals because they have less moisture than women do. This is because they do not take in aliment superfluously and do not live in such leisure or ease, and so the moistures that do exist in them are converted into the superfluities of horns, claws, and hair, as the Philosopher says.

11. Cf. Ar., *HA* 7.2 (582a34–583a13); A., *DA* 9.1.2.26 (*SZ* 1: 782–83).

12. Ar., *HA* 7.2 (582b28–583a4). Cf. A., *DA* 9.1.2.26, 15.2.6.101–12 (*SZ* 1: 782–83; 2: 1130–35).

1. To the first argument one must reply that a menstrual flow does not occur in the man for two reasons: namely, both on account of the abundance and excellence of his consuming heat and on account of the paucity of his moisture.

2. To the second argument one must reply that if superfluous blood in other animals is not converted into horns and claws, then this flux does occur in them, but this is against the course of nature. But the fact that they do suffer menses does not contradict the Philosopher since it occurs *per accidens,* since it arises from some illness in them and beyond the order of nature. This is why, etc.

Question 7: Whether the moon has dominion over the menstrual flow.

Further one inquires whether the moon has dominion over menstrual flow.[13]

1. And it seems not, because if it did have dominion, then the menstrual flow should increase when the moon waxes. But this is not true, since it grows more at the end of the month, according to the Philosopher.[14]

2. In addition, the moon and every heavenly body act by the power of light. Now, it acts the most when it is most illuminated. If then it had dominion over the flow of the menses, then it would occur on the fourteenth day more than on the thirtieth day.

3. Moreover, the moon does not have dominion over the emission of seed or of other waste products or over the flow of hemorrhoids. Therefore, etc.

The Philosopher says the opposite.

One must reply that the moon does have dominion over the menstrual flow and over everything moist and cold. Nevertheless, one must understand that two things are required for the menses to flow: their collection and their expulsion. The menses are collected when the moon waxes. Thus at the full moon their

13. Cf. Ar., *HA* 7.2 (582a34–583a13); A., *DA* 9.1.2.21–23 (*SZ* 1: 781–82).
14. Ar., *HA* 7.2 (582a34–b3); cf. A., *DA* 9.1.2.21–22 (*SZ* 1: 781).

collection is greatest. But after collection, a disposition for expulsion is required, and they remain in this disposition until the end of the month, and this is why they particularly flow then.

1–2. In this way a solution is apparent to the first and second arguments.

3. To the third, one must reply that sperm is hotter and more aggregated, and this is why it has less regard for the moon, because it is more globular than the menses, which are fluid and watery. Likewise, the other waste products, like urine and things of this sort, are things that are left behind after nature's operation, whereas the menses are something more intended by nature, and this is why the moon has more dominion over the menses than over other superfluities.

Moreover, hemorrhoids are caused by a superfluity of thick blood, because, when such blood is abundant in the body, it descends down below to where there are many veins, namely, in the womb, and then often one or two veins burst, and the blood flows sometimes on account of the opening of these veins. Therefore, this occurs by nature particularly among those living on gross and salty nutriment, like the Jews.[15] And because this blood is thick and has an earthy nature, the moon does not have dominion over its flow in the way it does over the menses.

Questions 8–10: Whether the menstrual flow is sudden. Whether such a flow causes infection in the eyes. Whether it accelerates the aging process.

Further one inquires whether the menstrual flow is sudden.

1. And it seems not. For every sudden occurrence is injurious to nature, but according to Avicenna that which occurs gradually is agreeable to nature. Since, then, the menstrual flow is gradual, it will be neither sudden nor injurious to nature.

2. In addition, nature acts and proceeds against disease in

15. The suggestion that the Jews' diet rendered them especially subject to hemorrhoidal bleeding helped introduce the medieval myth of Jewish male menses. See I. M. Resnick (2000).

three ways: namely, by digesting, dividing, and expelling matter, according to Galen in his *Aphorisms*. But the first two operations are not sudden, and therefore neither will be the third.

The Philosopher says the opposite.[16]

Second, one inquires whether such a flow causes infection in the eyes.

And it seems not. For that which is a source of cleansing is not also a source of infection. But the menstrual flow cleanses a woman. Therefore, it does not cause infection.

The Philosopher says the opposite.[17]

Third, one inquires whether the emission of semen and of the menses diminishes the memory and accelerates the aging process.

1. It seems not. The purification of the spirits contributes to memory, and the expulsion of crude and undigested superfluities postpones or retards the aging process. But the spirit is purified by the emission of semen and of the menses, and the aforementioned superfluities are expelled. Therefore, it contributes to memory and retards the aging process.

2. In addition, according to the Philosopher in the twelfth book of this work, fat blocks sensation and shortens life.[18] Therefore, in a contrary way, being thin will be a cause of their opposites. But the emission of seed and of the menses makes a person thin; therefore, etc.

The Philosopher says the opposite.

To the first question one must reply that the menstrual flow can be natural or unnatural. The natural flow arises from the strength of nature and the obedience of material, but an unnatural flow arises along with weakness of power and from recalcitrance [*inobedientia*] of the material. But every sudden action that arises from a strong power in an obedient material comes, with respect to the action that is arising, from an op-

16. Ar., *HA* 7.2 (582a34–583a13).

17. Ar., *De insomniis* 2 (459b28); A., *De somno et vigilia* 2.1.6.

18. Ar., *Part. An.* 2.5 (651a36–b11); cf. A., *DA* 12.1.7.88, 12.2.2.108 (*SZ* 2: 927; 934).

posite cause, and is not sudden in an absolute sense. And this is why the Philosopher understands that the natural menstrual flow almost seems sudden, because it occurs over a short period of time, for it ends in three or four days.[19] But the flow is unnatural when it arises slowly and is completed slowly, for example, over four months' time, or when it ends in death.

1. And with this a solution to the first argument is apparent, because every sudden act arising against nature's inclination is injurious. But the menstrual flow does not arise in this way, especially when it is a natural flow.

2. To the second argument one must reply that the division and digestion of matter are related to expulsion in the way generation is related to alteration. And thus, just as alteration occurs gradually over time, generation, which is its end, occurs suddenly and in an instant. Analogously, then, the expulsion of noxious matter appears to be almost sudden compared to division and digestion.

To the second question one must reply that just as among those having an affinity [*symbolum*] a mutual transition is easier, so too among those not having an affinity a mutual infection or alteration is easier. Now, then, the eye is particularly in agreement [*convenit*] with the brain, because it is cold and moist; and because the brain is diminished by the emission of the seed and the menstrual flow, sight is also diminished, since what injures one injures the other. And this is why in those people who have very frequent intercourse the first telltale sign appears in the eyes. Similarly, the eye is a very passive member, and at the time of menstruation the menses flow particularly to the eyes and infect them. And this is why if the eyes' object is clean, like a polished mirror or the eye of another person, it will immediately be infected by the eye of the one menstruating, since that one's eyes infect the air, and this air infects other air, and this infects still other air, as far as the mirror.[20]

19. Ar., *HA* 7.1 (581b28–582a2); *GA* 1.18 (725b16–19). A., *DA* 9.1.1.13, 15.2.6.103 (*SZ* 1: 778; 2: 1131–32).

20. The tainting of a mirror by a menstruating woman was a widespread belief. Cf. *DA* 7.2.5.133 (*SZ* 1: 646).

Therefore, to the present question one must respond that although the menstrual flow and its expulsion cleanses a woman, nevertheless when it remains in the body it infects all the members and especially the eyes and, in the same way, their object.

To the third question one must reply that an emission of this sort weakens memory for three reasons, because inordinate desires cloud the senses, according to Aristotle in the *Ethics*.[21] But when sperm and the menses are emitted frequently, they cause desires to burn excessively in a person, and as a result they obfuscate the senses, and so too the memory.

Moreover, memory flourishes in the posterior part of the brain. But this part of the brain is particularly diminished by the emission, and as a result memory is weakened.

The sensitive nerves are also debilitated by this emission and, as a result, so too are the motive nerves, because sensation is the principle of motion, and it thus induces aging because the consumption and evacuation of the radical moisture is the cause of aging. But by their frequent emission not only is what is superfluous evacuated, but also that which is agreeable to nature is expelled along with it, and this accelerates aging. Still, one should make a distinction concerning the body, because if there are superfluities in the body that nature does not need, but these are, rather, contrary to nature, then such an operation and expulsion contributes to the health and goodness of the intellect, by purifying the brain and [the faculty of] fantasy, without which there is no intellect. And very often it contributes to growth, because the heat is dulled by superfluities like these, and heat is the effective principle of growth, and this is why the heat flourishes and grows more freely once they have been expelled. If however the body is well disposed, and does not have excess and superfluous spermatic moisture, whatever it may be, then an operation like this is contrary to the aforementioned things, or it introduces contrary dispositions.

21. Cf. Ar., *Eth. Nic.* 1.1 (1095a3–6).

Questions 11–15: On the signs of conception.

Further, one inquires into the signs of conception. And first, whether a woman may conceive soon after her cleansing.[22]

And it seems not, because menses are required for conception. But the menses are diminished through the cleansing. Therefore, after her cleansing she will conceive more slowly.

The Philosopher says the opposite.[23]

Second, one inquires whether after conception the interior opening of the vulva closes so much that a needle cannot enter the aperture.

1. It seems not. In each case when what is thicker can exit, then what is thinner can exit. But after conception the superfluities of the waste products and the filth exit. Therefore, so too do subtle things exit more forcibly.

2. In addition, the stomach does not close after it has received food, so that it cannot open later. This is made clear by both the food and drink that are taken in later and nevertheless enter the stomach. Therefore, the argument will be the same for the womb.

The Philosopher says the opposite.[24]

Third, one inquires whether goose bumps [*horripilatio*][25] occur naturally after conception.

And it seems not. Because stiffness and goose bumps are nothing but the spread of bad material upon the members and especially upon the sensible members. But in pregnant women this matter is not spread, but is enclosed in the womb. Therefore, etc.

The Philosopher says the opposite.[26]

Fourth, one inquires whether the urine ought to be thin and clear in a pregnant woman.

And it seems not, because darkness and cloudiness in the

22. Cf. A., *DA* 9.1.3–4.28–48 (*SZ* 1: 783–91).

23. Ar., *HA* 7.2 (582b9–14).

24. Ar., *HA* 7.4 (583b29–31).

25. "Goose bumps": or "bristling of the hair" or "shivers." Cf. *DA* 6.1.3.22; 7.3.4.172 (*SZ* 1: 535; 664) and *QDA* 4.10.

26. Ar., *HA* 7.4 (584a2–12).

urine arise from the commingling of the humors and of the earthy parts. But in conception parts like these are in motion, and after conception these parts are in flux. And therefore, etc.

Similarly one can inquire why women who have been impregnated have an appetite so irrational that if they are lacking something from their desire they frequently miscarry.

To the first question one must reply that it is an emptied member's nature to attract and a full member's nature is to shun, and likewise it pertains to one that has been evacuated to incorporate and receive but this is not so for one that has been filled. Now then, the womb is a concave member, and for this reason, while it is filled with menstrual blood and with other superfluities, it can neither attract nor incorporate the seed; but after cleansing, because it is emptied, it attracts it, and because it has been evacuated, it incorporates it, and this is why she conceives more easily after cleansing.

To the argument one must reply that before cleansing the menses are so superfluous that they cannot be retained by the womb, and they are required for conception because they can close the womb's opening, something that could not occur if it were not first dried out.

To the second question one must reply that the more something desires, the more it attracts, and the more it attracts, the more it retains what has been attracted. And because a woman who has been cleansed has a great desire, she therefore attracts a great deal, and as a result she strives especially to retain, and this is why after conception the opening of the womb is immediately closed. Therefore, conception is a sign that it has closed.

1. On to the arguments. To the first, one must reply that the opening of the womb is double: one internal and another external. The internal one is closed, and it is not opened naturally until the fetus is formed. But the external opening is open, and through it superfluities that she does not need (for example, urine and the like) can exit. Nevertheless, the superfluities that remain, like sperm and semen, exit only through the internal orifice.

2. To the second argument one must reply that the womb's appetite and ability to receive are greater than the stomach's,

and the womb's ability to receive and that of the stomach have a different intensity, and this is why the argument does not prove the conclusion.

To the third question one must reply that when a person lives in fear and stress, then the natural heat and spirit cross from the exterior parts to the interior ones, and then the exterior parts are abandoned just as if they were left deprived of nature's control, and this is why they experience goose bumps and fear. So the proposition is correct. After conception the heat returns to the interior for the digestion of the semen and for the fashioning of the embryo or the formation of the fetus, and then the exterior members seem to be deprived of nature's control and then very often goose bumps occur.

And in this way a solution to the argument is evident. Because goose bumps can arise in two ways: either from the spread of bad, putrid matter over the interior sensible members, as in those suffering fever, or owing to the flight of the natural heat to the interior, as in the proposition.

To the fourth question one must reply that after conception the mouth of the womb is closed, and then nothing exits from it. And this is why dryness around the womb's orifice is one sign of conception, because this is an indication that the womb has received and absorbed all the sperm and retains what it has received, and this is why the urine is rendered clear immediately after conception but at the end, when the fetus is struggling to exit and the menses begin to flow, then the urine will be more turbid and cloudy on the surface, since this cloud on the surface attests to the inchoate digestion of matter.

To the last question one must reply that after conception part of the menstrual blood is borne to the mouth of the stomach and moves the stomach's appetite, and this is why it creates irrational desires in her. For just as in insane people a vapor ascending to the brain makes them imagine irrational things, so in the same way the menstrual blood ascending to the stomach creates irrational desires in her. And in addition to this, a pregnant woman has an appetite not only for herself but also for the fetus, whereas a man or a woman who is not pregnant has an appetite only for himself or herself. And this is why pregnant women sometimes want to eat charcoal or potter's clay and desire it a great deal. And if they are without something they desire, then they begin to despair, and, on account of this, nature withholds aliment from the fetus, and this is why very often she miscarries.

Questions 16–17: Whether the male is formed more quickly than the female.

Whether on the fortieth day the fetus is of the same size as a large ant.

Further one inquires whether the male is formed more quickly than the female.

1. And it seems not. For to the extent something is more perfect, then with just so much more difficulty does it acquire its perfection. But the male is more perfect than the female. Therefore, etc.

2. In addition, moisture is easily bounded by something else. But the female is moister than the male. Therefore, she is more quickly bounded when she is in the uterus.

3. Moreover, "if the one born is a female, her purgation will occur over thirty days, but if a male, over forty days."[27]

The Philosopher says the opposite. For he says that the male moves after forty days, and the female does not move before ninety days.

Further, one inquires whether on the fortieth day the fetus is of the same size as a large ant.

And it seems not, because after conception an amount of the menses and sperm exceeding the size of an ant is enclosed in the womb. Therefore, the one generated in the womb is larger than this.

The Philosopher says the opposite.[28]

To the first question one must reply that configuration aris-

27. Ar., *HA* 7.3 (583a26–b7). Cf. Avic., *DA* 9.5 (fol. 43rbA); A., *DA* 9.1.3.30–34 (*SZ* 1: 784–85).

28. Ar., *HA* 7.3 (583b12–18). Cf. A., *DA* 9.1.3.35–36 (*SZ* 1: 786–87).

es quickly from the strength of the heat, and the heat is stronger in the male, and this is why he is shaped more quickly in the uterus. Therefore, the Philosopher says that if what is retained in the womb falls on the fortieth day into cold water, the thing will appear to be wrapped in a web about the size of a large ant, at least if it is a male. If it is a female, however, then it does not appear formed yet on the fortieth day and this is owing to nothing other than its weak heat. Nevertheless, one must understand that there is a good bit of diversity with respect to pregnancy. When the fetus is formed over a given period of time, it will move in double that time and it will be born in triple that time. And this is why wise physicians say that movement doubles the time from conception, but birth triples it.

In the same way there is another difference, because in the first six days the conceived semen becomes almost like milk, and then over nine days it is changed into blood, and after that, over twelve days, it is consolidated, and then over nine days it is given shape, and the time after that prepares it for birth. But at least according to the Philosopher's understanding, the male is formed more quickly than the female in the natural course of things. If the opposite should occur, it will be contrary to nature.

1. On to the arguments. To the first, one must reply that the more perfect a thing is, the more slowly it attains its ultimate perfection; nevertheless, it can attain another perfection more quickly.

2. To the second argument one must reply that there is something that is moist and fluid, and a thing like this is given shape quickly and then quickly loses it, just as water immediately receives the impression of a seal and immediately loses it. But there is something that is at the same time both moist and solid, and something like this is not so easily given shape.

Or one can say differently that although the semen is shaped according to the womb, nevertheless the shape of the animal cannot be quickly formed. Thus the moist is poorly bounded by its own boundary, and this is why the seed of the female, which is more moist, acquires its bounded state more slowly.

3. To the third argument one must reply that the Philoso-

pher says that if the one born is female, purgation will occur after thirty days. But it is different for its formation. This is why the argument does not prevail.

To the other thing asked, one must reply that the proportion of a child outside the uterus to one inside the uterus is the same as that of a fully-grown man to a child. Therefore, in the first month it has, as it were, its smallest size and later acquires a size similar to that of a kind of worm[29] since, although the composite or the semen may have a large size, the whole of it is not converted into the fetus's substance. Some is converted into the fetus's substance, and some is reserved for nourishing it, and some is expelled just like the feces and a superfluity.

With this a solution to the argument is apparent.

Question 18: Whether a male or a female can be generated by contrivance.

Further one inquires whether a male or a female can be generated by contrivance [*per artem*].

1. It seems not, for what is determined by nature cannot be changed by contrivance. But the seed is naturally determined to be either male or female. Therefore, it cannot be changed.

2. In addition, whatever cannot exert power on the effect cannot exert power on the cause, because the cause is more powerful than its effect. But contrivance cannot change a male into a female, nor contrariwise. Therefore, it will not be able to change nature.

To the contrary. It takes greater power to produce a substance than to alter the action of the substance. But one who is sterile can give birth by contrivance, or a fetus can be produced from one who is sterile. And male and female differ only *per accidens*. Therefore, the material disposed to become a male can all the more be shaped into a female by contrivance, and contrariwise.

29. "Worm": *vermis,* elsewhere for A. a more generic term and thus might be translated "insect."

To this, one must reply that contrivance can contribute to the generation of a male or a female, because a strong heat is required for the generation of a male. But heat can be fortified and introduced to the semen by contrivance and medicine. Therefore, if the semen that is sent forth is disposed to become a female, it can be disposed to become a male by medicine, and contrariwise.

In addition, that which can alter the cause can also alter the effect. But a cause of a male is the strength of the heat, [living in] a frigid area, a northern wind,[30] and the evocation of the sperm from the right testicle, as the physicians say, and all these can be achieved by art or medicine.[31] Therefore, etc.

1. On to the arguments. To the first, one must reply that a natural disposition is of two types: either in the species, or in an accident containing the species. In one way, the male and the female do not differ by species, but only in terms of an accident containing the species; and although the species cannot be changed by contrivance, nevertheless a change can be made to the accidents of the species by contrivance.

2. To the second argument one must reply that something exists in potency in relation to another in two ways: either proximately or remotely. Whatever is proximate in potency and exists as if especially disposed to one effect cannot be altered. But whatever is remote in potency can readily be altered. Seen in one way, the matter for the production of a male or female is remote in potency, and this is why it can easily be altered. But those accidents by means of which they[32] are present in act are present

30. Ar., *GA* 4.1 (765a3–5). Galen, *De semine* 2.5 (ed. Kühn 4: 627); *De usu part.* 14.4 (ed. Kühn 4: 154). Avic., *Can. med.* 3.21.1.12.

31. Elsewhere, Albert explains that sufficient heat in the father's sperm is necessary to produce male progeny. When the sperm is derived from the right testicle (which, like other organs on the right side of the body, is warmer), when the body compensates for a colder climate or for the effects of the north wind by contracting its pores to preserve its heat, or when it is produced during a period of life when the complexion typically is warmer (e.g., adolescence), the odds of producing a male greatly increase. See *DA* 6.3.2.109 (*SZ* 1: 577),10.2.3.66 (*SZ* 1: 853), 18.1.3.27 (*SZ* 2: 1293), 18.2.9.94 (*SZ* 2: 1328), and 22.2.1.128(88) (*SZ* 2: 1528–29).

32. The subject of "they" would seem to be "male or female."

inseparably and they ought not to be altered. For although an alteration cannot be made in inseparable accidents, when they are present in act, nevertheless, nothing prohibits an alteration from being made in them before they exist in act and when they are in an indeterminate potency. And this is why, etc.

Questions 19–23: On the birth of a woman and her circumstances.

One inquires further whether a woman has a determinate time for giving birth.[33]

And it seems so. For a greater and nobler perfection is due to a nobler species. Other animals have a determinate time for giving birth. Therefore, so much more so will a woman.

The Philosopher says the opposite.[34]

Second, one inquires whether a fetus brought forth in the eighth month is weaker than a fetus brought forth in the seventh month.

And it seems not. For that which is consolidated and arranged over a longer period of time does not seem to be weaker. But one brought forth in the eight month is like this, when compared to one brought forth in the seventh month. Therefore, etc.

The Philosopher says the opposite.[35]

Third, one inquires whether a woman labors more giving birth than do the other animals.

And it seems not, because the end ought to be proportioned to the beginning. But a woman takes particular pleasure in intercourse and in conception. Therefore, she will suffer proportionally less in giving birth.

The Philosopher says the opposite.[36]

Fourth, one inquires why a woman suffers a miscarriage more

33. Cf. A., *DA* 9.1.4.45–48, 9.1.5.57 (*SZ* 1: 789–91; 794).

34. Ar., *HA* 7.4 (584a33–b1). Cf. Avic., *DA* 9.5 (fol. 44raD); A., *DA* 9.1.4.45 (*SZ* 1: 789).

35. Ar., *HA* 7.4 (584b1–12). Cf. Avic., *DA* 9.5 (fols. 43vbC, 44raE); A., *DA* 9.1.4.45 (*SZ* 1: 789).

36. Ar., *HA* 7.4 (584b14–21). Cf. A., *DA* 9.1.4.47 (*SZ* 1: 790–91).

often in the beginning or the end of her pregnancy than during the middle.

Fifth, one inquires why, when a pregnant woman ingests salt, this should be the cause why the fetus is generated without nails.[37]

To the first question one must reply that in those inferior species, the nearer something is to matter the more uniform it is, and the more remote it is from matter the more diverse [*difformitate*]. This is clear from comparing an element to compounds or homogenous parts to organic parts. Therefore, since a woman has more form than do other females in other species, she is more diverse. And in the same way reason and will are not subject to the impressions of celestial bodies in the same way other bodily powers are. But the life of other animals is ruled by sensation, which is an organic power, whereas reason and the will rule the power and life of a person. Therefore, because the other animals are ruled in a natural way, this is why they give birth uniformly, but a woman can govern herself by her will, so much so that she may give birth at different times, namely, in the seventh, eighth, ninth, or tenth month. In this way a solution is apparent to the argument.

To the second question one must reply that Saturn is the first among the planets in superior realms, and next is Jupiter, and then Mars, and so on for the others. But Saturn is cold and dry, and it is a malevolent planet. Thus, one born under Saturn is given to evil. Jupiter is warm and moist, and it is a benevolent planet, for it makes a person lovable, and so on for the others. Therefore, if the child that is born should come forth in the seventh month, and is then born under the moon, the birth will be for the good. If, however, the fetus has not been adequately arranged for its exit in the seventh month, and its exit is delayed until the eighth month, it is a sign of weakness in the fetus, and in the same way its birth will depend on Saturn, and under this planet the birth is bad or leads to something bad. And if it should come forth in the ninth month, it is born under

37. Cf. Ar., *HA* 7.4 (585a26–28). Avic., *DA* 9.5 (fol. 44raE); A., *DA* 9.1.5.57 (*SZ* 1: 794).

Jupiter, which is warm and moist, and the birth will be for the good. Thus the Philosopher says that, according to the opinion of the astrologers, if a woman should give birth in the eighth month the fetus will die or, if it lives, it will languish.[38] And with this a solution to the argument is evident.

To the third question one must reply that pain is caused by the sensation of a contrary, according to Galen and to Aristotle in the fourth book of the *Ethics*,[39] whereas pleasure is caused by the sensation of something agreeable. But now it is the case that among all the animals woman has more pleasure in conception, and this is only because she has a more subtle power of sensation, and pleasure during intercourse is caused by the movement of the sperm and its passage over the sensible members. But where there is greater pleasure there will also be greater suffering when suffering occurs. This is evident in the individual members, since the greatest pain occurs in the eye because it has the subtlest power of sensation. And this is why a woman suffers more in giving birth. And generally those that experience more pleasure during intercourse suffer more in giving birth. And another reason is that the woman abounds in spermatic blood when compared to other female animals, and this is why her fetus is relatively larger and the size of the fetus is the cause of her suffering during childbirth.

To the fourth question one must reply that the fetus in the uterus is just like a fruit on a tree. Now when fruit is new it quickly falls due to the movement of the wind, because its tether is not well consolidated, and it is the same at the end when it has ripened, because it is tied to the tree by moisture, and at the end the moisture has evaporated. But in the intermediate period it is well consolidated, and there is no defect in this middle period, and this is why it falls only with difficulty during this middle period. It is the same for the fetus, because in the beginning the cotyledons[40] are weak and not well consolidated, and

38. Ar., *HA* 7.4 (584b1–3).

39. Cf. Ar., *Eth. Nic.* 4.2 (1121a2–3), 10.2–4 (1179b9–1175a17). Galen, *De sympt. causis* 1.6 (ed. Kühn 7: 115). Avic., *Can. med.* 1.2.2.19.

40. *Cottilidones:* While "cotyledon" is a botanical term, A. uses it frequently to refer to the point of attachment of the placenta to the uterus, perhaps what

in the end they are so stretched that they easily are broken, and this is why women easily miscarry both at the beginning and at the end of their pregnancies, but in the middle period they are well consolidated. This is why, etc.

To the fifth question one must reply that salt protects from putrefaction and consumes the viscous moistures, and nails are generated from these moistures. And this is why, if a pregnant woman uses salt, the moistures that should be converted into nails are consumed, and the fetus will lack nails.

Questions 24–28: On the fetus before and after birth.

Further one inquires whether a woman, once she has been impregnated, can be impregnated again before the birth.

And it seems not. Because after impregnation the opening of the womb closes, and therefore, etc.

The Philosopher says the opposite occurs in many cases.[41]

Second, one inquires why it is that twins often survive when both are male or both are female, but if they are of different sexes, often either one or the other dies.

Third, one inquires whether the fetus's head ought to exit first.

And it seems not. Because a lower part ought naturally to seek out the area below. But feet are lower parts. Therefore, etc.

The Philosopher <says> the opposite.[42]

Fourth, one inquires why, when the fetus is born, it immediately cries and places its hand to its mouth.

Fifth, one inquires why in the first age infants sleep a lot, almost all day and all night, and why they are particularly moved to sleep by lullabies and the use of a cradle.

---

we would call the villi (Hyrtl, 1880, 151–53). Cf. DA. 9.1.5.56,9.1.7.68,9.1.7.71; 10.1.1.7; 16.2.7.127–29; 17.1.3.16 (*SZ* 1: 794; 799; 800; 829–30; 2: 1224–25; 1244). Gray (1977, 1164) used the term "cotyledons" to refer to the upper surface of the placenta.

41. Ar., *HA* 7.4 (585a2–14). Cf. Avic., *DA* 9.5 (fol. 44raE); A., *DA* 9.1.5.52–54 (*SZ* 1: 792–93).

42. Ar., *HA* 7.10 (587a24–25). Cf. Avic., *DA* 9.5 (fol. 43vbC); A., *DA* 9.1.7.69, 9.1.8.74, 76 (*SZ* 1: 799–800; 801; 802).

Last, why convulsions are particularly generated in infants.

To the first question one must reply that both a woman and a mare abound in semen, because they use a more noble nutriment, and this is why they can be impregnated twice. This is why the Philosopher says that in many places it happens that a woman gives birth to one child and then three months later gives birth to twins.[43] This would not occur unless after the first impregnation there is a second.

To the argument one must reply that although the opening of the womb closes naturally after the first impregnation, owing to the very great pleasure that she has during intercourse it often happens that she opens it, and also it happens on account of the penis [*veretrum*] opening its gate, etc.

To the second, one must reply that a male is produced from a stronger heat than a female is, and this is why it is formed and fashioned more quickly in the uterus than a female, and because of its quicker formation its exit also follows more quickly. And this is why, when it happens that in the uterus there are both male and female at the same time, either the female blocks the male's exit or the male accelerates the female's exit, and this is why often it happens that one or the other dies. When, however, both are male, then they agree in their matter and in their active cause, and in such a case they are formed at the same time and consequently have an almost simultaneous exit, and for this reason they live longer, healthier, and better than do other twins. And because the matter is the same and resembles the agent [cause], this is why so often twins like this are alike in their bodily appearance and in many of their dispositions.

To the third, one must reply that the upper parts like the head are more quickly formed in the uterus than are the lower parts, and they are larger in the uterus relative to the lower parts, and as a result they are heavier, and every heavy object inclines downward. This is why when childbirth approaches, the fetus turns its head downward. And in addition to this is the cause noted by Avicenna, that the fetus struggles to exit on account of the number of its needs. For it desires food, which it

43. Ar., *HA* 7.4 (585a12–21); cf. A., *DA* 9.1.5.54–55 (*SZ* 1: 793–94).

does not have in great quantity, and it desires to draw in air, which it does not have, and in the same way it wants more space, which it does not have. But the pathway for food and air is in the head, and the head is larger than the other members, and this is why the head naturally turns to the opening of the vulva.

To the fourth, one must reply that everything that is unaccustomed is a source of pain and is as if contrary to nature. But the child [*partus*] or fetus in the uterus exists as if in a warm bath, and when it is born it is exposed to the cold air, which is just like something contrary to it, and this exposure is the cause of its cry. In like manner, it naturally desires food and desires to draw in air, and nature incites and almost persuades the fetus that its mouth is the pathway for food and respiration, and this is why it naturally places its hand to its mouth when it emerges.

To the fifth, one must reply that the cause of sleep is the elevation of vapors to the brain and their coagulation on account of the brain's coldness, because "sleep is the binding up of the senses" that are external, according to Aristotle in *On Sleep and Waking*.[44] And the vapors that are condensed in the head because of the brain's coldness block the pathways for the spirits, which administer the senses, and this is why the external senses immediately cease from their operations and sleep is induced, and vapors like this abound in infants. Nevertheless, not every vapor like this is a cause for a long sleep but only a particularly sweet and soothing vapor, and not a pungent one. For in the senses there are many vapors and fumes that are choleric and melancholic and very capable of penetrating the membranes [*miringae*] of the brain.

To the sixth, one must reply that things to which we are accustomed are pleasing, and the fetus in the uterus is constantly in motion because it is moved and carried along by the womb's movement. This is why movement pleases it when it is outside the uterus. But another reason is that one intense operation withdraws from before another. In this sense, the infant's movement causes an arousal of the heat, an elevation of vapor, and a weakening of the brain, and this movement is therefore a cause

44. Ar., *De somno et vigilia* 3 (458a28–29).

of sleep because it makes the external senses cease from their operations.

And in addition an internal pleasure deflects the external powers from their operations, and a gentle motion and lullabies cause pleasure inside the infant. Therefore, the powers withdraw with heat and spirit to the interior and, as a result, the exterior parts are abandoned and are almost deprived of their influence [*a suo regimine*]. Therefore, they sleep, etc.

To the seventh, one must reply that a spasm is caused by a weakness in the nerves.[45] And this can occur from too much dryness, just as it is evident that a piece of parchment placed close to the fire is quickly curled up and becomes wrinkled owing to the moisture's consumption. Or it can occur owing to an excessive moisture and softness in the nerves, and this is the cause in infants because they live on a moist nutriment and use it almost continuously, like brute beasts. This is why spasms occur in them quite often, owing to the weakness and softness of their nerves. If this were universal throughout all their parts, "they would die within seven days," as is said in the text.[46] If, however, "they escape the seventh day," they are frequently cured, but they suffer more on the fourteenth day because the moon, which has dominion over all that is moist, has a greater effect on the fourteenth day.

45. Cf. A., *DA* 9.1.8.80 (*SZ* 1: 803–4).
46. Ar., *HA* 7.12 (588a8).

BOOK TEN

Question 1: Whether the mother or the father is more responsible for an obstacle to generation.

"PERHAPS IT HAPPENS to certain men," etc.[1] Here a determination is made concerning an obstacle to generation. A question is first raised concerning this issue in the tenth book, namely, whether the mother or the father is more responsible for an obstacle to generation.

1. And it seems that it is the father. Because the more noble a thing is, the more things are necessary for its operation. But the male is nobler than the female, and therefore more things are required for its activity. But that which requires more things can also be impeded by more things; therefore, etc.

2. Moreover, the fetus has various configurations and lineations. But the man's sperm is more viscous and globular, and as a result it is more antithetical to a linear shape. Therefore, etc.

3. Moreover, sperm is a superfluity of the nutrimental moisture. But the male has less of this moisture than the female, and there is less of it in thin people than in fat people, and women have more fat than men, and therefore, etc.

But the Philosopher says the opposite.[2]

One must reply to this that more obstacles are owing to the mother than the father, because the man is the cause of generation through the separation of his semen, whereas the woman is a cause of generation to the extent that she condenses and leavens the [separated] sperm and protects the fetus. And this is why she can impede generation in several ways.

Moreover, an appropriate disposition of the womb is required

1. Ar., *HA* 10.1 (633b12f.). Cf. A., *DA,* 10.1.1.1 (*SZ* 1: 827).
2. Ar., *HA* 10.1–5 (633b12f.). Cf. A., *DA,* 10.1.1–2.1–25 (*SZ* 1: 827–37).

for the production of a fetus. But the womb is a noble member, and this is why it is very prone to injury. It is a nerve-filled member, and as a result it frequently suffers spasms, which are an affliction of the nerves. Moreover, it often abounds in bad humors, and this is why sometimes it suffers from wind, because when bad matter is concentrated in some member, then nature returns to that place to expel that matter utterly, if it can. And if it cannot, it dissolves the matter into windy vapors. The infirmity comes to an end if the pathways for expulsion are made available for these vapors. If, however, the pathways are not made available for them, then they cause great distress and are very much in motion. For it is the same for vapors that are shut up in the belly of the earth and are unable to come forth. These vapors cause an earthquake as a result of their enclosure and their agitation, but, if an exit is available for them, they do not cause an earthquake but rather pass forth into the wind in the air, or into something else. It is just this way for vapors enclosed in some member, for they then cause giddiness or something else.[3] Thus there are many causes stemming from the man that are responsible for an inability to become pregnant: It can be either from a defect in the semen—if it is too hot, or too liquid, or too cold—or from a defect in the generative members, such as when the penis is too long or too short or cannot become erect or because the testicles are cold. And there are similar and even more causes stemming from the woman, because she not only attends to the semen but also conserves it, and there can be a defect stemming from conserving it. Therefore, etc.

1. To the first argument one must reply that although several things are required for the completed action of the man, nevertheless the woman requires many more for generation, and therefore, etc.

2. To the second argument one should reply that two things impede the semen: too much moisture or too much dryness, whereas a globular semen occupies the mean, and therefore, etc.

3. To the third argument one should reply that although the

3. Cf. A., *Meteora* 3.2.6.

woman has more superfluous moisture, she nevertheless does not have more spermatic moisture. An indication of this is that a man can impregnate more frequently than a woman can be impregnated; therefore, etc.

Question 2: Whether the retention of the sperm is more harmful to men than is the retention of menses in women.

It is asked, further, whether the retention of the sperm is more harmful to men than is the retention of menses in women.[4]

1. And it seems that the retention of menses is more harmful. This is because the more superfluities there are that are more impure and unclean, and the more they are withdrawn from the work of nature, so much more harmful will their retention be. But such are the superfluities of the menses in comparison to the sperm, and therefore, etc.

2. Moreover, when nature is deprived of a normal operation, then it is more distressed. But nature has ordained specific times for the flow of the menses, but not for the emission of sperm. Therefore, the retention of menses is more harmful than the retention of sperm.

On the contrary, the nobler something is, the more harmful it may be when it departs from the course of nature. But sperm is nobler than the menses. Therefore, the retention of sperm, etc.

To this it must be said that sperm and menses can be considered in two ways: either as they exist in their natural disposition—and thus the sperm is cleaner and purer and its retention causes more injury than the other's—or they can be considered as they exist beyond the disposition of their nature, and in this way the menses is even worse. But then the sperm is held in greater contempt, and the reason for this is that the nobler something is, the more nature hates its corruption and the more it is damaged by its injury. This is why a fever stemming from blood is worse and more dangerous than one stemming

4. Cf. A., *DA* 10.1.1.12–16 (*SZ* 1: 831–33).

from phlegm, bile, or melancholy. Nevertheless, blood is, of all the humors, nature's best friend, yet when it putrefies it is hated more for the same reason, etc. Thus it is in the first premise, etc.

1. In this way a response to the first argument is clear: that although the menses is more impure, its retention nevertheless injures nature less than the unnatural retention of the sperm.

2. To the second argument one must reply that the determination or indeterminacy of a set period for release is not the reason why the retention of sperm is more harmful, but its own bad quality when it is beyond nature's disposition.

Question 3: Whether the emission of semen weakens the male more than the female.

Further one asks whether the emission of semen weakens the male more than the female.[5]

1. And it seems that it weakens the female more. This is because where there are more emissions, there will be greater weakness. But the male only emits semen; the female, however, emits sperm and menses. Therefore, etc.

2. Besides, the one who takes greater delight in intercourse will be weakened more, according to Avicenna.[6] The reason for this is that, as a result of pleasure, one emits both the necessary and the unnecessary, but the female takes the greater pleasure, and is therefore weakened more.

On the contrary, semen is a superfluity of moisture. But weakness stems from the consumption of moisture and heat. Since, however, there is less moisture in the male than in the female, the male is weakened more, and therefore, etc.

One must reply that greater weakness occurs in the males. The reason for this is that that which is separated from the male is purer and is nearer to a conversion of its substance, and the further such a thing is separated from something, the nearer

5. Cf. A., *DA* 10.1.2.21 (*SZ* 1: 835).

6. Avic., *Can. med.* 3.20.1.11.

it is to weakness, and this is why the weakness is greater in the male.

1. On to the arguments. To the first, one must reply that although the woman has two emissions, the single one separated from the male outweighs these two.

2. To the second, one must reply that the pleasure is greater in the woman in a quantitative sense, but in a qualitative sense it is greater in the male, because the male's pleasure is more ordered than the female's. And in this way one can reply to the first argument that the female emits more in a quantitative sense but the male emits more in a qualitative sense. And this is why, etc.

Question 4: Whether a fetus can be generated from one part of the sperm.

Further one asks whether a fetus can be generated from one part of the sperm.[7]

1. It seems not. Sperm is separated from individual members, and this is why the fetus has individual members. Therefore, if some part were removed from the sperm, then a part like the one that has been removed from the sperm will not exist in the fetus. Therefore, etc.

2. Avicenna makes this clear, saying that one who has been mutilated generates one that is mutilated, and this occurs only because there is no part in the sperm corresponding to the mutilated part.[8]

The Philosopher says the opposite in the text.[9]

One should reply that a perfect fetus can be produced easily from part of the semen, which is quite apparent in the case of twins, since neither one of them is produced from the semen in its entirety. And the reason for this is that when the semen flows from individual members to a specific place, that semen is homogeneous [*unigeneum*], and this is why (provided it is present

7. Cf. A., *DA* 10.1.2.22 (*SZ* 1: 835–36).

8. Avic., *Can. med.* 3.20.1.3.

9. Ar., *HA* 10.3 (637a8f.).

in the requisite quantity) there is a potency in it for the whole, and similarly there is a potency for the part. This is why the Philosopher says that when there is too much semen, "the womb draws to itself that which is sufficient,"[10] and it ejects the residue or casts it out beyond the womb's orifice. And this is why one often believes that a woman is pregnant when she is not pregnant.

1. To the <first> argument one should reply that since the semen is an aggregate, one part is not situated next to another part as it is in the whole animal. And this is why, even if one part is removed from the semen, the part corresponding only to the hand or only to the foot is not removed—for these parts are mixed in with the whole—but a part of the whole is removed, and this is why the fetus is diminished.

2. To the second argument one should reply that it is one thing to remove a part from the semen flowing from a whole body and another thing to cut off semen from a mutilated body, since a potency for the whole can be present in the first semen even if a part is removed, but in the second semen the power that forms that particular member is absent because there is no other power derived or separated off from the mutilated part.

Question 5: Whether a mole [molyn] *can be generated naturally in a woman from semen emitted through pollution in sleep.*

Further one asks whether a mole [*molyn*] can be generated naturally in a woman from semen emitted through pollution in sleep.[11]

10. Ar., *HA* 10.3 (637a3–5).

11. "Molyn" derives from the accusative of the Greek, *mylē* (literally, "millstone") and was thus translated into Latin as *mola*, "millstone." The hard formation in a woman's uterus was often called called a "mooncalf" in English, but conventionally the Latin *mola* has been simply transliterated as "mole." While that convention is followed here, this word is unfortunate in that it conveys to modern ears the sense of a minute, surface blemish. Mole is discussed in *DA* at 10.1.4.31–36. and 18.2.6.82–83. (*SZ* 1: 839–41; 2: 1321–22). Cf. Ar., *HA* 10.7 (638a5f.); Avic., *Can. med.* 3.21.2.18–19; *DA* 11, fol. 44vb; A., *DA* 10.1.4.31–36 (*SZ* 1: 839–41).

1. It seems not. Because nothing is produced naturally unless it is intended by nature, since nature does not act without an intention for an end; but *molyn* or mole is not intended by nature, and therefore, etc.

2. Moreover, pollution during sleep results from the imagining of something pleasurable, like a beautiful young girl perceived through the senses or imagined in a waking state. But the imaginative power is an animal power, and the cause of a mole is a natural (that is, generative) power. Therefore, a mole will not be able to be generated from semen emitted through pollution.

3. Moreover, such generation occurs only in birds and <not> in a woman, because a hen conceives and produces a wind egg through pollution, as is said in the text.[12] But as often as she experiences a pollution, she produces a wind egg. Therefore, if a mole were produced in this way, as often as she experienced a pollution she would generate a mole. But this is false, and therefore, so too is the premise.

The Philosopher says the opposite.[13]

One should reply that a mole is a certain mass of flesh generated only from the semen of the female without the male's sperm, because while sleeping a woman frequently emits semen as far as the neck of the womb and the womb will sometimes then draw that semen into its depths and deal with that semen with its natural power, to the extent that it can. But because of the absence of the male sperm, in which there is a power that produces the sensitive soul, this mass lacks a sensitive form. Nevertheless it is nourished and increased by the woman's power and her nutriment to the extent that the woman is believed to be pregnant. This often lasts for a year and frequently for two years. We have seen a woman who had it for four years and after this she died, since no physician had the power to apply a cure. And very often this condition lasts until death, since a cure for it is difficult or impossible even for the most accomplished physicians, because it adheres so strongly to the womb, and because

12. Ar., *HA* 10.6 (637b33f.); 10.7 (638a27f.).
13. Ar., *HA* 10.7 (638a5f.).

it is connatural to it and is deeply rooted at its base. It is, however, commonly cured more effectively by dysentery or by diarrhea than it is by any medication. Thus I have observed skilled physicians treating it [in this way] and it was effective. The reason for this is that in dysentery there is a pronounced movement of the humors to the intestines—especially of the choleric humors—which eat away at and penetrate the intestines, and as a result the cotyledons [*cottilidones*] containing or anchoring the mole or *molyn* are ruptured or scraped away by the bile as if with a razor, and as a result it flows out. In a like manner a mole can be cured by a menstrual flow, because this mass is nourished by the menstrual blood, and thus when that blood or nourishment is lacking, the mole's cotyledons become attenuated and are made thinner, and, as a result, they are easily ruptured.[14] This explains likewise why a pregnant woman, if she is bled too much or suffers a menstrual flow, often experiences a miscarriage, because, once the blood has been diminished, the fetus is weakened, and the cotyledons by which the fetus is anchored to the womb are weakened, and then, having been destroyed, the fetus comes forth.

The premise, then, is correct.

1. To the first argument one should reply that something can be intended by nature in two ways: in a primary or secondary way. In a primary way, nature does not intend to produce a mole when it holds back the semen, but rather a fetus. This is why the woman emits semen at the neck of her womb, to meet the male's sperm, dreaming that she is having intercourse with a man. But since the male sperm is absent, nature is frustrated with respect to its first intention and thus intends to make what it can, and it is better to make a mole—that is, this fleshy mass—than nothing at all, or so it seems to nature. And this is why, etc.

2. To the second argument one should reply that pollution sometimes occurs simply from imagining a thing that is especially desired, but nevertheless this pollution only occurs along with the mediation of a natural expulsive power, since then a

14. On the cotyledons see above, *QDA* 7.33–39; 9.19–23.

natural power is obeying an imaginative power. But pollution often occurs when awake, without imagination, as has been found among nuns. Thus one must reply to this argument that the emission of semen does not occur in sleep without the operation of a natural power, like the expulsive power, and this is why, etc.

3. To the third, one should reply that the wombs in birds are always positioned uniformly, namely, next to the diaphragm, and this is why in them any semen emitted through pollution is always directed to one place, whether in sleep or in the waking hours. But the wombs in women are not always positioned so uniformly, since, as physicians are aware, they may rise up at one time and descend at another. This is why they are not always properly disposed to receiving or conceiving a mole, in the way that birds always conceive a wind egg from pollution, etc.

One should know further that a wind egg has a shape and disposition similar to that of a real egg, with the difference that it is not fit for producing a chick. This is not the case for the mole and the true fetus. And the reason for this is that birds are generated through a medium—namely, through the material of the egg—and do not form a fetus in the uterus. Rather, the fetus is formed outside the uterus. This is why the egg is the remote cause of its fetus, and this is why a hen can easily produce an egg like this without the assistance of the cock, but owing to the absence of the cock's semen, which forms the fetus, the egg is not fit for generating a chick. But a woman gives birth without a medium, for her fetus is formed in the womb and not outside of it. This is why, just as a chick is not produced outside the uterus in the case of the wind egg, neither can the mole's material be rendered in the likeness of a true fetus in the uterus. But in some locales, as it is said and as I have heard, a toad is always produced with the mole, since its matter is better suited for the production of a toad. Wherefore, as the story goes, the mole is generated by means of pollution, from sperm sent to the opening of the vulva and then drawn back again and anchored to the base of the womb along with the uncleanliness of the womb, and there it grows.

BOOK ELEVEN

Question 1: Whether there is a double mode of proceeding in science.[1]

"In every noble opinion," etc.[2] In this eleventh book one makes a determination regarding the scientific process. This is why in this eleventh book one inquires first whether there is a double mode of proceeding in science: one, descriptive,[3] and the other by assigning causes.

1. And it seems not. Because a demonstration is a syllogism that creates knowledge [*scire*]. Since, therefore, every science creates knowledge, then every science will be capable of leading to demonstration, and as a result none will be descriptive.

2. In addition, "to know is to recognize a thing's cause."[4] Whoever implants knowledge [*scientia*] implants knowledge [*notitia*] of the cause. Therefore, there will not be a descriptive process in any science.

3. In addition, all of our cognition arises from sensation. But

1. "In science": *in scientia,* which can also mean "in knowledge," but with this book A. proposes to examine the scientific method or process by which knowledge is acquired. A. employs several various terms, however, which can be translated as "knowledge." Whenever feasible, we will attempt to alert the reader as to A.'s shifts in vocabulary, either through notes or by leaving the Latin in brackets, while frequently using "knowledge" or "know" for a Latin word such as *scire* (and *scientia*), "cognition" or "come to know" or "recognize" for *cognitio* or *cognoscere.* No scheme will be perfect, however, and the interested reader is directed to the original.

2. Ar., *Part. An.* 1.1 (639a1f.). Cf. Avic., *DA* 11 (fol. 44vb); Averroes, *Part. An.* 1.1; A., *DA* 11.1.1.1–4 (*SZ* 2: 857–58).

3. "Descriptive": *narrativus,* i.e., through a narrative or descriptive approach, rather than by an investigation of causes.

4. Ar., *Anal. Post.* 1.2 (71b9f.), cf. Boethius's translation (PL 64: 713D). Cf. A., *Anal. Post.* 1.3.6; *Phys.* 1.1.1.

sensation is able to recognize accidents but not causes. Therefore, in no science is there recognition of causes.

The Philosopher says the opposite and argues by reason.[5] According to the Philosopher in *On Sophistical Refutations* [*Elenchis*], the wise person has two functions: one is "not to tell falsehoods about things he knows," and the other "to be able to expose one who tells falsehoods." But the first occurs by describing, and the second by assigning causes. Since, then, the functions of a wise person are necessary in every science, there will be a double process in every science.

One must reply that there is a double process in science. And this is clear both with respect to the object [*res*] and with respect to us. With respect to the object, because something is presupposed in every science, as a foundation for that science, and it is from this foundation that the causes of the consequences are subsequently assigned. But suppositions are conveyed only by describing them, whereas consequent conclusions are only conveyed by assigning causes. Therefore, etc. With respect to us, each process is required, since description pertains both to teaching and to learning, when one is in doubt, to seek the causes. And this is why the Philosopher, as the one most wise and expert in the sciences, proceeds in science first by describing and second by seeking after and assigning the causes of the things that have been described, and in so doing points out or hints at what we ought likewise to do.

1. On to the arguments. To the first, one must reply that demonstration is of two kinds: namely, *propter quid* and *quia*. A demonstration *propter quid* occurs through the cause, but a demonstration *quia* is considered through the effect or from a supposition.

2. To the second, one must reply that the Philosopher defines the most powerful knowledge in that spot when he says, "To know is to recognize a thing's cause," etc.

3. To the third, one must reply that although our knowledge [*cognitio*] arises from sensation, nevertheless our intellect

5. Ar., *Soph. El.* 1 (165a23f); cf. Boethius's translation (PL 64: 1009B).

can [achieve knowledge] of many things by discursive thought about which things sensation is ineffective. Therefore, etc.

Question 2: Whether the descriptive process is more necessary than the process that assigns causes.

One inquires further as to which process is more necessary.

And it seems to be the descriptive process. Because principles and suppositions are considered through description, and conclusions are considered through the assignment of causes. Since, then, principles are more necessary than conclusions, a knowledge [*cognitio*] of principles is more necessary than a knowledge of conclusions; this is why, etc.

In addition, "it is necessary that the one learning believe," according to the Philosopher in the *On Sophistical Refutations*.[6] But this is only the case if the descriptive mode is necessary, and therefore, etc.

The Philosopher says the opposite.[7] For he says that the descriptive process is for the sake of teaching or for the sake of the one teaching, but the process that assigns causes is for the sake of both. Therefore, the second is more necessary.

To this, one must reply that in one way the first process is more necessary, and in another way the other is. The second process is more necessary for an advanced audience but the descriptive process is more necessary to those who are less advanced. In fact, the natural process is from cause to effect, and for learning about a particular thing the natural order is first to propose a conclusion and second to prove it, and this is why, speaking naturally, the descriptive process is prior, and the other process is second. And because a first thing can exist without a later thing, the first process is therefore more necessary, since knowledge can be introduced through description without knowledge of the cause; but knowledge of the cause cannot exist without some description, because it cannot be introduced without a supposition of the principle, and this is why, etc.

6. Ar., *Soph. El.* 2 (165b3f.); cf. Boethius's translation (PL 64: 1009C).
7. Ar., *Part. An.* 1.1 (639a2f.).

Question 3: Which cause the natural philosopher ought to consider more.

On the supposition that the natural philosopher considers all causes, one next asks which cause he should consider more.[8]

1. And it seems it should be the material cause. Because the natural philosopher has especially to consider motion, and matter is the principle of transmutation. Since, then, matter is especially the principle of transmutation, he especially has to consider matter.

2. In addition, "logic and physics define things differently," according to the Philosopher in the book *On the Soul,*[9] because logic defines through form, and physics defines through matter.[10] But one should particularly consider the cause by means of which one defines. Therefore, etc.

On the contrary. According to the Philosopher in the second book of the *Metaphysics,* the principles of being and of knowing are the same.[11] But form is more the principle of being, and also, therefore, of knowing. But physics has to consider a principle like this more, and therefore, etc.

Moreover, the Philosopher proves that he has to consider the final cause first, because each thing depends on its end.[12]

One must reply that physics has to consider all causes, although it has to consider diverse causes in diverse respects. For if a thing is considered with respect to its generation, then physics especially has to consider the material and efficient causes because these causes are presupposed for generation. If, how-

8. Cf. Ar., *Part. An.* 1.1 (640b4f.); A., *DA* 11.1.3.27–43 (*SZ* 2: 867–74); *Phys.* 2.1.9.

9. Ar., *De anima* 1.1 (403a29f).

10. "Physics": *physicus.* In fact, the Latin does not denote the discipline, but its practitioner—not physics, but the natural philosopher engaged in the study of physical nature. Here and later, however, we will regularly translate these and related nouns (*physicus, metaphysicus,* etc.) as denoting the discipline, in order to clarify the differences between these divisions of philosophy. Likewise, *logicus,* a logician, is translated as "logic."

11. Ar., *Metaph.* 2.1 (993b28f.).

12. Ar., *Part. An.* 1.1 (639b3f.). Cf. A., *DA* 11.1.2.14–16 (*SZ* 2: 861–63).

ever, a thing is considered with respect to its essence, then physics most especially has to consider the formal cause. If, however, a thing is considered with respect to its operation and property, then physics especially has to consider the final cause. And because in this book the Philosopher inquires into the properties and operations of animals, he therefore principally considers the final cause here.

1. To the first argument one must reply that the principle of transmutation can be either active or passive. Thus, although matter is particularly the passive principle of transmutation, it is not the active principle, and this is why, etc.

2. To the second argument one must reply that logic defines by means of the parts of the definition [*ratio*], that is, by means of genus and *differentia,* and each of these is received from the form. And this is why it is said to define according to the form. But the natural philosopher defines by means of the parts of the essence, and considers something only in terms of its relation to matter. Therefore, in the book *On the Soul* he makes a determination about the soul only insofar as it is the act of a physical body.[13] And this is why he is said to define by means of the matter, because he defines by means of the parts integrating the essence.

Questions 4–5: Which cause is naturally prior. On the number of causes.

One inquires further into which cause is naturally prior.[14]

And it seems to be the material cause. Because what is ungenerated and incorruptible is prior to that which is corruptible and able to be generated. But matter is of this sort, and form is not, and the same holds true for the efficient and the final cause, and therefore, etc.

It seems to be the efficient cause. Because that which is the cause of all other things appears to be prior. But the efficient cause is like this, and therefore, etc.

13. Ar., *De anima* 2.1 (412a19f.).

14. Cf. A., *DA* 11.1.3.27–51 (*SZ* 2: 867–76); *Phys.* 2.2.5.

It seems to be the final cause. Because that which moves first is the first cause. But the final cause is like this, and therefore, etc.

It seems to be the form. Because *ens* and *esse* precede all other things. Therefore, that which is the first cause of being [*esse*] is first among the causes. But the form is like this, and therefore, etc.

And the same sort of inquiry can be made into the number of causes.

To the first, one must reply that something is prior to something else in two ways: either in completion or in generation. With respect to completion, the final cause is prior and the efficient cause is second, while the form is third, and the matter is last, because the efficient cause is called a cause only because it moves. But it moves only because it is moved earlier by the final cause. Thus, the final cause [*finis*] is the reason that there is an efficient cause. In addition, one generating generates one like itself. Therefore, the reason that the form of the one generated is as it is, is that the efficient cause was as it was. Therefore, the efficient cause is the reason why a form such as this is introduced. But matter is for the sake of form. Therefore, the final cause moves the efficient, and the efficient determines the form, and the form determines its matter. But the path of generation is just the opposite, because in order for something to come to be, first an agent is required, and, second, something on which it may act is required, and once these are present, it produces an effect. But it acts only by intention, and this is why the intent follows last, and why the efficient cause is first, and then the matter (as one undergoing some change [*patiens*]), and third is the form that is introduced, and last is the end which is intended, because the end is that which follows as a consequence.

In this way a solution to the arguments is evident.

One must reply to the second argument in a variety of ways, because each thing's cause is either internal or external. If it is external, then it is either the "cause by which," and this is the efficient cause, or it is the "cause for the sake of which," and this is the final cause. If it is internal, either it is act, and this is the form, or it is potency, and this is matter.

Or, in another way: Every cause is either a cause of being or a cause of becoming. If it is a cause of being, since being [*ens*] is divided into act and potency, then it is either a cause of being in act—and thus form is as a prior thing—or it is a cause of being in potency—and thus is matter. If it is a cause of becoming, then it is either an operating cause, and this is the efficient cause, or it is that for the sake of which it operates, and this is the final cause.

The Philosopher solves this in another way in the second book of the *Physics* where he says that "there are as many causes as there happen to be ways of inquiring why [*propter quid*] something occurs."[15] But it happens that there are four ways in which one can inquire why something happens, and this is why there are four [causes], etc. So much for this point.

Question 6: Whether the Philosopher ought to treat the mode of dividing here.

One further inquires whether the Philosopher ought to treat the mode of dividing here.[16]

1. And it seems not. For the Commentator, in the second book of the *Metaphysics,* says it is inappropriate to seek to teach and to inquire into the mode of teaching at the same time.[17] But in this book the Philosopher divides the animals, and as a result treating the mode of dividing is not part of this book's task.

2. Moreover, it is one thing to build a plough and altogether something else to use a plough, according to the Philosopher in the second book of the *Physics.*[18] Therefore, for the same reason, it is one thing to treat the mode of dividing and another thing altogether to use division.

15. Ar., *Phys.* 2.7 (198a14f.). Cf. A., *Phys.* 2.2.22.

16. Cf. A., *DA* 11.2.1.52–57 (*SZ* 2: 877–79). On the process of "division," see especially Michael W. Tkacz (1993), 130–59.

17. Averroes, *Metaph.* 2.15; Ar., *Metaph.* 2.3 (995a13f.).

18. Ar., *Phys.* 2.2 (194b5f.); A., *Phys.* 2.1.11. The word for "plow," *temo,* more accurately refers only to its central beam.

The contrary is apparent if one follows the determination made by the Philosopher.[19]

One must reply that some divisions follow after properties of form and others follow after properties of matter. The first division pertains to logic and to first philosophy, that is, metaphysics, but the second division pertains to physics.[20] This is why the Philosopher does not use the first division here but only the second mode of dividing, for he divides "animal" into blooded and bloodless and into winged and not-winged. And these are material divisions, and this is why they can be pertinent to the natural philosopher. Generally one can say that there is a threefold order of proceeding in science, namely, an order of division, definition, and synthesis:[21] Division is with respect to the genus, definition with respect to the species, and synthesis with respect to the properties and accidents. And these modes pertain to logic as far as teaching goes, but with respect to their actual use they can belong to other branches of philosophy. Therefore, the fact that the Philosopher treats here the art of defining[22] is not with respect to the physical world but with respect to form, and here he dons and arms himself with the habit of the logician, and this is why, etc.

Question 7: Whether the final differentia *is convertible*[23] *with the species.*

One further inquires whether the final *differentia* is convertible with the species.

1. And it seems not. Because the definition is said to be the boundary [*terminus*] of the thing. Beyond the boundary there is nothing, because nothing of the essence is beyond the definition [of the thing], but whatever belongs to its essence is with-

19. Ar., *Part. An.* 1.2 (642b8, 15, 19f.).

20. See *supra,* n. 10.

21. "Synthesis": *collectivus,* which suggests a process of gathering and collecting data.

22. Ar., *Part. An.* 1.1 (639a23f.). Cf. A., *DA* 11.2.1.52–72 (*SZ* 2: 877–85).

23. "Convertible": *convertibilis,* i.e., interchangeable with or capable of taking the place of the species.

in it. Since, then, several *differentiae* are placed in many definitions, if the final one were convertible [with the species], then the others would be superfluous or outside its essence.

2. In addition, the part is not convertible with the whole. But the *differentia* is part of the species, because the species is constituted from genus and *differentia,* and this is why, etc.

3. Moreover, each part of the definition is present plurally [*in plus*], whereas the whole is present equally [*in aeque*], according to the Philosopher in the second book of the *Posterior Analytics.*[24] But this would not be so were the final *differentia* convertible [with the species], and this is why, etc.

The Philosopher says the opposite.[25]

To this, one must reply that individual and specific *differentiae* have to be derived from the specific forms of things, and as a result a comparison of the form itself to the form considered essentially and a comparison of the *differentiae* to the species considered rationally will be the same. And therefore, just as the specific form does not extend beyond the thing formed—namely, that which was formed in the subordinates [*in suppositis*]—because the form is not found beyond the thing formed nor does the thing formed exist without form (for one does not find a cow apart from the form of cow nor the form of cow apart from a cow, and the same is true for the ass and others like this). And it will be the same for a specific *differentia* with respect to the species, [namely] that that *differentia* does not go beyond the species in subordinates, nor does the species go beyond it. Nevertheless, a *diffferentia* is not convertible in this way with the species, because then it would include the entire understanding of the species within itself. And the *differentia* alone is therefore not sufficient for a definition, but rather the genus is joined to it, and to another *differentia* in many cases. Sometimes there are several forms in one and the same thing, or there is one form which has several operations, and then the *differentia* derived

24. Ar., *Anal. Post.* 2.13 (96a33f); cf. Boethius's translation (PL 64: 755A). Cf. A., *Anal. Post.* 2.4.1.

25. Ar., *Part. An.* 1.2–3 (642b5f.). Cf. Averroes, *Part. An.* 1.4; A., *DA* 11.2.1.52–53 (*SZ* 2: 877); *Metaph.* 7.4.3; *Anal. Post.* 2.4.1.

from one form does not include an understanding of a *differentia* derived from another form, just as the form does not [include the understanding of another] form. Thus the intellective [form?] does not include the understanding of the sensitive form just as a rational [body?] does not include the understanding of the *differentia* of what is subject to sensation.[26] And this is why if "rational substance" or "rational body" is said, it would not adequately express the understanding of a human, as it is expressed when he is called a "rational animal," because "animal" includes the understanding of the sensible.

Nevertheless, one must understand that forms of natural things are hidden from us, and this is owing to their union with matter. And this is why, when inquiring into forms, the process of natural philosophy proceeds from effects, because the sensible effects are better known to us than the forms themselves, and because the forms are hidden from us we receive few *differentiae* from the forms, but rather receive almost all of them from the accidents. For something rational is received from reason, which is a power of the soul, and the same is so for the sensible. And this is why in our works few *differentiae* are mentioned that are convertible with the species unless they are accidental *differentiae,* like "capable of laughter"[27] or "capable of whinnying," because the accidents are known to us. But if the *differentiae* were named, they would be convertible with the species, just as has been said.

1. On to the arguments. To the first, one must reply that when a later *differentia* does not include a prior one, then the name is not applied to the whole composed from the genus and the superior *differentia,* and then it is necessary to posit several

26. The Latin is difficult here because of shifts in the gender of the adjectives employed. *Unde intellectiva* [feminine] *non includit intellectum sensitivae, sicut nec rationale* [neuter] *includit intellectum differentiae huius quod est sensitivum.* Our translation is one attempt, but surely not the only one possible, to make sense of the passage by anticipating the use of *corpus* in the next sentence.

27. What distinguishes a *differentia* like "capable of laughter" [*risibilis*] is that it is understood by most philosophers to be a proper predicate of the human, i.e., one that can only be predicated of the human being, and therefore in some sense it is convertible with the species.

differentiae in the same definition. Therefore, if the name "animal" were not applied to "animated sensible body," but only to "animated body," it would be necessary to posit in the definition of "human" both "sensible animal" and "rational animal." And this is why, because there is not one name applied to animated substance, "animated" and "sensible" are posited in the definition of animal.

2. To the second argument one must reply that something can be a part in many ways, according to the Philosopher in the fourth book of the *Metaphysics*.[28] Moreover, the *differentia* is a part of the definition [*ratio*], and thus, as it is a part, so is it not convertible with the species. Therefore, it is not converted in the species according to the definition, because it is a part of the definition, but it is converted in the subordinates so that it is not found in any subordinate in which there is not a species in which the final *differentia* is not present, etc.

3. To the third argument one must reply that in the second book of the *Posterior Analytics* the Philosopher accepts general *differentiae*[29] and each *differentia* is present in ones such as these plurally [*in plus*].

Question 8: Whether there is more pleasure in the knowledge [cognitio] *of superior or inferior things.*

One inquires further whether there is more pleasure in the knowledge of superior or inferior things.

1. And it seems that there is more in the knowledge of inferior things. For pleasure is caused by the union of the beloved with the lover. Where there is greater union or knowledge, there will be more pleasure. We have, however, more knowledge of inferior things (because they are near to us) than we do of superior ones, and therefore, etc.

2. In addition, just as a proportionate thing that is knowable pleases the cognitive power, so too a thing that is excessive and

28. Ar., *Metaph.* 4.25 (1023b12f.).

29. "General *differentiae*": *differentiae communes,* i.e., shared, common, or universal.

disproportionate corrupts that power. This is according to the Philosopher in the second book of *On the Soul*.[30] But superior things surpass the cognitive power more than lower things do. For he says in the third book of *On Heaven and Earth* that the more distant they are from us in terms of spatial distance, the more distant they are from us in terms of knowledge [*cognitio*].[31]

The Philosopher says the opposite.[32]

One must reply that greater pleasure consists in a modest knowledge of superior things than in a greater knowledge of inferior things, as the Philosopher says.[33] And the reason for this is that something acquired with greater difficulty is possessed with greater pleasure. But we acquire knowledge of superior things with greater difficulty than we do knowledge of inferior ones. Therefore, knowledge of them, once acquired, is possessed with greater pleasure.

In addition, the soul is, in some way, all things, for, as the Philosopher says in the book *On the Soul*, sense is the sensibles and "intellect is the intelligibles."[34] But there is a natural desire in the intellective soul with respect to all intelligible things. Therefore, the more intelligible something is, the more desire there is in the soul with respect to it. But superior things are more intelligible in and of themselves, because "each thing is intelligible when it is separable from matter," and superior things are more separable from matter than inferior ones.[35] But the more something is desired before it is possessed, the more it is loved once it is possessed, and pleasure is a sort of rest in the thing loved. Therefore, there is more pleasure in the apprehension of superior things, etc. And this is why the Philosopher says in the text that we love to know a little bit about superior things more than we love to know a lot about inferior ones.

30. Ar., *De anima* 2.12 (424a28f.); A., *De anima* 2.4.1.

31. Ar., *DC* 2.12 (292a14f.). Cf. A., *DCM* 2.3.13.

32. Ar., *Part. An.* 1.5 (644b31f.). Cf. Averroes, *Part. An.* 1.5; A., *DA* 11.2.3.82–83 (*SZ* 2: 888).

33. Ar., *Part. An.* 1.5 (644b34f.).

34. Ar., *De anima* 3.8 (431b22f.), 3.4 (429b30f.).

35. Ar., *De anima* 3.4 (429b21f.).

1. On to the arguments. To the first, one must say that there is more pleasure where there is a greater union of the thing loved with the lover, unless there is a another that is loved more, because there is a greater pleasure in a brief union of something greatly loved with the lover than there is in a long union of the lover with something loved less.

2. To the second argument one must reply that superior things are less well known to us the more distant they are from us, and yet we more greatly love to know something about them, and they do not so far exceed our intellect as to corrupt it. Because sense differs from intellect in this respect: namely, that sense, as a result of perceiving a sensible that exceeds it, fails to perceive other sensibles. Indeed, it is sometimes corrupted. But this is not the case for the intellect, because the more the intellect understands things that are more intelligible and more excellent, so much more, then, does it understand these lower or lowest things. And the reason for this is that it is not a corporeal power. Therefore, if sense were not an organic power, it would not be corrupted by a sensible that exceeds it; etc.

Question 9: Whether our intellect can achieve knowledge [cognitio] *of the first cause.*

One inquires further whether our intellect can achieve knowledge of the first cause.

1. And it seems not. Because nothing is in the intellect that was not first in sense. But the first cause does not fall under sense, and therefore, etc.

2. In addition, "just as the eye of the night owl is to the light of the sun or of the day, so is our intellect to those things which are the most evident in nature," according to the Philosopher in the second book of the *Metaphysics*.[36] But it is not possible for

36. Ar., *Metaph.* 2.1 (993b9f.). Cf. A., *Metaph.* 2.2. The night owl, *nocticorax*, is not the same as an owl (*noctua*) for A. Cf. *DA* 23.130(84) (*SZ* 2: 1639–40). Complicating matters is the confusion of two names. Isid., *Orig.* 12.7.40–41, uses *noctua* more generally and says he specifically equates the two (*nicticorax ipsa est noctua*), a statement repeated by ThC 5.92.2 (*nocticorax* is A.'s spelling). Rabanus Maurus, *De universo* 8.6, clearly states that some call the *noctua* a "sea

the night owl's eye to sustain and perceive the light of the sun or of the day without being injured or corrupted, and therefore, etc.

3. Moreover, the infinite is not intelligible by our intellect. The first cause is infinite, and therefore, etc.

The Philosopher says the opposite in the text and argues according to this reasoning: Nature does nothing frivolously or in vain.[37] Since, then, the first cause is the first intelligible, if it is unintelligible to us its intelligibility will be in vain, just as a visible which cannot be seen will be made in vain.

Moreover, in the third book of *On the Soul* he says that the soul is all things—sense is the sensibles and intellect is the intelligibles.[38] Since, then, the first cause is intelligible, therefore, etc.

To this, one must reply that our intellect cannot achieve knowledge of the first cause with respect to its essence. And the reason for this is that the object is proportioned to the cognitive power and the cognitive power is threefold. One exists in matter and uses an organ. Another does not exist in matter and does not use an organ. And the third exists in matter but does not use an organ. The first is the sensitive power, the second is a separate intellect, just as of the first cause or separate substance, and the third is the human intellect, which is in matter (the human body) but does not use an organ. But there is a proportionality between these three and their objects. The material object corresponds to the first, in that it exists in matter. An object that is absolutely separated from matter corresponds to the second. And an object that exists in matter corresponds to the third, but not in that it exists in matter. Therefore, properly speaking, the object of our intellect is the quiddity of the substance or a ma-

---

crow" (*corvus marinus*) but that it is not the same as a *bubo,* and later in the same passage he tells us that the *noctua* was rendered as *glaucus* in the Septuagint. Hugh of St. Victor 1.43 equates *noctua* and *nycticorax*. Cf. Vinc. 16.111. George and Yapp (1991, 148–50) discuss at length the *nicticorax/noctua* problem. On the *nyktikorax* cf. *GB* (207–9) and the additional work of Oliphant (1914, 61–62).

37. Ar., *Part. An.* 1.5 (645a1f.). Cf. A., *DA* 11.2.3.85 (*SZ* 2: 889).

38. Ar., *De anima* 3.8 (431b22), 3.4 (429b30f.).

terial thing, because it is in matter. But because it does not use an organ, it does not apprehend this quiddity materially; rather, it apprehends immaterially that which is in matter and it apprehends universally that which is in the individual.

Nevertheless, it must be understood that some things can be known in many ways. For some are known through their essences, and thus the first cause understands itself *per se* and, by knowing itself, it intuits all other things just as if in a mirror that depicts all things to itself. For it is itself a mirror bearing all things in itself. Other things are known through their species. And this is how we know and how intelligences know, but we know through acquired species, and intelligences do so through innate species. And still others are known by their effects, and this is how we know the separate substances, because we infer [*concludo*] the nature of the intelligences through motion, which is a common sensible. Something is known in a fourth way, by privation, and this is how we know a point and indivisibles. Thus we can achieve knowledge of the first cause in the last two modes, namely, by privation, that is, by understanding that it is incorporeal, immaterial, impassible, incapable of being generated, without fixed magnitude, and in the same way for things found to be akin to it.

The first [cause], God—the most true, most sweet, most powerful from eternity forever and ever and reigning through boundless ages—can be known in another way, that is, through his effects. And the Philosopher says this in the premise.[39] For he says that it should not be difficult for us to inquire into the natures of vile animals because knowledge of them and investigation into them present such admirable pleasures to us, just as (as he says later) in children who are beginning to learn. This is because the artifice of the first creator is reflected in them just as the handiwork [*opus*] of the sculptor [*statuifex*] is reflected in his statue and that of the blacksmith in a knife and that of the physician in preserving health and curing the sick. Therefore, from the knowledge of these vile animals we can ascend to the knowledge of the first cause just as we ascend from effect to

39. Ar., *Part. An.* 1.5 (645a15f.). Cf. A., *DA* 11.2.3.85 (*SZ* 2: 889).

cause. But this is the knowledge "that it is" [*quia est*] and not a knowledge of "what it is" [*quid est*], which is more knowable and is prior absolutely and with respect to itself [*secundum se*] but not with respect to us, but the first [is], namely, "that it is," etc.

1. On to the arguments. To the first argument one must reply that although the first cause does not fall under sense as far as its own essence is concerned, nevertheless it does fall under sense as far as its effects are concerned, and nothing is understood if it does not fall under sense in some way, either in itself or in another.

2. To the second argument one must reply that just as the night owl's or the bat's eye cannot endure the light of day, so neither can our intellect gaze at the essence of the first cause in itself, although it can gaze at its effects, etc. Nevertheless, it will be different for one stripped [of the body] after death, because then he will be able to do this, etc.

3. To the third argument one must reply that the infinite is of two types. One is infinite with respect to matter, and the Philosopher has enough to say about this type of infinite in the third book of the *Physics.*[40] This one is intelligible only through a privation of finitude. Another is infinite with respect to form and is said to be infinite because it cannot be bounded by form, and one like this is intelligible *per se,* etc.

To the first argument one must reply to the objection [*ad oppositum*] that although the first cause is not known by us through its essence, it is nevertheless not unknowable on account of this. So too, the sun is not visible in vain just because the night owl does not see it. Thus, even though we do not know the first cause through its essence, we nevertheless know it well enough through its effect, etc.

40. Ar., *Phys.* 3.4–8 (202b30–208a23).

BOOK TWELVE

Question 1: Whether homogenous [consimiles] *parts are generated immediately from the elements.*

"WE HAVE ALREADY stated above that the members," etc.[1] In his twelfth book the Philosopher establishes the composition of parts. Therefore, with respect to this, the twelfth book, we first inquire whether similar parts are immediately generated from the elements.

1. And it seems they are. This is because generation is analogous to decomposition [*resolutio*], although they differ either according to their causes or their ends, since that "which is first in generation is last in decomposition"[2] and contrariwise. But homogenous parts can be immediately decomposed into elements; therefore, they can also be immediately generated from them.

2. Besides, just as official parts come from homogenous ones, so too do homogenous ones come from the elements. But official parts are generated immediately from the homogenous ones. Therefore, homogenous ones are immediately generated from the elements.

On the contrary: One and the same thing is both the principle of nutrition and of generation, because "we exist from and are nourished by the same things."[3] But homogenous parts do not take their nourishment immediately from the elements; therefore, neither are they generated from them.

To this, one must respond that homogenous parts are not generated immediately from the elements, for although prime

1. Ar., *Part. An.* 2.1 (646a8f.). Cf. Avic., *DA*, 11 (fol. 44vb); Averroes, *Part. An.* 2.1; A., *DA*, 12.1.2.1–7 (*SZ* 2: 894–96).

2. Ar., *Eth. Nic.* 3.5 (1112b23f.).

3. Ar., *DG* 2.8 (355a10f.).

matter is in potency to every form, it nevertheless cannot receive every form immediately. Thus the Philosopher says in the eighth book of the *Metaphysics* that wine does not come from vinegar immediately nor does a living being come from a corpse immediately, because wine is generated on a vine.[4] Thus in order for wine to come from vinegar, it would first be necessary that the vinegar be changed into a nutriment for the vine. And, in the same manner, nature proceeds through several intermediaries when producing an animal, since the semen comes from a superfluity of nourishment, the blood from the semen, and the embryo from the blood. Thus, in order for the homogenous parts to come from the elements, it is first necessary that a mixture suitable to nutrition come forth from the elements, and then that mixture is converted into food, and then into chyle (that is, into a humor) and then it can nourish.[5]

1. On to the arguments. To the first argument one should respond that generation and corruption occur differently. For generation has being as its goal, and corruption has non-being, and this is why nature proceeds in an ordered way when generating, but proceeds as if in a disordered way in corruption.

Or one can state it in another way, namely, that when corrupting something, nature uses all the things acquired through generation as if for a single end, and that is why the elements come forth immediately from the homogenous parts, but not contrariwise.

2. To the second argument one should respond that it is not the same for the official parts in respect to the homogenous ones or for the homogenous parts in respect to the elements. For official and homogenous parts are produced from one nature and are informed by one common form, and so too that does not come from the homogenous parts and the elements, and this is why, etc.

4. Ar., *Metaph.* 8.5 (1044b31f.). Cf. A., *Metaph.* 8.2.3.

5. Chyle is the semi-liquid contents of the digestive tract during the first stages of digestion. See *DA* 1.3.3.563; 3.2.3.100, 3.2.4.125; 20.1.4.16, 18 (*chymus nutrimentalis*); and *DA* 23.77(40) (*SZ* 1: 262–63; 393–94; 405–6; 2: 1366–67).

Question 2: Whether the elements exist in act in homogenous parts.

Further one inquires whether the elements may exist in act in homogenous parts.[6]

1. And it seems that they may. For alteration is a change in quality, with the subject remaining the same. But "mixture is a union of altered mixables [*miscibilium alteratorum*]."[7] Therefore, the mixables remain in substance, changed only in quality.

2. In addition, according to the Philosopher in *On Generation* [*and Corruption*], a mixture differs from generation and corruption.[8] But if the elements were entirely corrupted in the homogenous parts, then a mixture would not differ from generation and corruption, because there would be a corruption of the mixable and a generation of the mixed.

3. In addition, power does not exist without substance, according to *On Heaven and Earth*.[9] But in the mixed "the power of the mixables is preserved."[10] Therefore, so too is their substance.

On the contrary. One thing does not really come from several things in act. Therefore, if the elements remained in act, the mixed would not really be one thing.

To this, one must respond that the elements do not remain in act either in homogenous parts or in any mixed things. The reason for this is that contrary forms cannot at one and the same time perfect the same matter, nor, according to the Philosopher in *On Generation and Corruption*, is any part of the mixed itself mixed.[11] If, then, the elements were present in act in the mixed, they would be present in act in some part of the mixed. But the forms of the elements are contraries, and, as a

6. Cf. A., *DCM* 3.2.8.
7. Ar., *DG* 1.9 (328b22f.).
8. Ar., *DG* 1.9 (327b6f.); A., *DG* 1.6.2.
9. Ar., *DC* 1.7 (275b5f.).
10. Ar., *DG* 1.10 (327b30f.); Cf. A., *DG* 1.6.3.
11. Ar., *DG* 1.9 (328a15f.).

result, contraries will exist at the same time in the same part, which is false, and therefore, etc.

Moreover, the heat of fire is one thing, against or beyond which the form of fire is not preserved, and the heat of air is something else, beyond which the form of the air does not extend. If, then, the elements exist in act in a mixed thing, there would be two heats of different degrees in the same part. But this contradicts what the Philosopher says in the third book of the *Physics.* For he says there that the same thing cannot be changed simultaneously by two whitening agents.[12]

Furthermore, if the elements were present in a mixed thing in act, then either some part of one will exist at the same time as, or right next to, another part of the other. If it is at the same time, since an element has its own dimensions, then two dimensions and, as a result, two bodies would exist at the same time, which is false. If it is next to it, then the mixture will only be a juxtaposition. But this is false, since, according to Avicenna, true mixture does exist when the very least part of one touches the very least part of the other.[13]

Nevertheless, the Commentator proposes that elemental forms exist in act in a mixed thing under a certain refraction, remission, and tempering, because their qualities are pushed back from a state of excess.[14] But in order to sustain this view he proposes many false ones, namely, that the elemental forms strive forward and are then pushed back. For he himself proposes that the elements are midway between substances and accidents, which is impossible, because then there would be a mean between contradictories, namely, between being in a subject and non-being in a subject.

1. On to the arguments. To the first, one should respond that for a mixture, it is necessary that there first be an alteration in the mixables, to the extent that the forms of the mixables are expelled and the form of the mixed is introduced. Thus, as long as the mixables are present during the process of alteration,

12. Ar., *Phys.* 3.3 (202a34f.); cf. ibid. 5.4 (227b8f.). See A., *Phys.* 3.1.8; 5.3.1.
13. Cf. Ar., *DG* 1.10 (327b33f.); A., *DG* 1.6.7.
14. Averroes, *DCM* 3; cf. A., *DCM* 3.2.8.

they remain, but once the form of the mixed has been introduced, the alteration ceases, and it is unnecessary for the mixables to remain. Thus a mixture is "a union of previously altered mixables." And if you were to say that only those things that are present in act can be united, and that a mixture is a union, and that they therefore are present in act, one must reply that if you understand by "mixture" a total transformation that occurs beforehand and is accompanied by a boundary, then, with the transformation going on beforehand, the mixables themselves approximate one another, and this is their union. But if we were to understand by "mixture" the final transformation, then that union ought not be referred to the mixables but to their powers.

2. To the second argument one should respond that alteration is a kind of generation. And likewise mixture is a kind of generation. Nevertheless, mixture is to be distinguished from generation properly speaking, because in the proper sense of generation the form of the one overcoming, corrupting, or generating is introduced. But this is not the case in a mixture, and if the four elements, equally disposed to be active or passive, come together for a mixture, then they act or are acted upon, one to the other, so that in the end all will be absent at the same time and one form having the power of all will then be introduced into their materials.

And in this respect there is another difference between a mixture and generation, because the power of the one corrupted is not found throughout the one generated, as the power of water is not present in fire, if fire should come from water.[15] But in a mixture the power of the mixables is preserved throughout, and this is why a mixture differs from generation.

3. To the third argument one should respond that every power reveals its own substance. Thus, if the power of the mixables is in the mixture, this power is not the ultimate, as it is in a mixable per se. Thus this power of a mixed thing is the power of a subject, but that of a mixable is like that of an efficient cause. Thus the power of a mixable is in the mixture just as the power

15. Cf. *QDA* 5.1 and note.

of the white is in the red and the power of water placed in wine, etc.

Question 3: Which elements predominate in homogenous parts.

Further, one inquires which elements predominate in homogenous parts.

1. It seems that the light elements do. Now, life depends on heat and moisture. Therefore, those things that are warm and moist especially predominate in living things.[16]

2. Moreover, when many things come together or toward some one other thing, and these things have greater power, they gain dominance. But light elements have greater power than heavy elements. Therefore, they predominate in a mixture.

To the contrary. Earth and water hold dominion in those things that are generated around a midpoint. But all living things are generated in this way, and therefore, etc.

One must respond that earth and water predominate in the homogenous parts of all living beings. And this is the reason: if several things come together, some will have greater power than others, so it is necessary that the weaker ones be increased in quantity in order to bring the whole thing into balance. For if one stronger in power were greater in quantity, it would immediately corrupt the other [weaker] one and convert it into its own nature. But heavy elements have less power for acting than do the light elements, and this is why it is reasonable that they are more abundant in quantity. An indication of this is that a mixed [body] is changed by the movement of the dominant element in it, and the movement of any mixed body whatsoever, is downward, to the extent that it is mixed. Therefore, the heavy elements predominate in it.

1. To the arguments. To the first, one should respond that heat is of two kinds: namely, elemental, and a certain tempered or proportionate one. Life is not supported by the first type, since this is repugnant to life, but rather life is supported by the

16. Ar., *De long. et brev. vitae* 5 (646a18f., b21f.).

second type. And such a heat can exist in a thing in which the heavy elements abound or predominate materially.

2. To the second argument one should respond that if light elements were either equal to or greater in quantity, they would attract or convert the others to their nature. But this is not the case, as has been said.

Question 4: Whether official parts are composed from homogenous ones by means of continuity.

It is asked further whether official parts are composed from homogenous ones by means of continuity or of contiguity.

1. It seems not. An end is not composed from those things which exist for the end, because the end is an extrinsic cause. But homogenous parts exist on account of official parts, as is said in the text.[17] Therefore, they do not compose them.

2. In addition, something is continuous whose motion is one and in which there is uniformity in the parts. But the motion of an official part is not always one nor always uniform in every part. Therefore, it is not composed from these by means of continuity. Nor is it by means of contiguity, because things are contiguous whose extremities exist simultaneously, but not as one.[18] But this cannot be said of homogenous parts. For if this were the case, then they would not share in one form.

The Philosopher says the opposite.[19]

One must reply that official parts are composed from homogenous parts. Now, official parts have various operations: consider that the tongue discerns tastes, and forms sounds [*voces*], and softens or grinds up nourishment, and so too for the others. Likewise, the hand receives and retains, extends and contracts. And this is why it is formed from various parts.

17. Ar., *Part. An.* 2.1 (646b11f.).

18. Cf. Ar., *Topica* 4.2 (122b28f.), cf. Boethius's translation (PL 64: 945A). Cf. A., *Topica* 4.1.4.

19. Ar., *Part. An.* 2.1 (646b30f.); cf. A., *DA*, 12.1.1.7–10 (*SZ* 2: 896–97).

1. To the arguments. To the first, one should respond that an end is of two types: intrinsic and extrinsic. An intrinsic end is a thing's form and it is properly the end of generation, since once this form is achieved generation ceases. An extrinsic end belongs to the very thing generated. Moreover, an intrinsic end is of two types: near and remote. The near end, however, is the proper operation of a homogenous part, because each part exists for its own particular work. But the remote end is the official part. This is why in all things that are generated in an ordered way the first one generated exists for the sake of the last.

Or, on the contrary, one may say correctly that the official part is the intrinsic and near end of the homogenous parts, because the perfective form of the official part is the perfection of a homogenous part, just as the form of the whole is the perfection of each part. The proper operation is the more remote end, because it is a secondary act. The extrinsic end of these homogenous parts, however, is the human, since all these lower things exist on account of the human and the human exists on account of the first cause. Through this a response to the first argument is clear.

2. To the second, one must respond that something can be continuous in two ways: either in nature or in quantity, so that if a line is drawn out straight, half of which is white and half black, that surface can be continuous with respect to its quantity, although it will not be continuous or uniform in terms of color. It is the same for homogenous parts. They are continuous in terms of location and in quantity, but they are discontinuous or contiguous respecting their own natures.

Question 5: Whether operations occur through similar or dissimilar parts.

Further one inquires whether operations occur through similar or dissimilar parts.[20]

And it seems first with respect to the nutritive power that they occur through similar parts. For certain animals in which

20. Ar., *Part. An.* 2.1 (647a5f.); A., *DA,* 12.1.1.12–16 (*SZ* 2: 898–900).

organic parts are not present have a vegetative soul. Therefore, their operation will be through similar parts.

The same seems true with respect to the vital power, since certain beings live which are lacking in organic parts, and therefore, etc.

The same seems true with respect to the sensitive power. For one element is predominant in an organ of any sensitive power whatsoever, just as water is predominant in the organ of sight, etc. But that one in which one element predominates is a similar part. Therefore, the sensitive power operates through similar parts.

Moreover, according to Aristotle in the second book of *On the Soul,* the nature of a medium and of an organ is one and the same. But the medium in the senses is similar to that in the parts; therefore, also the organ will be similar.[21]

The same seems true with respect to the motive power. Now in any thing, whenever dilation occurs on account of the appearance of something delightful, or contraction occurs on account of the appearance of something noxious, then in that same thing a motive power must be present. But these occur in similar parts, and therefore, etc.

One argues the opposite, first with respect to the vegetative power. For a vegetative operation occurs through that without which the vegetative operation cannot be completed. But a vegetative operation is not completed without dissimilar parts. This is clear in those that have blood, for blood only nourishes if it is digested in the liver and then carried to the individual members through the veins. Therefore, the nutritive power operates through dissimilar parts.

Second, the same thing seems true with respect to the vital power, which is in the heart, which the arteries, the lungs, and the homogenous members serve. But these are official members, and therefore, etc.

Third, the same thing seems true for the sensitive power, because for vision a pupil is required, or the glacial humor and the hollow nerve. And members like these are dissimilar, and therefore, etc.

21. Ar., *De anima* 2.7 (418b7f.), 11 (423b6f.).

The same thing seems true for the motive power, since, according to the Philosopher in the third book of *On the Soul,* in the first organ of motion there exist both something moving and something resting.[22] And everything like this in the parts is dissimilar, and therefore, etc.

To this question one should respond that some operations are due to homogenous parts and some to dissimilar parts. Now the operations of the nutritive power and the sensitive power are first due to similar parts, but the operations of the vital power and the motive power are due to dissimilar parts. For the natural power's operations are: to attract, digest, retain, and expel, and each of these occurs through some homogenous part. Now although these occur through many parts, because flesh attracts and bone attracts as do vein and nerve, nevertheless each part taken *per se* is homogenous. And this is why we say that natural operations first occur through homogenous parts, and secondarily through dissimilar parts that are made up of similar ones.

It is the same for the sensitive operation. For touch is the first sense, and touch is in every part of the body according to Avicenna, who says that God armed the animal everywhere with touch lest it suffer some ill unawares.[23] But nothing occurs in every part unless it is a homogenous part, and this is why the operation of touch occurs first through a homogenous part. In the same way, various parts come together for the operations of the other senses. Consider that in sight the glacial humor receives the appearance of a thing [*species*] and the hollow nerve remands it to the first visual principle. But these parts, taken individually, are homogenous, and this is why the sensitive power universally operates through similar parts first. Nevertheless, the operation of the external senses is only completed by means of dissimilar parts.

But the vital power exists principally in the heart; however, the heart is a dissimilar part, and other parts serve it. For the heart infuses the vital power, and the other parts receive it. Thus, in those breathing or having blood it is clear that the vi-

22. Ar., *De anima* 3.10 (433b14f.); 3.12 (434b32f.).

23. Avic., *Can. med.* 1.1.4.2; *De anima* 2.3.

tal power is kept safe in the heart only if the heat of the heart is tempered by the reception of something extrinsic to it. And this is why animals that walk breathe, and breathing occurs only by means of a dissimilar part.

But it is clear regarding the motive power that wherever there is a motive power, there is something moving and something moved, and these are dissimilar.

Or one can put it another way. For all operations occur through similar parts and dissimilar parts, but some occur first through similar parts and only secondarily through dissimilar ones, and some do the opposite.

As a result, the arguments conclude as true in part.[24]

Question 6: Whether a more intense heat is found in the element of fire, or in a mixture.

Further one asks whether a more intense heat is found in the element of fire, or in a mixture.[25]

It seems that it is more intense in the fire. According to the Philosopher in the second book of the *Metaphysics,* that which is in a category [*genus*] first especially has the characteristic of that category,[26] because that which first has this characteristic is a cause for all such things because they have this same characteristic. But fire is first hot; therefore, it especially has this characteristic.

Moreover, that thing is hotter which is more deprived of or more lacks the restraining action of a contrary; but the heat in fire is like this, and therefore, etc.

To the contrary. Operation makes the form known. To burn is heat's operation. But heated iron burns more actively than fire; therefore, etc.

Besides, that which conserves heat longer is hotter. But heated iron conserves heat longer than a flame in the air, once the combustibles have been removed. Therefore, etc.

24. One possible translation of these cryptic words: *Unde rationes pro parte verum concludunt.*

25. Ar., *Part. An.* 2.2 (648a19f.); A., *DA,* 12.1.3.32–49 (*SZ* 2: 906–11).

26. Ar., *Metaph.,* 2.1 (993b22f.).

To this, one must reply that one thing can be hotter than another in more than one way: either *per se* or through something else. If *per se,* then among all things fire will be hottest because in itself it has the cause of heat *per se.* For heat cannot be separated from the form of fire, just as a capacity for laughter [*risibilitas*] cannot be removed from a human.[27]

Something is hot through something else in another way, and such a thing can effectively be hotter than one that is hot *per se.* And heated iron is just such a thing, since the iron itself is hot through something else, because it participates in the heat of fire and is effectively hotter. And the reason for this is that every united power is stronger than it is when it is dispersed.[28] Now, however, the proper material for fire is rarified [*rara*], and the iron's material is more compact and solid. Thus the fiery heat incorporated into the iron is more unified, and for that reason is effectively hotter.

And, besides this, the more something resists an agent, the more strongly the form of the agent is impressed upon it, if it obtains victory over the thing being acted on. But the more solid the matter is, the more strongly it resists a hot agent, and this is why, when it overcomes the matter, the power of heat is more strongly impressed on it, etc.

Moreover, fire is easily divisible, and this is why if someone extends or reaches his hand into the fire, he can remove his hand without injury because the fire itself is easily parted. But heated iron is solid and not easily parted, and this is why it sticks more firmly to one touching it and, as a result, burns him more.

Through this the answers to the arguments are clear. Now, fire is hotter *per se* and *in se* and formally; nevertheless, heated iron is hotter accidentally and effectively, owing to the compact nature of its matter. This is because a quality impressed on a solid object is stronger than one impressed on another, just as moisture is stronger in water than in air, even though it may be in the air *per se* and in the water only through another agency. Thus to this one, etc.

27. That is, just as laughter is a proper predicate of the human, so too is heat a proper predicate of fire.

28. Ps.-Ar., *Liber de causis* 16 (Bardenhewer, 1882; reprint, 1970), 179.

Question 7: Whether the first digestion is in the mouth.

One asks further whether the first digestion, or the first digestive power, is in the mouth.

1. It seems so.[29] For, according to Avicenna, "grain placed and chewed in the mouth precipitates a sore there."[30] And this does not occur if the grain is first ground up by some instrument. Therefore, it seems that the grain is somewhat digested in the mouth, and therefore, etc.

2. In addition, "the root of a plant is analogous to an animal's mouth," as the Philosopher states in the second book of *On the Soul*.[31] But the nutriment's digestion occurs in the root; therefore, it will also occur in the mouth.

The Philosopher says just the opposite.[32]

In response to this, one should say that there is a disagreement over this between the Philosopher and many physicians. Now, some physicians posit a first digestion in the mouth, and a second in the stomach. The Philosopher posits a single digestion in the mouth and the stomach. Thus, according to the Philosopher, mastication—which occurs in the mouth—is in preparation for the stomach's operation, with the mouth disposing and preparing the food so that it may be more easily digested in the stomach. Thus one must say that digestion, properly speaking, is not in the mouth because the appetitive and attractive powers precede digestion. But the first appetite flourishes in the upper part of the stomach; therefore, there is no digestion before the nutriment reaches that part.

1. To the arguments. To the first argument one should respond that grain masticated in the mouth heats up in the mouth while it is being ground and is tempered by the saliva, and it is this heating and mixture with the saliva that precipitates sores,

29. A., *De anima* 2.2.8.

30. "Precipitates": may also be translated as "brings a sore to a head." Avic., *Can. med.* 1.1.4.2; *DA* 12.4 (fol. 44vb).

31. Ar., *De anima* 2.1 (412b3f.).

32. Ar., *HA* 1.4 (489a27f.); *Part. An.* 2.3 (650a8f.); cf. Avic., *Can. med.*, 1.1.4.2; *DA* 12.4 (fol. 45vb–46ra); A., *DA* 12.1.6.75–84 (*SZ* 2: 922–25).

because saliva is salty and pungent and penetrating. Thus, it precipitates a sore not because the grain has been digested but because it has been altered proportionally.

2. To the second argument one must respond that a tree's root is not only analogous to the mouth but also to the stomach, because a plant receives nourishment through the root just as an animal does through its mouth, but the nourishment is not dispatched from the root to another member in which it is digested, as happens in the animal when it is dispatched from the mouth to the stomach. Rather, one and the same root serves each operation, that is, both the reception and the digestion of the nutriment. And this is why in trees no member is set aside for superfluities because, as is said, they do not need any. And, besides this, the plants' nourishment is somewhat digested in the ground before they receive it. And so it is not the same, and this is why, etc.

Question 8: Whether the first digestion, which occurs in the stomach, changes the nutriment into a different species.

One asks further whether the first digestion, which occurs in the stomach, changes or alters the nutriment into a different species.[33]

1. And it seems not. For just as the nutriment is converted into the animal's substance, so too is it converted into the plant's substance. But in the first digestion the species of the plant's nutriment are not changed, and therefore neither are they changed in the first digestion of the animal.

2. In addition, digestion changes the species only by the power of heat. But the stomach is cold by nature, because it is nerve-filled. Therefore, it cannot change the species.

To the contrary. The species of the nutriment is changed in the second digestion, which occurs in the liver, because [there] it is converted into blood. Therefore, for the same reason, this occurs in the first digestion.

33. A., *DA* 3.2.3.100–101 (*SZ* 1: 393–94).

One must respond that the first digestion, which occurs in the stomach, when it proceeds by a natural path, changes the species of the nutriment. And I say "by a natural path" because if the natural power were outside its natural disposition, for example, if the retentive power were too weak and the expulsive power too strong, then it could happen that the food would be excreted under the same species in which it was consumed. But when it proceeds by a natural path, it happens otherwise, because in the beginning the nutriment becomes dissimilar, and in the end, similar. But this can happen only if it is changed into a contrary disposition. Thus the nutriment in the stomach is segregated into several parts by the power of the natural heat; of these parts, the chyle is said to be purer, and the waste product is said to be more impure. This chyle, however, is changed into blood during the second digestion, and in the third digestion this blood is changed into the members. Thus its species is changed.

1. To the arguments. To the first argument one should respond that the first digestion in a plant changes the species of the nutriment, as is the case in an animal. But that digestion that is in the ground does not occur through the mediation of any part of the plant. Thus, if this digestion is compared to the digestion that occurs in an animal, the argument does not succeed.

2. To the second argument one must respond that the stomach is cold and dry in its upper part, but it is warm in its lower part. Nevertheless, it does not have its own heat adequate for digestion; rather, it participates in the heat of the heart, the lungs, the spleen, and the liver. This is why the stomach is placed over them, just as a pot is placed over a fire. Thus, just as the pot is in itself cold, nevertheless by means of the heat of the fire surrounding it, it cooks the food. And so too, the stomach is in itself cold, but nevertheless it digests by the power of these other members, etc.

Question 9: Whether the four humors are generated from the same nutriment.

One inquires further whether the four humors are generated from the same nutriment.

1. And it seems not. Now, an agent is of two types: higher and lower. A higher agent acts according to the exigency of its power, but a lower agent acts according to the disposition of matter. Since, then, the agent in digestion is one and the material is uniform, it will generate a single thing.

2. Besides, sometimes the food is proportional to one humor, because sometimes it is choleric—like pepper and others like this—and sometimes it is phlegmatic. Therefore, since in those that have an affinity the transformation will be easier, the food is converted into only a single humor.

The Philosopher says the opposite.[34]

One must say that the four humors can be generated from the same nutriment because, whatever the nutriment is, however uniform it is, it is a mixture because, according to the Philosopher in the first book of *On Generation* [*and Corruption*] and in *On Sense and the Sensed,* simples do not nourish.[35] Thus the power of the four mixables is in every nutriment. And this is why the four elements can be separated from every nutriment, and in like manner the four humors, which correspond to the four elements, can be separated from every nutriment. For just as there are four substances in milk, and, with the addition of a coagulant three of the substances can be separated from it, one of which—like cheese—corresponds to earth, and another—like butter—corresponds to air, and a third—like whey—corresponds to water or, according to some, to fire. So too, diverse humors can be separated from nutriment through a natural power, because the earthy parts are more converted into melan-

34. Ar., *Part. An.* 2.3–4 (649b9f.). Cf. Avic., *DA,* 12.5 (fol. 46ra); A., *DA,* 12.2.1.97–105 (*SZ* 2: 930–33).

35. Ar., *DG* 2.8 (335a11); *De sensu et sens.* 4 (441b26f.); A., *DG* 2.2.17; *De sensu et sens.* 2.6.

choly, the watery into phlegm, the airy into blood, and the fiery into choler.

1. To the arguments. To the first argument one should respond that although the first matter of nutriment is one, nevertheless it itself has a varied disposition and has power under various powers. Therefore, because it is the material of a mixture, and is acting on a single thing that is nevertheless disposed in various ways, it can produce diverse effects. Natural heat acts one way when using the power of the soul on earthy parts, and another way on airy parts. This is because natural heat first assembles homogenous parts, as when lead and silver are joined together at the same time, and fire separates one from the other. This is how a natural heat acts when digesting nutriment, etc.

2. To the second argument one should respond that although some kind of nutriment may correspond more to one humor than to another, nevertheless it exists in potency toward any of them, and this is why although any humor can be generated from it nevertheless that humor will be generated most from the one to which it corresponds, etc.

Question 10: Whether blood is the last food for the members.

One inquires further about blood. And, first, whether blood is the last food for the members.

1. And it seems not. That which is generated during the first digestion changes its species in the second. Therefore, what is generated in the second will change its species in the third. But blood is generated in the second digestion, and therefore it changes its species in the third. And, as a result, something coming after the blood is food for the members.

2. In addition, the veins are the site for blood. If, then, blood were the last food for the members, no member would be nourished other than the vein which contains the blood.

The Philosopher says the opposite.[36]

One must say that the last food for the members is of two types: common and particular [*appropriatus*]. If we are speaking of the last common food, the blood is the last food. If, however, we are speaking of the last particular food, then some moisture comes after the blood. For blood that is borne to any member undergoes a further digestion in that member and is changed into a moisture. This moisture is in a proximate disposition to the member. Nevertheless, the Philosopher takes this in the first way; the physicians speak of it in the second.

1. By this a response to the first argument is clear, that although blood may change its species in the third digestion, it is nevertheless the last common food.

2. To the second argument one should respond that blood seeps through the medium of the veins after the fashion of sweat,[37] and thus reaches the parts surrounding the veins, and for this reason it nourishes not only the veins, but the exterior parts as well.

Question 11: Whether thick blood is more nourishing than thin blood.

Further one asks whether thick blood is more nourishing than thin blood.[38]

1. It seems that the thick blood is more nourishing. This is because nutrition is the complete assimilation of the nutrient to the one nourished. But thick blood is better assimilated to the members than thin blood. Therefore, etc.

2. In addition, thicker and more globular semen is better suited for generation, as the Philosopher said above.[39] Therefore, for the same reason, thick blood is better suited for nutrition.

36. Ar., *Part. An.* 2.3 (650a33f.). Cf. Avic., *DA* 12.6 (fol. 46rb); Averroes, *Part. An.* 2.4; A., *DA*, 12.2.1.101 (*SZ* 2: 931–32).

37. Cf. *QDA* 13.8.

38. Cf. A., *DA* 12.2.1.102–5 (*SZ* 2: 932–33).

39. Ar., *HA* 7.1 (582a30f.).

To the contrary. Things that are more easily converted are more suitable nourishment. But thin things are more quickly converted. Therefore, they nourish better.

Besides, that thing nourishes better which prepares the member as more suitable for its operation. But thin blood is such a thing. For those having thin blood have better sensation, intellect, and motion, which is the case only because a thin thing nourishes better. Therefore, etc.

To this one must respond that something is said to nourish better than another in two ways. In one way, it is because it remains longer in the members and is more difficult to digest. And in this way thick and hard things are said to nourish better, and beef and eel are said to nourish better than the meat of a chicken or a ram, and the like. Something is said to nourish better in another way, because it is converted more easily into the members' substance, and in this way thin things nourish better because they are easily converted and remain in the members only a short time, like the meat of a chicken or a capon and the like.

The Philosopher speaks in the first way when he says that thick blood nourishes better.[40] And his argument is that the second digestion proceeds by thickening and coagulating. And for this reason the thicker it is, the better it is for or the closer it is to nutrition. The physicians understand things in the second way, when they say that thinner things are better nourishment. The reason, however, why those having thin blood have better sensation, intellect, and motion is that where the blood is thin and pure, there the spirits are thinner. Now, the spirits serve and minister to the sensitive and motive powers. The thinner these spirits are, the better adapted they are to motion or for moving the senses and the organs. And this is why such animals as these have better sensation and motion, and even have purer phantasms. They therefore have a good intellect. For our intellect "either is a phantasm, or does not exist without a phantasm," as is said in the third book of *On the Soul,* and the purer the phantasms are, the better, etc.[41]

40. Ar., *Part. An.* 2.4 (650b14f.).
41. Ar., *De anima* 3.8 (423a12f.); A., *De anima* 3.3.12.

Question 12: Whether zirbus *and fat have a complexion that is naturally hot or cold.*

Further one asks about *zirbus* and fat [*pinguedo*],[42] and whether they have a complexion that is naturally hot or cold.[43]

It seems that it is cold. This is because that which is more abundant in cold things is more of a cold nature. But *zirbus* and fat are more abundant in cold things, since they are more abundant in women than in men and more abundant in winter than in summer. Therefore, etc.

Moreover, whatever is dissolved by cold is coagulated by heat, and vice versa. But fat and *zirbus* are dissolved by heat, and therefore they are coagulated by cold. And if this is true, then, taken by themselves, they have a cold nature, just like ice, etc.

To the contrary. Blood is hot and moist. A thing that is more digested is all the hotter for it. But *zirbus* and fat are nothing other than well digested blood. Therefore, they are hotter.

Besides, whatever things that are found only in animals that have blood are hot; but *zirbus* and fat are things like this, and therefore, etc.

To this, one must reply that *zirbus* and fat can be compared two ways. One way is to a heat having the power to convert the nutriment into the members, and, seen in this light, they have a cold nature. For on account of the fact that nature cannot convert an oily blood or any other such humor into the substance of the members, it leaves the blood outside the members, owing to its volume. But it nevertheless acts on it by digesting it beyond the nature of blood, and in this way, with respect to the members, they have a cold nature. If, however, these are compared to an undigested nutriment, then they are hot, because whatever is watery in them is dissolved and consumed, and this is why it is more easily converted into fire than is blood, since

42. Var. *zyrbus, zirbum, zyrbum.* In technical usage, it is the omentum. It often seems, however, to refer to any sort of fat that protects an internal organ and, distressingly, sometimes seems to be a mere synonym for another type of fat (see, e.g., A., *DA* 1.2.6.196 [*SZ* 1: 119]).

43. Ar., *Part. An.* 2.5 (651a20f.); Avic., *DA*, 12.6 (fol. 46rb); Averroes, *Part. An.* 2.5; A., *DA*, 12.2.2.106–9 (*SZ* 2: 933–35).

blood, owing to its wateriness (and not its oiliness) extinguishes fire. But zirbus and fat, on account of their airiness, feed the fire and strengthen it even more. Thus, these are neither intensely hot nor intensely cold, but occupy a mid-ground. Nevertheless, they are said to be cold because they are generated only from a defect of heat, as when natural heat cannot convert the blood entirely into the members, and this is owing either to its weakness or to the quantity of the nutriment.

Responses are then clear to arguments in this way, because they proceed along their own lines.

Question 13: Whether the marrow is the nutriment for bones.

Further one asks about the marrow, whether it is the nutriment for bones.[44]

1. It seems not. For everything that is nourished has pathways through which it receives nourishment. But bones, since they are solid and hard, do not have such pathways through which the nourishment might be introduced; therefore, etc.

2. In addition, the marrow is related to bones in the same way that fat is related to flesh. But fat does not nourish the flesh because it is outside the flesh; therefore, neither does the marrow nourish bones.

3. In addition, blood is the last food for the members. But blood is located in the veins. Yet there are no veins in the bones. Therefore, the bones are not nourished by blood, and, as a result, neither are they nourished by the marrow.

The Philosopher says the opposite.[45]

One must reply that marrow is the nutriment for the bones, for the bones exist as a support for the flesh, which is a soft part. But, now, it is true that a contrary approaching its contrary necessarily acts on it. Bones are cold and dry, which is evident from their heat. Flesh, however, is moist and hot because of blood. Thus, if flesh were nourished and not the bones, the

44. Ar., *Part. An.* 2.6 (651b20f.); Avic., *DA*, 12.6 (fol. 46rb); Averroes, *Part. An.* 2.6; A., *DA*, 12.2.3.109–12 (*SZ* 2: 935–36).

45. Ar., *Part. An.* 2.6 (652a2f.).

bones would quickly fail, and this is why it is necessary that the bones be nourished just as the flesh is.

But one must understand that although bones appear solid and hard, they are nevertheless porous bodies, even though while they are alive this is not apparent to the senses. But this is clear when they are broken. Thus moisture, which is the last nutriment of the members, seeps into the middle of the bones, and whatever is more earthy and dry is converted into their nutriment, because like is nourished by like; but that which is more refined passes over all the way to the hollow of the bones and is converted into marrow. Yet because bones are more solid than other parts, they cannot receive a nutriment sufficient for them as quickly as the other members do, and this is why nature—having foresight like the head of the household [*paterfamilias*]—places the marrow in the bones, and they are continually nourished by it. Yet nevertheless they draw more of what is earthy and dry into the nutriment and leave behind more of what is hot and moist. And this is why the marrow of melancholy animals such as a stag or ox is better and sweeter than the marrow of a phlegmatic animal or one of another complexion (for example, a sanguineous complexion), like a pig or a human, because in the first ones the earthiness of the marrow passes over into the bones and the airiness remains in the marrow. In the others, the contrary is the case, and for this reason, etc. This is because the harder the animal's flesh is, the sweeter and better its marrow is, because each is nourished by one like itself. Thus, when flesh is hard, it assimilates to itself that which is hard and melancholy in the food, and what is more refined and airy it leaves behind. And the opposite is true for flesh that is refined and soft. And for the same reason a broth or soup made from a melancholic animal with a hard flesh is better than a broth made from an animal with soft flesh, so that the broth of a stag or hare is better than a broth made from a pig, and a cow's is better than a ram's. Because when the flesh is hard, that which is refined and airy alone leaves it when it is cooked. For this reason, etc. But when the flesh is soft, as in a pig, a great deal of phlegmatic and watery moisture leaves it, which then is mixed in the broth. As a result, this does not become as sweet or as well suited to nourish a human nature as the first one. Thus one must reply that marrow is the nutriment for the bones; and in what way has been stated.

1. On to the arguments. To the first argument one must respond that bones are nourished because in truth they have porous pathways through which nourishment passes, although these are not clearly seen owing to their solidity. But these are evident when they are broken, and they are visible to the senses in the bones of the dead.

2. To the second argument one must respond that just as fat, while it surrounds the flesh, does not nourish it, so too the marrow, when it is within the bone, does not nourish the bone, although at such a time its potential for doing this is remote. But just as natural heat can convert the fat surrounding the flesh into nourishment for flesh, so can it convert the marrow into nourishment for bones.

3. To the third argument one must respond that although veins do not pass through the bones, nevertheless the veins' ends and their orifices are joined to the bones, and the moisture seeps out from these ends and is absorbed by the bones' pores. For this reason, etc.

Question 14: Whether all bones are nourished.

One asks further whether all bones are nourished.[46]

1. It seems not. For some bones are not hollow, like the bones of feet and hands. Therefore, no marrow, which is the nourishment for the bones, is found in ones like these. Therefore, etc.

2. In addition, whatever can be nourished can also grow and, as a result, extend. But bones, since they are dry, hard, solid, and compact, cannot be extended. Therefore, etc.

On the contrary. Everything participating in life is nourished. But every bone participates in life, and therefore, etc.

46. Cf. Avic., *Can. med.* 1.1.5.1; Averroes, *Part. An.* 2.9; A., *DA*, 12.2.2.111–12 (*SZ* 2: 935–36).

To this, one must respond that every bone is nourished. And an indication of this is that when the body grows the bones are proportionally increased in size, either in length or in thickness. But this can only be if the bones are nourished. And besides this, just as the heat in the parts surrounding a single bone acts on it, so too is this the case in the part surrounding any given bone. Just as that one needs nourishment, so too does each one.

1. To the arguments. To the first argument one must reply that although certain bones are solid (except for their hollow area), they nevertheless remain porous and therefore can be nourished.

2. To the second argument one must respond that bones can exist in two conditions. They are sometimes perfectly hard, and other times imperfectly hard. In the first condition they neither grow nor extend, but in the second condition they can grow, just as other members do.

Question 15: Whether the marrow is hot.

One inquires further into the complexion of the marrow, and whether it is hot and moist.

1. It seems not. For what is derived [*rescinditur*] from something else bears witness to it, according to Galen in his *Tegni:*[47] but marrow is derived from the bones' nourishment, which is cold and dry. Therefore, marrow will be cold and dry.

2. In addition, every effect bears witness to its principle. But the first marrow is the marrow of the brain and the second is the spinal cord [*nucha*] crossing through the vertebrae. But these marrows are cold and moist, and therefore so will be all the others.

The Philosopher says the opposite.[48]

One must reply that marrow is of two types: one is ordered to nutrition, and this one is generated from spermatic blood or from

47. Galen, *Ars* 4 (Kühn 1.315–17).

48. Ar., *Part. An.* 2.7 (652a27f.); cf. A., *DA* 12.2.3.113 (*SZ* 2: 936).

the nutriment, and this marrow is situated in bones. The other marrow is ordered to sensation and towards influencing motion, and this kind of marrow is the marrow of the brain and of the vertebrae. But the first is hot and moist, because provident and learned nature always has ordained that one contrary be next to another in the body, so that they might temper one another. And nature situated the first marrow in bones, which are cold and dry. It is necessary, then, that this marrow be hot and moist.

Moreover, the first marrow shares the nature of blood, and an indication of this is that it changes its heat in accordance with a change in the blood, which has one heat in youths and a different heat in old people. But blood is hot and moist. For this reason, the marrow is hot and moist. But the second marrow is cold and moist. And the reason for this is that sensation consists in being acted on and being altered. Therefore, that which serves the senses must be especially capable of being acted on and hardly active at all. But the cold is less active than the hot, and the moist is more passive than the dry, and this is why the marrow of the brain and vertebrae is cold and moist. But nevertheless the spinal cord, which is the marrow for the vertebrae, is hot in comparison to the brain, but cold in comparison to the other marrow, because it is nearer to the brain and for this reason receives coldness from it. But it is nearer the heart than the brain, and this is why it is hotter than the brain itself.

1. To the first argument one must reply that something may be derived from something else in two ways. It may be according to the path of division, and something thus derived bears witness to the one from which it is derived. Or it may be according to the path of separation. By this path, the four elements can be derived from the same mixed body, and one derived in this way does not always resemble the one from which it is derived. It is in this way that marrow is derived from the bones' nourishment.

2. To the second argument one must reply that not every marrow takes it origin from the brain or from the spinal cord, but the bones' marrow takes its origin from the spermatic blood, and this is why it has a yellowish color, whereas the spinal cord has a white color.

Question 16: Whether the brain is cold owing to the heart's tempering action.

One asks further whether the brain is cold owing to the heart's tempering action, just as the Philosopher himself says.[49]

1. It seems not, for that which is the principle of sensation is not without heat, because sensation does not sense without heat. But the brain is the immediate principle of three of the senses. Therefore, it is not without heat.[50]

2. In addition, whatever floats is hot because it is the nature of a hot thing to rise up. But the brain floats, and therefore it is hot.

3. In addition, every part exists for the sake of its own operation, as the Philosopher says at the end of book eleven.[51] But the heart is not an operation of the brain. Therefore, it is not necessary to posit the brain's coldness for the sake of the heart.

The Philosopher says the opposite.

One must respond that the brain is cold for the sake of tempering the heart, for the heart has a superabundant heat, and life requires that tempering occur in that in which it exists, and this is why nature ordained some member, in which cold is superabundant, by which the heat of the heart might be tempered, so that a middle disposition might be brought about. And for the same reason nature arranged the brain opposite the heart so that they might temper each other. And it arranged that the brain be above the heart because heat's motion is from the center to the circumference, and cold's motion is from the circumference to the center. This is why the heart is in the middle, so that it may cause heat to flow to the parts on the circumference, and especially upwards. But the brain exists in an elevated part so that it might cause its motion to flow from the circumference to the center, and especially downwards. Thus, if the heart were above and the brain below, the one would not

49. Ar., *Part. An.* 2.7 (652b19f.); cf. Avic., *Can. med.* 1.1.3.1; *DA* 12.6–7 (fol. 46rb–47va); Averroes, *Part. An.* 2.7; A., *DA* 12.2.3.116 (*SZ* 2: 937).

50. Cf. *QDA* 1.43; Ar., *De anima* 3.1 (425a6).

51. Ar., *Part. An.* 1.5 (645b14f.).

influence the other, because the brain would move downward where the heart would not be, and the heart would move upward, where the brain would not be, so that neither one would stand in the path of the other's influence. This is why provident and wise nature, trained to obey its Creator, thus ordained that the one would assist the other, etc. The brain's operation is an indication of its coldness. Now, when vapors reach the brain, they grow thick—often, more than is necessary—and then, according to the Philosopher, a catarrh results from the brain's coldness. This is just as occurs in the macrocosm when rain is caused by a great condensation of vapors due to the coldness of an intervening gap in the air [*medii interstitii aëris*].[52] And on account of this, the brain's big size is the cause of somnolence, and this only arises due to its coldness, by which the vapors are thickened.

1. To the arguments. To the first argument one should respond that something can be hot in two ways: either by complexion, like fire, pepper, and wine, or by participation (that is, through something else) like heated water. Now then, the brain has a cold complexion; nevertheless, it has as much heat as it needs, or as much as it needs for sensation, owing to the heart's influence.

2. To the second argument one must respond that floating can occur from two causes: in one way, from heat, and this is why a candle burns in water and always rises up until it is consumed, especially if something is removed from the lower part of the candle so that a little bit of the wick extends about a finger's length above the tallow or wax. The reason for this is that the flame floating on the water moves upward, and draws the candle with it. This is why the candle's body always rises up and floats.

In another way something rises up owing to its viscosity, and this is why oil and mucous and things like this float since, owing to their viscosity and oiliness they cannot break up or divide the water and its parts and descend; or, because such things are airy,

52. Ar., *Part. An.* 2.7 (652b33f.); cf. A., *DA,* 12.2.3.116–17 (*SZ* 2: 937–38).

they have to float. And this is the reason why the brain floats.

3. To the third argument one must reply that ultimately the brain exists for the sake of its own operation. Nevertheless, it has the complexion it does for the sake of tempering the heart. Thus ultimately it exists for the sake of the heart; this is why, etc.

Question 17: Whether sensation is in the brain.

One asks further whether sensation is in the brain.

1. It seems that it is. The first in each genus is, according to the Philosopher in the second book of the *Metaphysics,* exemplary of the genus.[53] Thus, because fire is the first heat, therefore of all things it is especially hot. But the brain is the principle of the sensitive power because the sensitive nerves arise from the brain's web. Therefore, sensation thrives especially in the brain.

2. In addition, touch is the guardian for the entire bodily mechanism, as both the Philosopher and Avicenna say.[54] If, then, touch thrives in any part of the body, then it thrives, as a result, in the brain.

The Philosopher says the opposite. For he says that sensation is not in the brain, and it is not in any other moisture.[55]

To this one must respond that sensation is not in the brain, taken absolutely. The reason for this is that sensation does not occur without heat. The brain, however, is a cold member, for it is composed of earth and water. This is why, when the brain is cooked, it resembles earth, according to Isaac [Israeli] in his *On Diets* [*Dietae*]. And this is why the brain cannot be the instrument of sensation.

Furthermore, sensation does not occur without blood or something analogous to blood, as is said in the text.[56] But there

53. Ar., *Metaph* 2.1 (993b24f.).

54. Cf. Ar., *De anima* 2.3 (415a3f.); 2.11 (423b29f.). Cf. Avic., *Can. med.* 1.1.6.5; *De anima* 1.5; A., *DA,* 12.3.2.173 (*SZ* 2: 960).

55. Ar., *Part. An.* 2.7 (652b2–6); cf. Averroes, *Part. An.* 2.7; A., *DA* 12.2.3–4.113–34; 12.3.1.165–71 (*SZ* 2: 936–45; 956–59).

56. Ar., *Part. An.* 2.7 (652b9f.).

is no blood in the brain, and therefore neither is there sensation. Notwithstanding the fact that there is no sensation in blood, according to the Philosopher, nevertheless without blood or something analogous to it there is no sensation. This is because something generated from blood or served by the blood is the first instrument of sensation.[57] It is the same for the brain, because certain things are generated from the brain's power—for example, the animal spirits—without which there could be no sensation. Now, the vital spirits are generated in the heart and flow through the arteries to the brain, and there, owing to the brain's coldness and the narrowness of the veins and the opposition to their motion, these become animal spirits. And then they are sent to the particular senses, and this is why the animal spirits immediately proceed from the brain only to three senses: namely, hearing, smell, and sight. For this reason, it is said that the brain is the principle of the senses, although in a formal sense sensation is not in it.

1. To the arguments. To the first argument, that that which is first in a genus is exemplary of that genus, it must be said that this is true of those that share the same name and cause, but not of other things. For although the sun is the principle of heat, nevertheless it is not exemplary of the genus in act because there is no elemental quality in the sun.

Or, it can be stated otherwise that just as the sun is the principle of heat and nevertheless is not, formally, the first hot principle, thus too it can be said that even though the brain is the principle of sensation, nevertheless it is not the first sentient.

2. To the second argument one should respond that touch is the guardian of the whole bodily machine, but nevertheless it is not present organically in every part of the body. Rather, there are certain parts in which touch does not exist, and this is why these parts are surrounded by other parts, in which touch is present. Such parts are the humors, and this is why, etc.

57. Ar., *Part. An.* 2.10 (656b20f.).

Question 18: Whether the vital power has something analogous to it and whether it exists in some organ.

It is asked further whether the vital power has something analogous to it and whether it exists in some organ.

1. It seems not. A universal power corresponding to the entire body does not have a part as an organ. This is evident from the power of growth [*virtus augmentativa*], because every part grows. This is why life is not in an organ, but rather every part participates in life. This is why it is not in a part as if in an organ.

2. In addition, organs depend on their functions, and the functions proceed from the powers. An organ does not correspond to that act that proceeds immediately from substance but only to that which proceeds from a power. But life proceeds from substance or immediately from the soul. Therefore, life is not in some part as if in an organ.

On the contrary. Every organic operation proceeds through some mediating bodily part. But in these lower regions life is a bodily operation.[58] For we live in the soul and the body, as is said in the second book of *On the Soul*.[59] Therefore, it occurs through some mediating part.

It must be said that life is present in a given part as if in an organ, because the soul is one and the parts of the body are many. Since, then, it is necessary for there to be some order among the parts in one thing having several parts (otherwise, there would be no unity there), it is necessary that there be one part of the body on which other parts depend. And it is necessary that the first seat of the soul be in this part, and from this part the powers of the soul must flow to the other parts. This part is the heart, because the heart is the first one generated in the animal, and the first to live and the last to die, which would not be so if life were not based in the heart; and this is why the heart is situated in the middle [of the body] just as the prince is in the middle of his kingdom.[60] Thus perhaps one can say that

58. Whereas, in superior forms, like angels, life does not require a body.
59. Ar., *De anima* 2.2 (414a16f.).
60. Ar., *De motu animalium* 10 (703a30ff.); A., *De princ. mot. proc.* 2.8.

the soul is not in every part, just as the prince is not in every part of the kingdom, but rather dwells more in a castle or in the middle of the kingdom.

1. On to the arguments. To the first argument one must respond that life is in every part of the body, but it is in the heart in a radical way and in other parts by virtue of participation and power, just as a king living in the middle of the realm lives in all parts of the realm in power and by virtue of participation. And this is why the heart is the proper organ of life.

2. To the second argument one must reply that although the soul is the first principle of life, nevertheless it is not the immediate principle, because life does not extend from the soul except through some mediating power, like the sensitive or vegetative power, since sensation is a sort of life; and this is why, etc.

Question 19: Whether one element is dominant in any given sense organ.

Next, one inquires whether one element is dominant in any given sense organ.

1. It seems not. For that which consists in a mean proportion of certain other things, possesses none of them predominantly. But a mean proportion of tangibles is present in touch. Therefore, in touch there is no element that dominates.

2. In addition, if one element is dominant in some sense organ, earth and water will be dominant in any sense organ, because these prevail in all mixed bodies, according to the Philosopher in the second book of *On the Soul* and in *On Sense and the Sensed.*[61] But the one in which water is dominant is sight. Therefore, there are not several senses.

The Philosopher says the opposite in the text.[62]

One must reply that, according to the Philosopher in the sec-

61. Ar., *De anima* 3.1 (425a3f.); *De sensu et sens.* 2 (438b16f.); A., *De sensu et sens.* 1.14.

62. Ar., *Part. An.* 2.1 (647a9f.); *De sensu et sens.* 2 (438b18f.). A., *DA* 12.1.1.12 (*SZ* 2: 898).

ond book of *On the Soul,*[63] the nature of the organ and of the medium and of the object are one and the same. But they are the same in diverse ways, because the nature of the organ and of the medium is the same as if of two things receptive of the same thing, but the nature of the medium and of the object is the same just as the nature of act and potency, because the object assumes the role of the agent, and the organ and the medium the role of the one undergoing change. This is why the nature is said to be the same, because it is analogous, since that which is the organ and medium in potency, is the object in act. But in every object some one element is dominant, so that in smell the hot and the dry dominate, and this is the same nature as fire. And the clear and the bright is in color, and this belongs to the nature of air, and so too for the others.

And this is also apparent with regard to the medium, because whatever things perceive the same thing received that thing through some common nature. But the organ and the medium receive the sensible; therefore, they do so through some common nature. But the medium receives only through something that is found in the element; therefore, the organ does not. This is because taste and touch are changed in a material way—touch by being warmed, and taste by becoming moistened—and therefore in their organs something earthy is especially dominant, for example, earth. But sound changes with local motion mediating the change, and this is why a light element predominates in the organ perceptive of it, and air is of this sort. And likewise for sight. And in smell the hot and the dry dominate, because it is changed by altering, since it is not changed by warming.

One must understand, nevertheless, that this dominance can be understood in two ways: materially and formally. If materially, then the various elements are not dominant in the diverse organs, since earth and water dominate in every organ. If formally, then the proposition is true because for an element to dominate in some organ means nothing else than that the organ has a certain disposition toward an element that is especial-

63. Ar., *De anima* 2.11 (424a1f.).

ly analogous to the object. Through this disposition the organ is receptive of the object, just as on account of clearness and brightness the eye is receptive of something visible.

1. To the first argument one must respond that, formally speaking, touch consists in a mean proportion of the elements, for otherwise it would not perceive the extremes. With respect to their material, however, earth and water dominate, just as in every mixed body.

2. In the same way the response to the second argument is clear. For this reason, etc.

Question 20: Whether the organ of touch and taste is in the heart and whether the organs of the other senses are in the head.

One inquires further whether the organ of touch and taste is in the heart and whether the organs of the other senses are in the head.

1. It seems not. What is common to all is proper to none. But the heart is common to all the senses; therefore, it will not be specific to any one of them.

2. In addition, if the organ of taste were in the heart, since the heart infuses its power into all parts, then it would not perceive a taste more through one part than through another but rather would perceive a taste through every part, just like touch. But this is not true, because taste thrives in the tongue and not in the other parts, and thus, etc.

3. In addition, a more noble location is due to a more noble power. But sight is a more spiritual sense. Therefore, it demands a more noble location, and this is the heart.[64] Therefore, it has to be situated in the heart, more than does the organ of taste and touch.

The Philosopher says the opposite.[65]

One must respond that principally and naturally taste and

64. Cf. Ar., *Metaph.*, 1.1 (980a23f.).

65. Ar., *Part. An.* 2.10 (656a27–35). Cf. Avic., *DA* 11 (fol. 44vb); Averroes, *Colliget* 2.11.20; Averroes, *Part. An.* 2.12; A., *DA* 12.1.1.13–15, 12.1.7.91–96, 12.2.8.148–49, 12.3.1.168–71 (*SZ* 2: 898–99; 928–29; 950–51; 958–59).

touch must be next to the heart, but the other senses must be in the head. And the reason for this is that nothing is sensed without heat. But earth is especially dominant in the organ of touch and taste, and as a result that organ, left to itself, is cold and is incapable of sensation, and this is why nature, wise in its wisdom (which, according to Galen in *On Critical Days* [*De criticis diebus*], is boundless),[66] ordained that the organ of touch and taste be next to the heart. This is so that they may receive heat, which they do not have by virtue of their complexion, from the heart, which is the source of heat, taking it by participation in and by the influence of the heart. And an indication of this is that when any part of the body is injured by something tangible, the pain passes immediately to the heart. The other senses, however, are more spiritual. Thus, they would be impaired more by the excess heat of the heart if they were situated there. So if the organ of sight were next to the heart, the glacial humor, in which sight thrives, would be easily dissolved, and the organ of smell would be burned by too much heat. And likewise the air, connatural to hearing, would be overly moved by the excessive heat of the heart. This is why these three senses are situated in the head. For nature is provident and on its own would never provide defective coverings for these parts, like the eyelids for the eye, and for smell the protective covering within the nostrils, and for hearing the [outer] ears as well as the twists and turns within the ears. Nevertheless, these exist in diverse ways among animals, because some animals have a refined sense of smell and have a covering over it, and these do not smell unless the covering is lifted up, and they smell only while breathing.[67] Others have a stronger olfactory organ that is not easily damaged, and ones like this do not have a covering nor do they smell [only] while breathing. It is the same for the eyes. Those having tough eyes, like fish, do not have eyelids but those that have sharp eyes do have eyelids. And if the eyelids are made of a soft, light flesh, then the animal closes its eye with the up-

66. Galen, *De diebus decretoriis* 2.2 (Kühn 11.843–44).

67. This seemingly senseless statement may refer to a situation that exists in certain animals who submerge beneath water. Hippopotamuses, seals, walruses, and amphibians have just such a mechanism in their nostrils.

per eyelid. If, however, the eyelid is tough, then it closes its eye with the lower eyelid, like the hen or birds, or with a membrane stretched out from the corner of the eyelids, like birds of prey, like the sparrow hawk [*nisus*] and others like it, because animals such as these cannot move the upper eyelid without straining their eyes.[68] And such are the animals that have a rough flesh or a rough and spiny hide owing to the insertion of feathers, and the serpents, which have a rough and scaly hide.

1. On to the arguments. To the first argument one must reply that although the heart is a member common to each sense, nevertheless one sense may require the heart's influence in a more immediate way than another sense, such as those that exist for the sake of life's soundness. Those, however, which are for cognition do not require as much.

2. To the second argument one must respond that touch perceives tangible qualities and is the guardian of the entire body, and this is why tangible qualities are perceived in every part of the body, through the power of touch. But taste perceives only the nutriment, which is taken in from outside, and because the entrance of the nutriment from outside is not apparent in every part of the body, the power of taste does not flourish in every part of the body, but only in the mouth and on the tongue. Thus the tongue receives the species of flavor and entrusts it to the heart through the nerves.

3. To the third argument one must respond that something which is nobler in itself nevertheless is not nobler in comparison to something else. The eye is one of the most noble parts of the body. Nevertheless, it would not be fitting for the entire body to be just like the eye, because then it would be liable to injury. Thus, although a middle position is the most noble, nevertheless it would not be the most suitable for the eye, because vision is especially spiritual, and material, active qualities especially flourish in a middle position, and extremes of heat would impair the spiritual nature of vision. An indication of this is that sight enjoys a tempered cold more than heat, since the eye is in-

68. The membrane is the nictitating membrane. On the *nisus,* see *DA* 23.24.129(83) (*SZ* 2: 1639).

jured more by fire and by looking at it than by water, because fire dissolves the tender substance of the eye, whereas cold water congeals the visual spirits and strengthens them, according to Avicenna.[69] And this is why, etc.

69. Cf. Avic., *Can. med.* 3.3.1, 1–4.

BOOK THIRTEEN

Question 1: On the complexion of the heart.

"IT FOLLOWS NOW to speak about the nature of the teeth," etc.[1] In this thirteenth book one should first inquire about the heart, because earlier there was an investigation into teeth and nails. And first one inquires into the complexion of the heart.[2]

1. And it seems that the heart is hot and moist. For the one generating and the one generated are alike. But the heart is the principle of the blood's generation, and the blood is hot and moist. Therefore, the heart is hot and moist.

2. Moreover, like is nourished by one like itself. But the heart is nourished by the thinner blood, and therefore, etc.

3. Moreover, life depends on the hot and moist. But the heart is the principle of life. Therefore, etc.

To the contrary. According to the Philosopher, the heart is opposite to the brain, both in its disposition and its location.[3] But the brain is cold and moist. Therefore, the heart, by its opposition, is hot and dry. The proof is here in the *Topics:* "When you have two contraries, if some property is in the one, the contrary property will be in the other."[4]

To this, one must reply that the heart can be considered in two ways: either materially or formally. If materially, then the

1. Ar., *Part. An.* 3.1 (661a34f.). Cf. Avic., *DA* 12.14 (fol. 50rb). Averroes, *Part. An.* 3.1; *Colliget* 1.4; A., *DA* 12.3.6.210–20 (*SZ* 2: 975–80).

2. Cf. Ar., *Part. An.* 3.4 (665a27f.). Avic., *DA* 12.2 (fol. 45rb). Averroes, *Part. An.* 3.4; A., *DA* 13.1.4.25–28 (*SZ* 2: 995–97).

3. Ar., *Part. An.* 2.7 (652b26f.).

4. *Si oppositum in opposito, et propositum in proposito.* Cf. Ar., *Topica* 7.3 (153a33f.), and Boethius's translation (PL 64: 990D). The phrase also appears earlier at *QDA* 4.1–2.

heart's complexion is melancholic[5] because the heart is very hard and compact, and this is why it nourishes most poorly, according to Isaac [Israeli] in his *Diets* [*Dietae*].[6] If it is considered formally, then the heart is choleric with a hot and dry complexion. It is clear that it is hot because the natural heat and spirits flourish in the heart. It is also clear that it is dry, because heat cannot be preserved long in a moist material. But heat is preserved in the heart for life's duration. And this is why it is necessary for it to be in a dry material.

Moreover, the heart is a principle of motion and is not susceptible to injury, according to the Philosopher in the first book.[7] Therefore, it is necessary for it to have the power to resist noxious things, and the moist cannot do this as the dry does. This is why, etc.

Therefore, briefly one can say that the heart has a melancholic complexion as a result of its mixture, whereas insofar as it is the principle of the generation of blood and the spirits it has a choleric complexion. Thus, when properly speaking about the heart, just as now one inquires about the heart, it is hot and dry; nevertheless, it becomes moist by virtue of the blood contained in it, which is delegated to it from the liver.

1. On to the arguments. To the first argument one must reply that what is generated resembles the one generating in acts of generation taken univocally. But the heart is not related in this way to the blood.

2. To the second argument one must reply that although the nutriment is similar in the end, in the beginning it is dissimilar.

3. To the third argument one must reply that life depends on the hot and the moist, whose principle is in the heart, because the radical moisture and natural heat take their origin from the heart. Therefore, life depends on these just as it does on proximate principles. But it does not necessarily follow from this that the heart is hot and moist formally, but it suffices for it to be hot and moist effectively.

5. That is, cold and dry.

6. Isaac Israeli, *Lib. diet.* part 5. (*Opera omnia,* Lugd. 1515 fol. 138va).

7. Ar., *HA* 1.17 (496a4f.); *Part. An.* 3.4 (667a33f.). Cf. *QDA* 1.55. A., *DA* 1.3.4.583; 13.1.4.37 (*SZ* 1: 268; 2: 1000–1001).

Question 2: Whether an animal that has a large heart is fearful.

One inquires further whether an animal that has a large heart is fearful, or whether a large heart is a sign of fearfulness, as is said in the text.[8]

1. And it seems not. Males are generally larger than females, but males are bolder than females. Therefore, a large heart corresponds more to boldness than to fearfulness.

2. In addition, boldness is caused by the natural heat and by the strength of the power and spirits. But these can be contained better in a large heart than a small one, and therefore, etc.

The Philosopher says the opposite.

One must reply that a large heart may arise from two causes: either from an abundance of matter alone or from the intensity or strength of the active power [*virtus agens*]. The first type of largeness is not a good sign, because power is much weaker in a dispersed matter than in a compacted matter, just as the Philosopher says, that "fire warms a large house less than a small one."[9] The second type of largeness is a good sign, however, because if the heart is large and the heat and spirit are proportionate to it, then a large heart can do more than a small one. Therefore, if the heart is large and the other parts are proportionate to it, this is a sign of boldness. If, however, it is large and the power of these others is not proportionate to it, then this is a sign of fearfulness. And in this case a small heart is better than a large one. And for this reason an animal with a small heart may be bolder than an animal with a larger heart, just as the lion is bolder than the horse, and the dog is bolder than the cow.

1. On to the arguments. To the first argument one must reply that males are bolder than females not only because they surpass the females with respect to the size of their material mem-

8. Ar., *Part. An.* 3.4 (667a11f.). Cf. Avic., *Can. med.* 3.2.1.1; *DA* 13.3 (fols. 51rb–vb). Averroes, *Part. An.* 3.4. A., *DA* 13.1.4.35 (*SZ* 2: 999–1000).

9. Ar., *Part. An.* 3.4 (667a24f.). A., *DA* 13.1.4.36 (*SZ* 2: 1000).

bers but also with respect to the strength of their power and of the other spirits, etc.

2. To the second argument one must reply that if the natural heat corresponds proportionately to the large size of the heart, a large heart is better than a small one. But the Philosopher is not speaking about this; rather, he speaks by way of comparing the same heat to a large and a small heart. And therefore the effect [of the heat] has greater influence on a small heart than on a large one.

Question 3: Whether a softer heart is better.

One inquires further whether a softer heart is better.

It seems not. For every dry thing, the hotter it is, the harder it is. But the heart is dry, as has been said. Therefore, the hotter the heart, the harder. But the harder it is, the better it is, and therefore, etc.

In addition, that member is better disposed which can better resist harmful things. But the harder the heart is, the better it can resist harmful things. Therefore, etc.

The Philosopher says the opposite.

One must say that softness can arise from two causes: either from a defect in power and in natural heat that is unable to consume the superfluities and to harden a member, and this type of softness is in women when compared to men. Or, softness can arise from an abundance of blood, and this type of softness can be in one animal compared to another, and this type of softness is a sign of a good disposition, because in the second book of *On the Soul* the Philosopher says of this type that "those soft in flesh are sharp in mind."[10] But the first type of softness is a sign of mobility, inconstancy and defect, and this is why women are more inconstant and changeable than men. Therefore, the Philosopher understands in the second way that a softer heart is better, and not in the first way. And, just as a statement has been made about softness, so too by way of opposition a distinction must be made about hardness.

And in this way one can respond to the Philosopher.

10. Ar., *De anima* 2.9 (421a26f.). Cf. A., *De anima* 2.3.23.

Question 4: Whether the heart is hotter than the spirit.

One inquires further whether the heart is hotter than the spirit.

It seems not. In the genus of hot things those that are more rarefied [*subtiliora*] are hotter. But spirit is more rarefied than the heart. And therefore, etc.

In addition, the heart is situated opposite to the brain, so that the heart's heat will be tempered by the brain's coldness, and vice versa. But this is not the case for the spirits, since they are not situated opposite the brain, and therefore, etc.

To the contrary. That which is the principle of heat in something is hotter than is a derived heat [*principiato calore*] in something. But the heart is the principle of heat and of the spirits, and therefore, etc.

One must reply that one thing can be hotter than another in two ways: either formally or accidentally. Formally, the more rarefied are hotter than the dense, as air is hotter than water and fire hotter than earth. But accidentally, heated solids are hotter than rarefied things, seeing that the heat is incorporated into them more and is preserved longer. This is why heated iron heats more than the hottest water, owing to the subject's solidity. And this is why, formally speaking, the spirits are hotter than the heart, but speaking accidentally the heart is hotter than the spirits because it is more solid than they are, and this is why the heat is incorporated into it more and is retained longer.

By this, a solution to the arguments is apparent.

Question 5: Whether a heart so flows to the other parts that it receives nothing from them.

One inquires further whether a heart flows to the other parts in such a way that it receives nothing from them.[11]

It seems not. Because everything acting physically undergoes something when acting and is moved when moving. But the

11. "Flows to": throughout this section, A. uses the verb *influo* without restraint. At any given time it can mean "flow to," "influence," or "infuse." Cf. A., *De princ. motus proc.* 2.6.

heart is a thing that moves physically and acts. Therefore, it undergoes something and is moved. But every thing that is moved and undergoes something receives something from something else. Therefore, etc.

Moreover, the heart is situated opposite the brain, so that it receives tempering from the brain's coldness. Therefore, it does not flow to the other parts in such a way that it receives something from others.

The Philosopher says the opposite.[12]

To this, one must reply that some members are only receiving members, like flesh and bone, and others are receiving members and also flow out to others, like nerve and vein, since what they receive from others they infuse into others, and then there is some member that so infuses that it does not receive. This is apparent from the Philosopher's argument in the eighth book of the *Physics*,[13] where he says that if two can be found joined and still another can be found that exists *per se*, and yet another remaining one can be found that exists *per se*, as if something is both moving and moved, and something else is so moved that it is not moving, then it will also be possible to find something that is so moving that it is not moved. So in his premise. But this rule should be understood with respect to things that are united accidentally.

Nevertheless, one must understand that reception is of two types: one is virtual and the other is material. And the material can be of many types, just as one of the members receives from another so that what has been lost is restored or so that its disposition will be rendered temperate. And the heart receives from the other members in these ways. For it receives nutriment from the stomach and from the liver and veins, for it is subject to loss just as are the other members. Moreover, it receives cooling through the lungs' drawing power, because the lungs serve as a fan for the heart, and likewise it receives cooling through the coldness that flows in from the brain. But virtual reception does not suit the heart in comparison to the other members be-

12. Ar., *Part. An.* 3.4 (667a34f.). Cf. A., *DA* 13.1.4.37 (*SZ* 2: 1000–1001).
13. Ar., *Phys.* 8.4 (254a7f.).

cause the heart is the first seat and first dwelling or domicile for the soul, infusing power to all the other members, and it does not receive power from them but only accidental or material dispositions.

In this way a solution to the arguments is apparent. For they prove that the heart receives something from other parts, but they do not show that power flows to it from other parts.

Question 6: Whether the heart is susceptible to illness.

One inquires further whether the heart is susceptible to illness.

1. It seems so. Because *syncopis* and *cardiaca* are afflictions of the heart. But these are illnesses, and therefore, etc.[14]

2. Moreover, nothing passes from one end to another except through a medium. But illness is a path to death, and the heart is subject to death, and therefore to illness.

3. In addition, every member that is capable of being extended or contracted beyond its requisite state can become ill. But the heart can be extended and contracted; therefore, etc., because a person dies on account of too much joy or sadness. Therefore, the heart can become ill.

The Philosopher says the opposite.[15]

One must reply that there are four times when each illness is subject to treatment, namely, at its beginning, its growth, its stasis, and its decline. The beginning is when, while the illness is imminent, the animal begins to cease its customary operations. Growth is when the signs of illness intensify. But stasis is when the signs of illness reach such a limit that it cannot intensify with nature remaining; the time of decline is when the person begins to convalesce.

Further one must understand that a member can suffer in

14. See *DA* 22.2.1.45 (*SZ* 2: 1472). *Cardiaca* can indicate many heart-related problems ranging from mere heartburn to more serious disorders involving either the heart or stomach. "Syncope" is a term still used for fainting.

15. Ar., *Part. An.* 3.4 (667a32f.). Cf. Avic., *Can. med.* 3.2.1.1, 3.2.2.6. A., *DA* 13.1.4.37 (*SZ* 2: 1000–1001).

two ways: either *per se* or through another. *Per se,* when there is some pustule or other illness on the member itself; through another, as when some illness exists in one member and another suffers along with it.

Therefore, one can reply to the question in this way, that if we are speaking about an illness or some suffering arising from the pain in another member, then in this sense the heart can suffer and become ill, because whenever another member suffers the heart suffers. If we are actually speaking about its own illness, then this sort of illness can exist in the heart in terms of the beginning and growth [of illness], but in terms of stasis this is impossible, because in all things that are properly ordered, when the first ceases then all subsequent things cease as well. But the heart is the first member, and this is why when the heart ceases its proper operations it is impossible for the other members to use their proper operations. Therefore, so long as the animal endures, the heart is always healthy.

1. On to the arguments. To the first, one must reply that *syncopis* and *cardiaca* are illnesses in other parts; nevertheless, they are called afflictions of the heart because the heart ceases from its operations during them.

2. To the second argument one must reply that death is not universally the end of illness. Now sometimes it arises from no previous illness, and if some illness does precede it is not principally an illness of the heart but of the other members.

3. To the third argument one must reply that the heart expands in joy. Therefore, it can expand so much that it cannot close itself, and then the animal dies suddenly and without any illness. But during sadness the natural heat, spirit, and vital power retreat to the interior of the heart, while abandoning the exterior parts. Therefore, in this instance the heart sometimes closes so much that it cannot open itself further, and then death occurs. But both these forms of death, whether they arise from joy or sadness, arise without illness and suddenly. This is why, etc.

Question 7: Whether the blood is generated first in the heart or in the liver.

One inquires further whether the blood is generated first in the heart or in the liver.

It seems that it is in the liver. For blood is generated in that part in which the other humors are generated. But this occurs in the liver and not in the heart, and therefore, etc.

Moreover, if food enters the windpipe,[16] very often suffocation occurs, as the Philosopher says.[17] Therefore, it is necessary that food be thinned and purified before it approaches the heart, because food passes through more members than air that has been breathed in, since food crosses from the stomach to the liver before it approaches the heart, and therefore, etc.

The Philosopher says the opposite.[18]

One must reply that blood can be generated in two ways: either from something moist, as when it is generated from phlegm, or not from something moist, as when it is generated from chyle. And this second generation too can arise from something in two ways: either virtually or materially. And both of these also happen in two ways: either with respect to the whole, or to the part. In the first way, blood can be generated in any member, as when it is generated from something moist. In the second way it can be generated both in the heart and in the liver. But with respect to the whole and materially, it is generated in the liver; but with respect to the part and formally or virtually, it is generated in the heart because the blood generated in the liver is thick and not completely digested, but the blood generated in the heart is thin and perfectly digested. Therefore, according to the physicians the first generation of the blood occurs in the liver, and according to the Philosopher the first generation occurs in the heart, because according to the Philoso-

16. "Windpipe": *arteria.* In *DA* A. regularly uses the term "artery" for the windpipe.

17. Ar., *Part. An.* 3.5 (668b6f.).

18. Ar., *Part. An.* 3.5 (667b6f.). Cf. Avic., *Can. med.* 1.1.4.1, 3.12.1.1, 3.14.1.1; *DA* 3.1 (fol. 31raA–vaB). Averroes, *Part. An.* 3.4. A., *DA* 1.1.5.72–73, 13.1.5.39–42 (*SZ* 2: 1002–4).

pher the heart is the principle and origin of the veins, whereas according to the physicians this is the liver. And this controversy between the Philosopher and the physicians can be solved by the distinction of "the prior" because "prior" is said in two ways: in terms of generation, and in terms of perfection, as a boy is prior to the man in terms of generation but is not prior in terms of perfection. Thus, in the premise, blood is prior in the liver in generation and in time but it is prior in the heart in terms of perfection.

In this way one can make clear the solutions to the arguments, for they proceed by their own paths.

Question 8: Whether some sweat can be bloody.

One inquires further whether some sweat can be bloody.

1. It seems not. For nature acts against illness in three ways: namely, by dividing, digesting, and expelling. Therefore, digestion precedes expulsion. Therefore, sweat that has been expelled was previously digested. But the third digestion, from which sweat proceeds or by which it is expelled, proceeds by whitening. Therefore, no sweat is bloody.

2. In addition, the matter that is expelled from the body is threefold, according to the medical authorities: namely, the coarse matter through the anus, the moderate through a nasal discharge, and the thin through sweat. Since, then, blood is a thin material, it is not expelled through sweat.

The Philosopher says the opposite.

One must reply that sweat sometimes is bloody, and then it has a reddish or purplish color. And sometimes it is phlegmatic, and then it has a white color. And sometimes it is choleric, and then it is yellowish in color. And sometimes it is melancholic, and then it is black. The fact that the sweat may be bloody arises from two causes with respect to the body, and two with respect to the soul. With respect to the body, these are the thinness of the blood and the permeability [*raritas*] of the veins, and in respect to the soul these are a lack of digestive power and sadness or fear, because during sadness or fear the exterior members are abandoned without control, and this is why blood flows hither and thither and frequently exits through sweat, and therefore the sweat is bloody.[19] And so too for the other humors.

1. To the first argument one must reply that in the normal order of things, digestion precedes expulsion, but nevertheless an irregular process often occurs or causes expulsion before digestion.

2. One must reply in the same way to the other, that a flow of blood that occurs from the nose is normal, but a flow of blood through sweat is irregular and contrary to nature; indeed, it is an illness.

Question 9: Whether venous blood is thicker than arterial blood.

One inquires further whether venous blood is thicker than arterial blood.

1. It seems not. The more digested are thicker than the undigested, according to the Philosopher in the fourth book of *On Meteorology*.[20] But blood in an artery is more digested, because there is more heat and spirit in an artery than in a vein, and these are the principles of digestion. Therefore, etc.

2. Moreover, it is the business of heat to thin and rarefy. But blood in a vein is hotter than in an artery, because the right side [of the body] is generally hotter than the left, and a vein proceeds from the right ventricle of the heart, whereas an artery proceeds from the left. Therefore, etc.

3. In addition, a person is bled from a vein and not from an artery. But this is only because blood from a vein can go forth with less pain and difficulty than blood from an artery, owing to its thinness. Therefore, etc.

The Philosopher says the opposite.[21]

One must reply that blood in an artery is thinner than in a

19. See Lk 22.44, where Jesus' "sweat became as drops of blood."

20. Ar., *Meteora* 4.2 (380a2f.); A., *Meteora* 4.1.15.

21. Ar., *Part. An.* 3.5 (668a4f.). Cf. A., *DA* 13.1.5.43–44 (*SZ* 2: 1004).

vein. And the reason for this is that veins are properly the vessels for blood, whereas arteries are vehicles for spirit and heat. Therefore, there is little blood in arteries (namely, just as much as suffices to warm the spirit and the vital heat) in comparison to veins, where there is a great deal of nutrimental blood, and this is also because motion and heat have a power for rarefying and thinning. Therefore, through the continuous motion of the vital spirit bearing the vital power or life to the individual members, and through continually attracting air and natural heat, blood in the artery is rarefied and diffused, and, as a result, is thinned. But there is a greater abundance of blood in a vein, and it is ordered for the nourishment of the parts of the body, and this is why this blood is thicker.[22] And therefore, the spirits are nourished from the first, and the solid members from the second. And in addition, it is the business of heat to consume thin things and to leave behind thick. But heat is greater on the right side than on the left, as was argued, and this is why what is left behind is thicker, because it is more completely digested.

1. To the first argument one must reply that blood in an artery is not more digested. Or it can be said, in another way, when digestion acts on something whole, that which is more digested is thicker. But where digestion does not act on the whole, but only on thin parts, then it is not necessary that what is more digested also be thicker. And so too in the premise, and this is why, etc.

2. To the second argument one must reply that it is the business of heat to rarefy and to thin. Therefore, the heat, which is on the right side, causes the fumes to evaporate from the blood contained in the veins. Thus, what is left is rendered thicker *per accidens*, through the path mentioned.

3. To the third argument one must reply that an artery has the same nature as nerve because the spirit contained in it is in continuous motion. And this is why it is necessary for the artery

22. The troublesome phrase *sicut membra spiritibus*, which ends this sentence in one manuscript and in the edition, has been omitted here for the sake of clarity. It bears every sign of a marginal gloss that has entered the text.

to be more resistant, and this is why an artery that has been divided or cut into is made whole again only with difficulty, and on account of this one is not bled from the artery. And in addition, because the pulse and motion of the spirits is in the arteries, if one were bled from an artery it could happen that the blood would flow out without restriction or that the blood flow could not be restrained owing to the continuous motion of the artery, which impedes the healing of a wound. And a third reason is that there is a greater abundance of blood in veins than in arteries. This is why one may be bled better from a vein, because an abundance of spirits may escape from an artery, and from their escape *syncopis* or even death could occur. Therefore, etc.

Question 10: Whether the liver is generated from semen or from menstrual blood.

One inquires further into the complexion and generation of the liver, whether the liver is generated from semen or from menstrual blood.[23]

1. And it seems to be from the blood. For the liver is coagulated blood. But coagulation does not remove the nature or species of a thing, as is clear in the case of frozen water and coagulated milk and, similarly, glass. Therefore, the liver is generated from blood.

2. Moreover, the complexion and accidents of a thing attest to its origin. But semen is white, as is said in the third book, and those that are generated from semen have a white color, like bones, veins, and the like.[24] But the liver is reddish and has the color of blood. Therefore, it is generated from blood and not semen.

To the contrary. A principal member is generated from semen and not from blood. But the liver is a principal member; therefore, etc.

23. Ar., *Part. An.* 3.7 (669b13f.); Avic., *Can. med.* 1.1.4.2, 3.14.1.1; *DA* 13.7 (fol. 52vb); Averroes, *Colliget* 1.27; *Part. An.* 3.7 A., *DA* 13.1.6–7.49–59 (*SZ* 2: 1006–11).

24. Ar., *HA* 3.22 (523a15f.). Cf. A., *DA* 3.2.9.179 (*SZ* 1: 432).

One must reply that the liver is not generated from blood but first from semen. Because there are four principal powers in the body, which the four principal members serve immediately: for example, the brain serves the animal power, the liver the natural power, the heart the vital power, and the testicles the generative power. If the other members are from semen, then these members arise from semen. And an indication of this is that a principal member that has been amputated does not regenerate. And so it is for the liver, so that if it is wounded it cannot be healed, and this is clear whether medicaments are applied or whether physicians treat it. Because in the first generation of the embryo two bubbles are found in the semen: the heart arises from one of these, and the liver from the other, and this is why one must say that the liver arises from the semen.

1. On to the arguments. To the first argument one must reply that the liver is not said to be coagulated blood, because it is blood, but because it resembles blood in color.

2. To the second argument one must reply that a similarity in color or even complexion does not argue for a unity of substance, and this is why, etc.

Question 11: Whether the motion of the lungs is natural.

One inquires further into the lungs. And first, whether the motion of the lungs is natural.[25]

It seems not. Natural motion occurs simultaneously in both the artery and the heart, because the artery's motion comes from a natural motion in the heart. But the lungs' motion does not occur at the same time with the motion of the heart. Therefore, it does not proceed from the natural power, which is in the heart.

In addition, natural motion is not subject to speeding up or slowing down. But the lungs' motion can be speeded up or slowed down, since a person can hold in his breath for a long time and, as a result, slow down the lungs.

25. Cf. Ar., *Part. An.* 3.6 (668b33f.); Avic., *DA* 13.2 (fol. 51ra–rb); Averroes, *Colliget* 1.23; *Part. An.* 3.5. A., *DA* 13.1.6.46–47 (*SZ* 2: 1005–6).

To the contrary. The lungs' motion is on account of the heart's motion. But the heart's motion is natural, and therefore so too is the lungs'.

One must reply that one motion in an animal is solely natural, and another is only animal, and still another is composed of both natural and animal. The motion of the heart and spirit in the arteries is only a natural motion. Animal motion is only a progressive motion. A composite motion is like the motion of urine and excrement. And the motion of the lungs is like this sort of motion. Now, if one should consider the first principle of this motion, this comes from the heart and this is why it is a natural motion; if, however, we should consider the mode, this is an animal motion because it can be speeded up or slowed down.

In this way a solution is apparent to the arguments.

Question 12: Whether a pulse exists in every animal.

One inquires further whether a pulse exists in every animal.

It seems so. For the pulse exists on account of the heart's motion, just as breathing does. But breathing is present in every animal having lungs. Therefore, so too is the pulse.

In addition, the branches are moved once the root is moved. But the arteries arise from the heart. Therefore, once the heart is moved, the arteries are moved. But the motion of the arteries is the pulse. Since, then, motion is present in the heart of every animal that has a heart and arteries, the pulse's motion will exist in every animal.

Moreover, the pulse is nothing other than the motion of the spirit inside the arteries. But this sort of motion is present in every animal that has arteries. Therefore, etc.

The Philosopher says the opposite.[26]

One must reply that a pulse can be understood in two ways: either on behalf of an extrinsic motion, and thus it is said in contrast to a pulling and twirling movement in the seventh book of the *Physics,* and this kind of pulse can be found in every animal.[27] Pulse can be understood in another way on behalf of

26. Ar., *Part. An.* 3.6 (669a19f.).

27. Ar., *Phys.* 7.2 (243b16f.). Cf. A., *Phys.* 7.1.3.

an intrinsic motion, and this type is double: either accidental or natural. An accidental pulse arises from "windiness," and a pulse like this can exist in any animal, and it is said to be a twitch in the members or a hiccough. If it is a natural pulse, it will be either in the lungs or in the heart. If in the heart, then again it can exist in every animal that has a heart. If in the lungs, then this can exist in two ways: either according to the customary course of nature—and so again a pulse exists in every animal that has lungs just as does movement of the lungs—or according to a non-customary course of nature, and this is nothing other than a sigh or a deep breath. And a pulse like this exists in an animal only as a result of fear or hope. Now when a human is afraid, he can see by his senses a quick and unaccustomed motion of his lungs. And the Philosopher talks about this sort of pulse. And because a human alone properly has hope for future goods and fear of evils, this is why the Philosopher says that a pulse like this exists only in a human.

Question 13: Whether the heat in an artery is natural or accidental.

One inquires further whether the heat in an artery is natural or accidental.

1. And it seems to be natural. Because for each thing it produces, nature provides something by which it may be preserved. Therefore, it does this too for natural heat. But this is not preserved anywhere other than in an artery, and therefore, etc.

2. In addition, the radical moisture is the food and fodder of the natural heat, whereas superfluous moisture extinguishes it. Therefore, it is necessary that it be preserved in something else away from the superfluous moisture. Therefore, etc.

To the contrary. The arteries arise from the heart. Therefore, that same heat that flows from the heart is preserved in an artery. But one such as this is accidental, and therefore, etc.

One must reply that the artery does not serve the natural heat, but more so the accidental and nourishing heat. And the reason for this is that a natural heat exists in every member. If, therefore, an artery were for preserving the natural heat, an artery would be in every member. But this is not true, and therefore, etc.

1. On to the arguments. To the first argument one must reply that natural heat is preserved in several ways, namely, through the expulsion of superfluities, through ventilation, and through a corresponding warming of the nourishment. And this is why it is not necessary to posit arteries for the preservation of natural heat, because it is preserved in other ways.

2. To the second argument one must reply that natural heat can be extinguished by superfluous moistures. This is why an animal body is porous, so that the superfluities and fumes can be expelled through sweat and through other wastes or products of digestion.

Question 14: Whether the diaphragm is necessary.

One inquires further about the diaphragm, and whether the diaphragm is necessary.

1. It seems not. A part that divides is not required between members that share their influences with each other. But the spiritual and nutritive members share their influences, because the heart infuses the liver, and the liver the heart, and the diaphragm is arranged between them. Therefore, it is not necessary, because it impedes these mutual influences.

2. In addition, spiritual members do not differ from nutritive members more than nutritive members do from sensitive or generative members. But no part dividing them from one another falls between the nutritive, sensitive, and generative members. Therefore, neither does one fall between the nutritive and spiritual members.

The opposite is clear from the Philosopher's intention.[28]

One must reply that according to medical authorities there are four distinct regions in the body of perfect animals. One be-

28. Ar., *Part. An.* 3.10 (672b10f.). Cf. Avic., *DA* 13.4 (fol. 51vb). Averroes, *Part. An.* 3.10; A., *DA* 13.1.9.69 (*SZ* 2: 1016–17).

gins at the top of the head and ends at the root of the tongue, and this is the region of the animal in which the brain rules. The second begins at the root of the tongue and passes through the windpipe to the lungs and heart, as far as the belly and intestines. And this region is called the vital or spiritual region, in which the heart rules. The third region begins at the root of the tongue and passes through the esophagus and the belly, liver, and intestines, and ends at the lower parts of the body. And this one is called the nutritive region, in which the liver rules. The fourth begins from the kidneys and the lower part of the vertebrae, and ends at the anterior parts. And this is called the generative region, in which the testicles rule. But it is the case that there must exist a division among these regions; otherwise, the parts of one would be damaged by the parts of another, since they lead to or are arranged for diverse organs. Therefore, the *uvula,* because it is a certain membrane that closes the opening to the windpipe, divides the animal and vital regions.[29] And similarly there is a certain membrane called the *syphac* that divides the generative from the nutritive regions.[30] And the diaphragm is between the vital and nutritive regions, but this one is stronger and larger and more evident than all of the other intermediate webs [*telae*].[31] And the reason for this is that undigested vapors rise up from the stomach, which, if they approached the heart immediately, would suffocate it or induce *syncopis.* And this is why, lest the heart suffer some flaw or injury or failure, either from the stomach or intestines, provident nature arranged a device that would both hold and hold in check, namely, the diaphragm. And this is why it is thicker and denser in the extreme parts at which it is joined to the liver and spleen and ribs, but it is thinner in the middle part which is near to the heart, lest it bear some injury to the heart or lungs on account of its thickness.

There is another reason for the diaphragm's necessity, since,

29. At *DA* 1.2.10.246–47 (*SZ* 1: 138) Albert identifies the *uva* or *uvula* as a "fleshy substance hanging over the upper part of the *epyglotis.*"

30. The *syphac* may have several meanings, but here probably refers to the peritoneum. For an identical usage, see *DA* 1.2.21.416 and cf. *DA* 1.2.24.461 (*SZ* 1: 203; 222).

31. For A., a *tela* is generally a finely textured membrane.

if there were no diaphragm, putrid fumes from the corrupted feces would reach the heart and kill it, as [indicated] before, and, moreover, we would always perceive a putrid odor in the nose and mouth, just as we do on the outside, if there were no diaphragm. And therefore, etc.

1. On to the arguments. To the first, one must reply that although the nutritive and spiritual members are divided by a web, nevertheless there are pathways in the web itself that lie open to their influences.

2. To the second argument one must reply that there is a division between the other two regions, just as there is between the spiritual and nutritive regions, and this is why, etc.

Question 15: Whether an affliction of the diaphragm disturbs the intellect.

One inquires further whether an affliction of the diaphragm disturbs the intellect.

It seems not. For the intellect is not subject to harm and is incorruptible, according to Aristotle in the third book of *On the Soul.*[32] Therefore, it is not moved or disturbed by an affliction in the diaphragm.

In addition, the incorporeal is not affected by the corporeal. But the intellect is incorporeal and does not use a bodily organ. Therefore, it cannot be affected by the diaphragm.

The Philosopher says the opposite.[33]

One must reply that an affliction in the diaphragm does move and disturb the intellect *per accidens,* but not *per se.* And the reason for this is that the diaphragm is an abundantly nerve-filled web and particularly sensitive. As a result, it shares in the nature of the webs of the brain, and because it is near the heart and is a sort of medium through which the heart infuses its powers into other members, this is why, when the diaphragm is blocked, of necessity the heart's influence is withdrawn from the other parts.

32. Ar., *De anima* 3.5 (430a17f.).
33. Ar., *Part. An.* 3.10 (672b28f.). Cf. A., *DA* 13.1.9.70 (*SZ* 2: 1017).

Further, although the intellect is not an organic power, nevertheless it needs other powers which do use organs, since nothing can be understood without a phantasm, as is said in book three of *On the Soul,* because the intellect either is a phantasm or does not exist without a phantasm.[34] And this is just what happens when the intellect's operation is impeded by an injury to the power of memory or phantasms, as is evident among the insane or those suffering lethargy [*lytargici*], not because the intellect uses some organ but because it abstracts species from sensibles and strips and despoils them of matter. Thus the intellect is disturbed by an injury to the diaphragm, because, once an injury has occurred, the heart too is damaged and the influences of the heart are withdrawn from the other members.

With this, one can respond to the arguments. For they prove that an affliction in the diaphragm does not disturb the intellect *per se,* but just as has been said.

Question 16: Whether touching the diaphragm is a cause of laughter.

One inquires further whether touching the diaphragm is a cause of laughter.

1. It seems not, because one and the same thing is not the cause of contraries. But it has already been said that touching the diaphragm is a cause of pain, and therefore it cannot be a cause of laughter.

2. In addition, laughter arises from the purity of the blood and the thinness of the spirits. But these do not arise from the diaphragm, and therefore, etc.

The Philosopher says the opposite.[35]

One must reply that the movement of the diaphragm is a cause of laughter. And the reason for this is that laughter arises from a perception of something agreeable and pleasant, just as, in a contrary fashion, weeping arises from the apprehension of something disagreeable. Now, however, when something touch-

34. Ar., *De anima* 3.8 (432a12f.).

35. Ar., *Part. An.* 3.10 (673a1f.). Cf. A., *DA* 13.1.9.71 (*SZ* 2: 1017–18).

es the diaphragm or some part near it without causing it damage, a certain warming and a certain tickle is caused by this motion in the diaphragm, and from its warming the blood in the arteries of the heart is thinned, and the cleaner the blood is the more agreeable it is to nature. This is why laughter arises from the warming of the diaphragm, because the mouth and the face are indicators of interior dispositions.

Something that contributes a great deal to this is the fact that whatever is very sensitive easily apprehends a contrary thing and, as a result, is easily oppressed by pain. Similarly, that same thing, because it is particularly sensitive, is easily subject to pleasure when it apprehends something agreeable. And the diaphragm is very sensitive, and heat is agreeable to nature, and this is why pleasure proceeds from the warming of the diaphragm.

1. By this a response is evident to the argument, that one and the same thing is not a cause of contraries in one and the same way, but nevertheless can be a cause of contraries in diverse ways.

2. A response has already been stated to the second argument, and so much for it. Therefore, according to some, when the diaphragm is touched by the heart, it sounds like a drum [*tympanum*]. In this way the sound of laughter is caused, etc.

Questions 17–18: On laughter.

One inquires further whether laughter is only proper to a human.

1. And it seems not. For laughter arises from touching the diaphragm. But the diaphragm exists in other animals, just as it does in the human. And therefore, etc.

2. Moreover, tickling the underarms and the palm of the hand and the sole of the feet causes laughter. But these parts, or ones analogous to them, exist in other animals. Therefore, etc.

The Philosopher says the opposite.[36]

In addition to this, one can inquire why touching or tickling

36. Ar., *Part. An.* 3.10 (673a6f.). Cf. A., *DA* 13.1.9.71 (*SZ* 2: 1017–18).

the underarms or the sole of a foot, rather than other parts, causes more laughter.

To the first, one must reply that laughter arises immediately from the cleanness of the blood and the spirits' thinness. Therefore, because the spleen is the receptacle for melancholy, which is the dregs of the blood, it is said to be the principle of laughter, because the spleen cleanses the blood when it attracts the melancholy. But in other animals the blood is very thick, and this is why laughter does not exist in other animals.

The reason this occurs more in the underarms than in other parts is that the underarms are network of pores [*emunctoria*] for the heart and partially adjoin the inner diaphragm. Therefore, an inordinate motion in them at once causes motion in the diaphragm and, likewise, warms it.

And this occurs in the human more than in other animals on account of the thinness of his skin. This arises on the sole of the foot and the palm of the hand because these parts are more nerve-filled and have a good sense of touch. This is why, etc.

Moreover, unaccustomed things are very quickly perceived, and the soles of the feet are not accustomed to being tickled or stroked, and this is why tickling the sole of the foot causes laughter more readily than tickling the hand or other parts of the body.

Moreover, a reason why a person can endure his own touch on these parts rather than that of others is this: namely, habituation and trust of nature. For when parts are based in the same root, the nature of one part trusts another just as it does itself. But when they are not based in the same root, an affliction of the part arises as if it were distrustful of the other, and this is why a person bears his own touch on many parts on which he does not bear the touch of another. And another cause is habituation, which is almost like another nature or which can be changed into nature, according to the Philosopher in the second book of the *Ethics*.[37]

Thus one must respond to the question that the cleansing of the blood in a human is the reason why laughter is properly present in him.

37. Ar., *Eth. Nic.* 2.1 (1103a19f.).

1. To the argument one must reply that merely touching the diaphragm is a cause of laughter only when it is accompanied by clarity in the blood, because the purer the blood, the more it is diffused to the parts and causes joy.

2. To the next argument one must reply that tickling can occur in other animals owing to an unaccustomed touch, or it may happen sometimes that some of them may have thin skin, etc., yet laughter does not occur, etc.

Therefore, it is well known that when a woman is stroked on her breasts and belly, this arouses her then for intercourse, and, although she may hide the fact, she burns then like fire. And this happens for this reason, that her testicles located in the womb hang down on certain nerves and cotyledons coming from the breasts and the belly.[38] Thus when the breasts are touched, the sperm is driven out of them and they pour it out into the womb. This effusion into the womb arouses a soporific warmth in her, just like water cascading over limestone. For this reason, women particularly desire intercourse then. Therefore, at that time, because of the vehemence of the desire, they secretly urinate and emit sperm and sometimes cleanse themselves. Therefore, crafty suitors, as I heard in confessions in Cologne, tempt women with this kind of trick and touch. Women who seem to reject such things, if truth be told, desire them all the more and intend to consent, but in order to appear chaste they deny such things, etc.

Question 19: Whether birds and fish ought not to chew.

One inquires further whether birds and fish ought not to chew.

1. It seems not. Because mastication is a certain preparation for digestion. But fish, since they are cold, have a weak diges-

38. In *DA*, the term "cotyledon" is used primarily to indicate lobules on the placenta that receive nourishment from the veins. It is also commonly used to indicate the point of juncture where a stem meets a piece of fruit. Some similar point of intersection must be envisioned here.

tion, and similarly the nutriment of birds—legumes, for example—is difficult to digest. Therefore, they need to chew.

2. Moreover, the reason why fish do not masticate is, according to the Philosopher, that then too much of the water that surrounds them would enter their bellies and suffocate them.[39] But this conclusion does not follow, because then it would work to prove the same for animals living in air.

3. In addition, the reason why birds do not masticate is that they lack teeth. But this does not suffice, because birds have hard beaks in place of a mouth and teeth. Therefore, just as others chew with teeth, so birds can chew with their beaks.

The Philosopher says the opposite.[40]

One must reply that birds and fish do not chew. There are many reasons for this with respect to fish. One relates to the water that surrounds them, because if they were to chew, water would enter excessively and would work to suffocate and drown them. Another reason is related to their nutriment, which is soft and easily convertible, and this is why they do not need to chew. A third reason is related to gluttony, since these animals are gluttonous and snatch their food so eagerly that they swallow their nutriment almost whole.

The reason related to birds is that they do not have teeth, and even if they had them they would get in their way because birds have curved beaks, or beaks that hardly follow a straight line. Therefore, if birds had teeth, their mouths would remain always open, and as a result curved beaks would be useless to them for retaining food.

1. On to the arguments. To the first argument one must reply that although chewing is preparation for digestion, nevertheless fish do not need to chew because their nutriment is soft and digestible. Nor do birds need to do this, because either their nutriment is easily digestible or they have some instrument in which their nutriment is prepared, and this is why, etc.

2. To the second argument one must reply that water is, acci-

39. Cf. Ar., *HA* 8.2 (592a19f.); *Part. An.* 3.14 (674b34f.).

40. Ar., *Part. An.* 3.14 (674b17f.). Cf. Averroes, *Part. An.*, 3.14.; A., *DA* 13.2.1.80–82 (*SZ* 2: 1022–23).

dentally, softer than air and there is a great deal of windiness in fish because it contributes to their swimming. Now those learning to swim and experienced swimmers alike place on themselves bladders and other light things that are not easily submerged in water. They do this so that they will be supported by them. So too, nature provides fish with a great windiness inside their body so that they may be supported better while swimming. This windiness would be easily corrupted by water entering excessively. And for this reason, while walking animals chew, air exits the same way that it enters, because the lungs are in continuous motion. But this is not the case for fish, because they cannot expel water through their gills as easily as it enters through their mouth, on account of the gills' narrowness.

3. To the third argument one must reply that although birds have beaks in place of teeth, teeth can exist for many reasons for which beaks cannot, because beaks are uniform, as it were, whereas teeth are diverse, because some are sharp for breaking up and some are broad for grinding, and this is why, etc.

Question 20: Whether birds ought to have a double stomach, like a liver and crop.

One inquires further whether birds ought to have a double stomach, like a liver and crop.

1. And it seems not. Because if a weaker thing has power over something, a stronger thing has power over the same thing, and even more so. But a crop is weak and tender, and a liver is strong and thick. Therefore, whatever a crop can do, a liver can do, too. Therefore, it does not need a crop.

2. In addition, what is natural is the same for all. But not all birds have a crop, as is evident among the heron and cranes. Therefore, it is not natural for birds to have a crop.

3. Moreover, possession of a crop is only necessary owing to a weak digestion. But digestion is weaker in fish than it is in birds. Therefore, fish need a crop more than birds do.

The Philosopher says the opposite.[41]

One must reply that many birds do need a crop, for many reasons. One reason is that the warmer they are during the third digestion, the more heat they lack in the first and second digestions. Therefore, generally the liver in birds is particularly temperate and this is why they need [a crop], so that the nutriment may be softened and prepared before it reaches the *iecur* [liver], that is, the *hepar* [liver]. Another reason is that they lack teeth. Now, because they lack teeth and their nutriment is hard—for example, whole grain and similar sand, stones, and things of this sort—and these are not broken up by teeth, therefore it is necessary that they be softened in something before they arrive at the liver, which is a solid member. A third reason is that many birds do not live as predators or acquire nutriment by violence, and this is why nature provides them with an instrument in which the nutriment can be preserved until the time when it is needed.

Nevertheless, not all birds have a crop. The nutriment for some birds is soft and easily digestible and convertible. Examples are those birds that live on worms,[42] or others that live off the fruit of trees, and still others that live off flesh. And in the same way some birds have short and straight intestines, and the food they receive quickly passes out of ones like this, and animals like this, for this reason, do not need a crop. But those animals do need one who are nourished on solid grains or stones or things of this sort. And the reason for this has been given.

1. On to the arguments. To the first, one must reply that the liver is ordered for digestion, but the crop is ordered for retaining and softening, and this is why the crop is tender whereas the liver is hard and solid. Therefore, although the liver is more powerful than the crop, nevertheless this is not entirely so, just as the intellect can do more than a sense, yet it still requires a sense, because a sense can act on something on which the intellect cannot act in the same way, just as sight can see color but

41. Cf. Ar., *Part. An.* 3.14 (674b17–675a21); A., *DA* 12.2.1.80–83 (*SZ* 2: 923–25).

42. "Worms": *vermes*. The word can also indicate insects or any of several ignoble creatures. Cf. *DA* 26 (*SZ* 2: 1739–64), where these are listed.

the intellect cannot. This is so in many other instances as well. Therefore, although the liver is stronger, nevertheless it cannot do what the crop does because it is not ordered for that, just as the intellect cannot see.

2. To another argument one must reply that it is already apparent that the nutriment of some birds is more convertible than the nutriment of others, and those who live more as predators live on cruder foods.

3. To the third argument one must reply that the nutriment for fish is liquid and easily convertible; they have short and straight intestines, and this is why they do not need a crop. Nevertheless, some fish that are not gluttonous live more on harder foods, and they have something like a crop, as the Philosopher says, and this is why, etc.[43]

Question 21: Whether blood is the nutriment for the stomach.

One inquires further whether blood is the nutriment for the stomach.[44]

It seems not. For each member is nourished by that which it digests. But the stomach does not digest nutriment in the blood but in the chyle. Therefore, it is not nourished by blood but by chyle.

Moreover, the first generation of blood is in the liver. But nutriment does not pass from the liver to the stomach, but vice versa. Therefore, etc.

To the contrary. "Blood is the last food of all the members," according to the Philosopher, and therefore also of the stomach, since it is a member.[45]

Moreover, the veins are vessels for the blood. But there are veins in the stomach. Therefore, etc.

One must reply that there are two webs or tunics in the stomach, namely, an exterior and an interior one. The exterior is nourished by blood, but the interior is nourished by chyle. But

43. Cf. Ar., *Part. An.* 3.14 (675a8f.).
44. Cf. A., *DA* 12.2.1.97–105 (*SZ* 2: 930–33).
45. Ar., *Part. An.* 2.4 (650a34f.).

because the Philosopher says that "blood is the last food of all the members," one can say that the stomach is nourished by blood with respect to each web or tunic. For even though blood is not generated in the stomach, nevertheless after the generation of the blood in the heart is completed, it is diffused through fine veins to all the members. Thus, sometimes the stomach is even warmed when it separates the chyle from the blood, since it is strengthened and grows warm and becomes stronger. Nevertheless, its perfected nutrition is received from blood; otherwise, its web would not be covered with thin veins.

One can respond to the arguments in this way, because each arrives partially at the truth. And this is why, etc.

Question 22: Whether several intestines are required beyond the stomach.

One inquires further whether several intestines are required beyond the stomach.

And it seems not. Because what nature can perform through one, it does not perform through several. But the pure is separated from the impure in the stomach. Therefore, the impure can adequately be expelled from the stomach without an intestine.

Moreover, if intestines are required beyond the stomach, then either they are required for further digestion or only for expulsion. If only for expulsion, then one would suffice. If for digestion, then they would be wider above and narrower below. But the opposite is apparent in ruminant animals and in others, like the human and the dog.

The Philosopher says the opposite.[46]

One must reply that intestines are necessary for the sake of the belly, since the stomach is not joined immediately to the anus, because then it would be excessively cooled and cold impedes digestion. And this is why it is separated from the anus by some medium, and that medium is the intestine. But although the first digestion occurs in the stomach, nevertheless digestion

46. Ar., *Part. An.* 3.14 (675b28f.). Cf. A., *DA* 12.2.1.97–105, 13.2.2.84, 86 (*SZ* 2: 930–33; 1025).

is not completed there, because, just as that which is pure and separated by the stomach is not absolutely pure but rather is further purified and digested in the members, so too the impurity expelled by the stomach is not absolutely impure. Rather, something pure can still be separated from it, and this is why there are several intestines beyond the stomach, in order to digest it further. But because digestion is of two types—of the pure and of the impure—for this reason some intestines serve to digest what exits from the stomach, when it is pure, and others serve to digest it when it is impure. But because there are three [stages] in digestion, namely, a beginning, a middle, and an end, for this reason three intestines serve each digestion, so that what is begun in the first is assisted in the second and completed in the third. And this is why in individual animals, for the most part, there are six intestines. Nevertheless, these are ordered differently in different animals, since in those that live on crude nutriment—like sheep, and oxen, and the like—the upper intestines are narrower and the lower ones are wider, because the more their nutriment is digested the cruder is that which remains behind. And because at the end it is especially crude, this is why the lower intestines are wider. But because at the beginning it is thinner and liquid, this is why the upper intestines are narrower. But it is just the opposite in the human and the dog, for a contrary reason, since the human and the dog are alike in many ways. They have a big appetite, and this is why they have a small stomach. If they had a large stomach, many would confound themselves with a desire beyond all bounds, because the diaphragm and, as a result, the spiritual members would be injured by the repletion of the stomach. A human and a dog live on almost the same nutriment, namely, what is thin and easily convertible, and this is why nature arranged for them narrower lower intestines, so that the nutriment would not exit too quickly. And an indication of this is that a dog does not defecate without pain.

Through this a solution to the arguments is apparent.

BOOK FOURTEEN

Question 1: Whether bile [fel] *is only a superfluity.*

"THE DISPOSITION, then, follows in this manner," etc.[1] Here the Philosopher makes a determination regarding the parts of bloodless animals. And this is why one inquires for the first time in Book Fourteen whether or not bile [*fel*] is only a superfluity.[2]

1. And it seems not. Because every member is generated from something like it, from which it is established in its complexion and essence. But there are many choleric members. Therefore, they are nourished by something choleric. But bile is such a thing, and therefore, etc.

2. Furthermore, everything that is produced as a result of nature operating in an ordered manner [*ordinate*] is produced for the sake of something. But four humors are generated in the stomach from chyle, which are necessary for living beings. But one of these humors is contained in the gall bladder. Therefore, bile is necessary.

The Philosopher says just the opposite and criticizes Anaxagoras for saying the opposite.[3]

One must respond that choler [*cholera*] is of two types, since one is necessary and the other is helpful. And the necessary is also of two types: universal and particular. The universal is for the sake of the choleric members, and the particular is for the sake of the gall bladder. The helpful sort is also of two types:

1. Ar., *Part. An.* 4.1 (676a22f.).

2. Throughout this book A. uses both *fel* and *cholera* where we would likely use only the term "bile." As an aid to the reader, we have sought to keep the two distinct.

3. Ar., *Part. An.* 4.2 (676a16f.). Cf. Avic., *Can. med.* 3.15.1.1; *DA*, 14.5 (fol. 53vb); Averroes, *Part. An.* 4.2; A., *DA* 13.2.4, 96–100 (*SZ* 2: 1030–33).

one is universal, and flows in the body with the blood, and the other is particular, and runs to the bottom of the stomach in order to promote digestion.

Thus, when one asks whether bile is only a superfluity, one must respond that it is, because that choler that mixes with blood through the power of nature that cleans the blood is sent to the proper receptacle, that is, to the gall bladder. Thus bile is a superfluity of the choler flowing with the blood.

(1–2).The first argument proceeds concerning the choler, which is outside the gall bladder and out of which the choleric members are nourished. And one must respond in like manner to the second argument, and the Philosopher does not speak about this. This is why, etc.

Question 2: Whether some of the bile or choler is naturally sent to the intestines.

One asks further whether some of the bile or choler itself is naturally sent to the intestines.

1. And it seems not. This is because not only expulsion is necessary to the digestive power, but retention is also necessary. Now expulsion is through the hot and dry, whereas retention is through the cold and dry. But none of the black bile [*melancholia*] is sent to the intestines for the sake of retention, and therefore neither is any of the choler sent for the sake of expulsion.

2. Moreover, a light humor naturally rises up. But of all the humors choler is the lightest. Therefore, it will not naturally descend nor will it be sent to the intestines.

The opposite is clear according to the Philosopher's determination.[4] For he says that the intestines of choleric ones are bitter. But this is due only to the admixture of bile.

One must respond that nature sends choler to the intestines because digestion does not advance unless the expulsion of impurities follows. Now, then, the expulsive power operates with

4. Ar., *HA* 2.5 (506a31f.). Cf. A., *DA* 2.2.1.86–89, 13.2.4.99–101 (*SZ* 1: 326–28; 2: 1030–33).

the mediating influence of the hot and the dry, and bile and choler are sent to the intestines in order to expel the impurities more easily.

1. To the first argument one must respond that the intestines do not require the retention of impurities, but rather their expulsion, and this is why the intestines do not need black bile. And, moreover, for this reason—because the intestines are naturally cold and dry—they therefore have sufficient power of retention on their own without the assistance of black bile.

2. To the second argument one must reply that nature works according to the intention of the end, and not according to the intention of the material. But now it is the case that heat is the principle of digestion, and the cold and the dry are the cause of appetite, and this is why black bile is sent to the opening of the stomach,[5] in order to promote or arouse the appetite, but choler is sent to the bottom of the stomach in order to promote digestion.

Question 3: Whether choler or bile is the cause of a short lifespan.

One inquires further whether choler or bile is the cause of a short lifespan.

1. It seems not. That which ignites heat and consumes impurities is more the cause of a long lifespan than a short one. But choler is such as this, and therefore, etc.

2. Moreover, the four humors are analogous to the four elements. But of all the elements fire is especially incorruptible, as is said in the fourth book of *On Meteorology*.[6] Therefore, similarly the humor also is analogous to fire. But such is choler, and therefore, etc.

The Philosopher says the opposite.[7]

One must reply that choler (that is, *fel*) causes a short lifespan more than a long lifespan, when speaking *per se*, because that

5. "Opening of the stomach": *os stomachi*, which, in *DA*, can sometimes mean the esophogus. Cf. *DA* 1.2.10.242 (*SZ* 1: 136).

6. Ar., *Meteora* 4.1 (379a14f.). Cf. A., *Meteora* 4.1.7.

7. Ar., *Part. An.* 4.2 (677a30f.). Cf. A., *DA* 13.2.4, 100–101 (*SZ* 2: 1032–33).

which consumes the radical moisture is not a cause of long life.[8] But choler is of this sort. If it is intense, it burns the blood and consumes the other humors. Nevertheless, if the choler is tempered, it causes a long lifespan just as if removing an obstacle, because through its heat it tempers the cold of black bile and through its dryness tempers the moisture of phlegm. Choler that is collected in the gall bladder, however, is only a superfluity. Thus it can easily overflow and mix with the nearby parts, and this is said to be the cause of a short lifespan.

1. On to the arguments. To the first argument one must respond that that choler flowing inwardly with the blood and that which serves as nutrition for the choleric members by itself consumes the moistures. But that choler that is in the gall bladder is not of this type.

2. To the second argument one must reply that although fire is incorruptible [*imputribilis*] in and of itself, nevertheless fire causes other things to putrefy [*putrescere*], and this is why, etc. Or it suffers putrefaction [*putrefit*], as is evident in a tertian fever.

Question 4: Whether serpents should have members that serve the motive power.

Next one inquires into the motive parts of animals. And first, whether serpents should have members that serve the motive power.

1. It seems so. This is because every bodily operation occurs through a mediating bodily organ. But progressive motion is a bodily operation. Therefore, it occurs through a mediating bodily organ.

2. Furthermore, every animal has some part for motion, by which it is held up when it is in motion. But that part is the instrument of motion, as is clear concerning the feet of animals. Therefore, there is such a part in serpents.

8. When A. says *cholera sive fel*—"choler (that is, *fel*)"—he seems to equate the two, another indication of the extent to which the two terms are confused.

The Philosopher says the opposite.[9] For he says, "Serpents do not have feet," nor do they have wings, and these are the parts dedicated to local motion.

One must respond that serpents do not have parts dedicated to motion. And the reason is that there is a double operation in an animal. One occurs in one part only, as the first digestion occurs in the stomach and the second digestion in the liver. The other is found in individual parts, like the third digestion. But now it is true that that which deals with all the parts in an essential way determines no part exclusively for itself. But the motive power of the serpent is in every part of its body, and this is why the motive power in serpents does not have an organic part dedicated to itself.

1. On to the arguments. To the first, one should respond that every bodily operation occurs either with a body part mediating it or with the entire body mediating it, and the motion of serpents occurs in the second fashion.

2. To the second, one must respond that the serpent is not held up on a part dedicated for its motion, but when it raises its head, it holds itself up by the rest of its body and by its tail, and when it holds itself up on its head portion, the tail draws itself back in, and this is why it is unnecessary for it to have feet. The Philosopher's argument, however, is that, from the fact that something has blood, it should have either two or four feet. But neither two nor four feet would be adequate for the serpent, owing to the length of its body, because they would either be distant from one another, or not. If distant, then the intermediate parts would be unsupported. If not, then the extremities would be unsupported.

Question 5: Whence the serpent's motion begins, as if from a principle.

One inquires further whence the serpent's motion begins, as if from a principle.

9. Ar., *HA* 2.14 (505b12); *De incessu animalium,* 4 (705b26–27). Cf. Averroes, *Part. An.* 4.10; A., *DA* 2.1.8.83, 14.1.1.1–6 (*SZ* 1: 325; 2: 1045–48).

1. It seems to begin from the head. Because the principle of motion is that from which the motive power proceeds and by which it is regulated. But this occurs from the head and the brain, and therefore, etc.

2. Moreover, that in which there is the greatest flexibility is the proper principle of motion. But this is in the head and the neck, because it can turn its head to the rear, as the Philosopher says.[10] Therefore, etc.

To the contrary. That is the first principle of motion which, when it ceases, causes motion to cease. But this occurs with the cessation of the heartbeat. But the heart is in the middle. Therefore, the serpent's motion begins from the middle.

And one must reply that the motion of any given animal begins from a principle that is situated in the middle, and it proceeds from there to the circumference. Thus the serpent would not move its head unless there were something in its middle setting the head in motion. Thus the first principle of motion is in the heart, just as the Philosopher says in the book *On the Movement of Animals,* and it flows from there to the other parts.[11] Now, the heart is in the middle of every animal. But in some animals it is located between the place for the entrance and for the exit of the food, and in others it is in the middle between the right side and the left side (as in those capable of motion).

1. On to the arguments. To the first, one must respond that two things are necessary for motion: namely, imagination and execution. With respect to imagination, the principle of motion is in the head and the brain; but with respect to execution, the principle of motion is in the middle, namely, in the heart, and this part sets the other in motion.

2. To the second argument one can respond in the same way.

10. Ar., *De incessu animalium* 7 (707b18f.).

11. Ar., *De motu animalium* 9 (702b12f.), 10 (703a14f.); A., *DA,* 13.2.7.122–23 (*SZ* 2: 1043–44).

Question 6: Whether the heart has a dedicated location in serpents, in ringed animals, and in bloodless animals.

One inquires further whether the heart has a dedicated location in serpents, in ringed animals, and in bloodless animals.[12]

1. It seems not. Something lives only through the influence established by a member in service to life. But parts of a serpent or a ringed animal live when they are cut off.[13] Therefore, they receive the influence of a member in service to life. But the heart is just such a member. Therefore, the heart will be in any given divided part.

2. Furthermore, everything that lives participates in that through which life is shared. Therefore, anything that lives with respect to a given part participates in that thing through which life is shared with respect to that given part. But the serpent lives in this way, and life is shared through the heart. Therefore, etc.

On the contrary. Whatever is generated first in an animal has a determinate location in the animal. But the heart is generated first in an animal. Therefore, etc.

One must respond that either the heart or something analogous to the heart has a determinate location in serpents and, similarly, in ringed animals. And the reason for this is that when many things are ordered to one thing and for the sake of one thing, if these are determined and have a determinate location, then how much more so will the principle be determinate. But there is a brain and there are lungs in serpents, and the lungs are the fan for the heart, and likewise the brain exists for the tempering of the heart. Therefore, since the brain and the lungs have a determinate location in serpents, how much more so will the heart.

1. On to the arguments. To the first, one must say that serpents and ringed animals are extremely uniform across their

12. Ar., *Part. An.* 4.5 (681b33f.); A., *DA* 13.2.7.122–23 (*SZ* 2: 1043–44).

13. "Ringed animal": Literal translation of *animal anulosum,* referring to the segmented bodies of insects. Akin, but not identical, to A.'s other term for insect bodies, *rugosus,* lit., "wrinkled."

parts, and this is why the parts participate in life in the manner of the whole, and why the parts that are cut off continue to live. This is evident when an eel has been cut up, not because there is a principal member of life in the parts or because there is a heart in that part after it has been cut off, but rather because, due to the uniformity with the whole, they retain the [heart's] previously established influence.

2. To the second argument one must respond that it is unnecessary for every living being to participate in a principal member through which life is shared; rather, it suffices for it to participate in a part that does the sharing or with which it is shared.

Or one can say, in another way, that the soul is in things in two ways: either radically, or in the manner of one exerting influence. Thus the soul exists in every part radically, but it is in the heart alone in the manner of one exerting influence. And it is for this reason that the parts, because they have the radical principle of life, continue to live even after they have been cut off. But because the influence from a principal part (like the heart) has ceased, they do not live a long time, etc.

Question 7: Whether parts that have been cut off live after they have been cut off.

One inquires further whether parts that have been cut off live after they have been cut off.

1. It seems not. For "soul is the act of an organic body," etc., and the soul is the principle of life.[14] Therefore, life exists only in an organic body. But the parts of such animals are not organic, and therefore, etc.

2. Moreover, animal parts are preserved by the form of the whole, and this is why an eye that has been plucked out is not an eye. But parts that are cut off are no longer preserved by the whole. Therefore, they do not live longer.

3. Furthermore, if parts that are cut off live, then they must be in some genus of life, and, as a result, especially in the first

14. Ar., *De anima* 2.1 (412af.).

genus of life, namely, the vegetative. But it is proper to vegetative life to generate one like itself and to grow. But parts that have been cut off do not grow, they are not nourished, and they cannot generate ones like themselves, and therefore, etc.

The Philosopher says the opposite, for he says in the text that the parts of animals such as these that have been cut off live just like the parts of trees.[15]

One must respond that the parts of ones like this do live after they are cut off. And the reason for this is that this is the way it is in lower regions,[16] that the more perfect the form is, the greater diversity it requires in matter, since the more perfect it is, the more operations it has, and when there are diverse operations, there is a corresponding diversity among the parts. And similarly the more imperfect the form is, the less diversity it requires in matter. And this is why in these lower regions the human body is more diverse and the body of an element is more homogenous. Likewise, the more the form of a mixed body approaches the nature of an element, the less diversity it requires in its matter. But among animate things the form of the plant is nearer to the form of an element, and this is why a part is homogenous with the whole in the species or in the particular [*in specie vel in re*], although not in name, because in the part of a plant the soul is similar in species to the soul of the whole plant, although the plant is not diversified.[17] And because such animals with ringed bodies lack blood, they approach very near to the nature of the elements, and this is why they require only very modest diversity among their parts in comparison to perfect animals. Rather, the kind of soul that is in the whole is the kind that can be in every part, because just as the form of fire does not require great diversity in its parts, so too the form of animals such as these requires only very little diversity in their parts, among the genuses of animals.

1. On to the arguments. To the first, one must respond that a perfected soul is "the act of an organic body" and all the parts

15. Ar., *Part. An.* 4.5 (682a4f.). Cf. A., *DA* 13.2.7.123 (*SZ* 2: 1044).
16. I.e., sublunar regions.
17. This is a most difficult passage.

of animals such as these are sufficiently organic if they are divided transversely, because a cuttlebone [*sepion*] or something in place of a spine is extended along the length of such animals, and the soul is rooted in this just as if in a subject.[18] And this is the reason why, with the part remaining whole, they can live for a time, namely, until the complexion of the part itself is dissolved by the action of its surroundings.

2. To the second argument one must reply that some parts are more uniform with the whole and others are not. The more uniform parts are preserved by the whole when they are united with the whole, and they are preserved by their own form when they are divided because, through the division, that which earlier was common to it now becomes proper to it. But this is not so for parts that are not uniform with the whole, because the form of such a one or of a whole animal demands greater diversity than is found in one part *per se*. And this is why such parts are preserved in species only while they are united to the whole.

3. To the third, one must respond that such parts live a double kind of life, because parts like these after their division still have motion and fantasy and, as a result, sensation, as the Philosopher proves in the second book of *On the Soul,* because if they are pricked with a needle they draw themselves in.[19] But they would not do this if they did not perceive, and sensation presupposes a vegetative [soul]. Therefore, they live with vegetative and sensitive life. But because every living thing requires nutriment, and there is a corresponding determinate organ for the reception of nutriment, and because these divided parts have been deprived of such an organ, for this reason they are not nourished, and, as a result, they do not grow and they do not endure in being for very long. Yet plants absorb nutriments from the ground through the root, and the parts of plants are not produced from spermatic matter but only from nutritional matter, and this is why when a part of the plant is put in the ground it can develop roots, because the formative power of the root exists as it were in every part of the plant, and its matter is

18. For *sepion*, cf. *DA* 12.2.8.152 (*SZ* 2: 952). Note that *spinae* can also mean small bones or the spines of animals like porcupines.

19. Ar., *De anima* 2.2 (413b19f.).

the nutrimental moisture, and this is why, etc. But this is not the case for animals, owing to their greater perfection, since in one part there is no formative power of another unless it belongs to that which is generated from the nutriment. Thus, when the serpent's tail is cut off, it regenerates because it arises from the nutrimental moisture, but this is not the case for the head, because it arises from sperm, just as is the case for every other animal, and this is why, etc.

Question 8: Whether a soul is one in number in some part that has been divided.

One asks further, whether a soul is one in number in some part that has been divided.

1. It seems that it is. For if a new soul were introduced into a part which was not there previously, either this would occur as an operation of nature or by means of an act of cutting-off alone. But it would not be as an operation of nature, because nature produces like from like, and in an act of cutting-off there is no soul acting on the whole to introduce form to the part. Nor would it be through the act of cutting-off alone, because generation is a boundary [*terminus*] for change and the act of cutting-off is not a natural change but rather a violent one. Therefore, there can be no principle for introducing a new soul.

2. In addition, no change occurs from one soul into another soul unless there is [first] a resolution into prime matter, because when the soul withdraws then death occurs, and a living being does not arise from a dead one, just as wine does not come from vinegar, unless first it is reduced to prime matter.[20] If then a new soul were introduced to the part that has been cut off, and which previously was under another soul, it would be necessary for it to be reduced to prime matter first. But this is not true, because there is nothing resolving it there.

On the contrary. It is impossible for something that is one and the same in number to perfect many materials at the same

20. See *QDA* 12.1.

time. Therefore, the soul cannot be the same in number in diverse parts.

One must reply that the soul is not the same in number in those diverse parts that have been cut off, but it is the same in species. For just as the form of a continuous and undivided fire is single in act and yet there are many forms in its related potential [*multae in potentia propinqua*], in order to reduce that potency to act, all that is required is the dissolution of its continuity. So too the soul in these undivided animals is one in act and many in potency because the soul is extended as far as the extension of the matter, and division alone is required for the reduction of this potency to act. But the many are not one in number, since they are opposed. Since, then, many souls arise through division, just as many fires arise from the division of one fire, it follows that the soul will not be one in number in diverse parts.

1. To the first argument one must reply that a soul that is absolutely new is not introduced into a part that is cut off, but rather that the same soul which earlier was in potency is now in act. And it arises in act through division alone because it existed previously in a potency so closely related that nothing was lacking for it to exist in act, except removing an obstacle to it. But the obstacle was its continuity, and this is why dissolving that continuity causes this soul to exist in act. Nor, however, does it follow that it is one in number with a soul of another part, because those things differ in number whose material is different.

2. To the second argument one must respond that a thing having one soul in act and in perfection, cannot assume a new one, once that one has been cast aside, unless there should occur some reduction to prime matter. But this is not what was proposed, because the part did not have the soul previously in act, but only in potency, and therefore, etc.

Question 9: Whether fleshiness on a person's head impairs the intellect.

One asks further whether fleshiness on a person's head impairs the intellect or whether a human head naturally ought not to be fleshy.

1. And it seems not. Because "sensation does not occur without heat," according to the second book of *On the Soul*.[21] But the senses are located in the head, and the brain, which is a very cold member, is in the head. Therefore, it is fitting that there should be a hot part on the head which may temper that cold. But flesh is such a part, because it is especially hot and moist. Therefore, etc.

2. In addition, the head's fleshiness is repugnant, according to the Philosopher's explanation, only when there is impairment of the intellect.[22] But the intellect, since it is not an organic power, cannot be impaired in its operation by a bodily part.

The Philosopher says the opposite.

One must reply that excessive fleshiness on the head does impair the intellect, and among animals only the human has a capacity for intellect, and this is why of all the animals the human head ought to be less fleshy in respect to its size. And the reason for this is that flesh, since it is moist and hot, contributes a great deal to digestion and, as a result, contributes to a natural power. But the natural and animal powers exist as contraries, as it were, and this is why that which contributes principally to one does not contribute to the other. Rather, just as flesh and bone have almost opposite complexions, and since flesh is hot, and is suited to the natural power, so too then bone is suited to the animal power. And this is why the natural members—like the liver, spleen, and heart—are fleshy and lack bones, whereas the head, in which the animal virtues are located, is on the contrary very bony and nerve-filled and less fleshy.

Moreover, fleshiness on the head makes the head heavy and weighed down. But the human naturally has an erect stature owing to the excellence of his heat. Therefore, fleshiness on the head causes a human to depart from his natural posture and impairs the direct flow that occurs from the heart to the superior parts. As a result, fleshiness on the head causes the impairment of the senses, and the operation of the intellect fails because of

21. Ar., *De anima* 3.1 (425a6).

22. Ar., *Part. An.* 4.10 (686a28f.). Cf. Averroes, *Part. An.* 4.10; A., *DA* 14.2.1.25 (*SZ* 2: 1056).

the weakness of the senses. Thus it is that children, when they have a head that is too fleshy, have poor discretion, and old people, likewise, since they have heads that are weighed down, do not have good powers of imagination or memory, because the spirits, which are the vehicles of the powers, cannot come there easily from the heart, and the animal spirits are not generated out of them. But digestion is poor, because the head's fleshiness draws the natural spirits from the liver to itself.

1. On to the arguments. To the first, one must reply that although the sensations do not arise without heat, nevertheless excessive heat impairs a sense, and this is clear by taking a look at places, seasons, and nutriment, for the intellect's operation cannot continue in places that are exceedingly hot nor does it occur as well in exceedingly hot seasons as it does in temperate places and seasons. Thus, if the head is too fleshy, the heat flowing up from the heart and likewise the heat conserved in the flesh displace the cold of the brain too much and hinder it from its proper operation.

2. To the second argument one must respond that although the intellect does not use an organ *per se,* nevertheless it does require other powers that do use organs, and these powers are hindered by too much fleshiness, and once they are hindered the intellect is hindered *per accidens,* because our intellect "is either a phantasm or does not exist without a phantasm," as is said in the third book of *On the Soul,* etc.[23]

Question 10: Whether the upper part [*of the body*] *in children is larger than the lower.*

Next one inquires whether the upper part [of the body] in children is larger than the lower, and whether the contrary is true for youths and those more advanced in age, and whether the contrary condition occurs in beasts.

1. It seems not. The upper part is the one through which a living thing receives nutriment. Thus the root is high up on a

23. Ar., *De anima* 3.8 (423a12f.).

plant.[24] But the human being receives nutriment through the mouth, which is situated on the head, and the head is smaller in children than the rest of the body, and therefore, etc.

2. If one were to say that the upper part does not consist of only the head but of the head and the trunk together, then it seems that just as this part is larger in children than the lower part, the same will hold for a youth. Therefore, this cannot be understood this way.

The Philosopher says the opposite.[25]

One must respond that among humans the upper part is larger than the lower during childhood, but during youth the opposite is the case, and the contrary to this occurs among brute animals. And the reason for this is that among all the animals the human has the largest brain in proportion to his body, and in childhood the brain is very cold and moist. And because nature produces the principal members first and sees especially to their nourishment, this is why in the beginning the upper part of the body is larger than the lower in the human. I understand the upper part to consist of everything from the top of the head to the lower part of the trunk. But over the course of time, heat flourishes in the human and begins to dissolve the superfluous moistures, which are in the upper part. Once these are dissolved, a freer pathway is open to the lower parts for nourishment, and this is why during youth the lower parts, like the tibias, the testicles, the penis, the anus, and the vulva, begin to grow larger. And an indication of this is that during childhood children do not walk erect, but they move and crawl like reptiles on their hands and feet, and this is owing to the weight of the upper part and the weakness of the lower part. Nevertheless, its larger size should not be understood simply in terms of quantity, but rather understood in terms of proportion, because the upper part is always larger quantitatively, but in the begin-

24. *Unde radix sursum est in planta,* which appears contrary to experience. Indeed the Aberdeen bestiary, ca. 1200, reminds us that a root, *radix,* is so called because it is fixed deep in the ground (Aberdeen University Library MS 24, fol. 78r).

25. Ar., *Part. An.* 4.10 (686b2f.). Cf. Averroes, *Part. An.* 4.10; A., *DA* 14.2.1.26–27 (*SZ* 2: 1056–57).

ning it is disproportionately larger. But during youth the contrary is so for the lower part, because the head is not proportioned to the legs.

In other animals, however, the opposite is the case since the lower parts are larger at first. Thus the legs of a foal [*pullus equinus*] are not much shorter than the completed length they will have later, or have at the end. And the reason for this is that the body of animals such as these is supported by the lower parts, and this is why it is necessary that such parts be solid and very earthy. But there is earthiness in the semen of brute animals, and this is why a good bit of it is converted into the lower parts at the very beginning. But over the course of time the nutriment can be more easily converted into more rarefied parts (namely, those that are above) than into the dense parts, and this is why the upper part increases in size then, and the lower part is proportionally diminished.

1. On to the arguments. To the first, one must reply just as has already been said.

2. To the second, one must respond to the contrary, that the upper part is always quantitatively larger, but nevertheless it is not always larger in terms of proportion, and this is why, etc.

Question 11: Whether hands should belong to the human alone.

Next one inquires whether hands should belong to the human alone.

1. It seems not. For the hand is the organ of organs. Therefore, those things that require organs equally, equally require hands. But brute animals need organs as much as the human does, since if these were lacking they would suffer a lack as much as would the human. Therefore, they need hands just as the human does.

2. In addition, nature always does that which is best. But it is better for the human to have hands than to have front feet. Therefore, by the same reasoning, it will be better for brute animals. Therefore, hands should not belong to the human alone.

The Philosopher says the opposite.[26]

One must reply that hands belong to the human alone because the hand is just like an organ of the intellect, with which a human being especially executes whatever he entertains in the intellect, because just as through the intellect a human has a capacity for all things intelligible, so too through the hand he has a capacity for all things capable of being performed. Just as the human has an almost infinite power through the intellect, since he cannot understand a given number of things without being able to understand still more, so too with the hand he cannot do a given number of things without being able to do still more, should a reason arise. And although all the members obey the intellect and reason, yet none do so as much as the hand. Thus, when someone attempts to express something that he understands intimately, he can hardly restrain his hands, since the hand so completely obeys the intellect that it naturally tends to manifest in an outward act what is conceived internally in the soul. Thus the hand should belong only to an animal that has intellect, and since the human alone has intellect, this is why, etc.

Furthermore, the hand bears witness to a refined sense of touch, so that among all the members the power of touch flourishes especially in the hand, and this is why it belongs only to that animal which has the best sense of touch, and the human is of this sort. Thus one must reply briefly to the question that the hand exists for the sake of the intellect and an excellent sense of touch, and this is why it belongs only to the human.

1. On to the arguments. To the first, one must respond that although all animals require organs, nevertheless none does just as the human does, because every animal has some kind of armament by nature: one has horns, another talons, or teeth, or feet for flight, but the human, on the other hand, is lacking in every type of armament. Thus if he had by nature every type of armament, his body would be terribly unbalanced. And this is why nature did not give the human some armament naturally,

26. Ar., *Part. An.* 4.10 (687a2f.). Cf. Averroes, *Part. An.* 4.10; A., *DA* 14.2.2.31–33 (*SZ* 2: 1058–60).

but gave him hands with which he can make for himself every type of armament, in accord with his intellect and reason.

2. To the second argument one must respond that not everything that is better for one is better for another. It is good for sight that the eye have the disposition it has (namely, clear); nevertheless, it would not be good for the whole body to be just like the eye. Thus it is better for brute beasts to have feet rather than hands, because if they had hands they would only be able to use them as they use feet, owing to their lack of intellect and reason. And this is why, etc.

Question 12: Whether a human should naturally have wings and fins.

Next one inquires whether a human naturally should have wings and fins or whether naturally he might swim or fly.

1. It seems so. For whatever is lighter can better be supported in something light. But a human is lighter than birds and fish. Therefore, he can be supported in the air better than a bird and in water better than a fish. The major [premise] is clear, and the minor [premise] declares that heat is the cause of lightness, and there is more heat in the human than in other animals.

2. In addition, a dead human floats in water. Therefore, much more so a live one, since a live one is lighter than a dead one, as is evident by experience [*per experimentum*] since a dead one is heavier than a live one for those carrying him, because four can hardly carry a dead man while two can carry a live one with ease.[27]

The opposite is clear.

One must respond that it is not natural for a human to swim or to fly, because through his reason and through his hands the human has the power to form instruments for himself, by means of which he can pass through water just like a fish, and this is why he makes a ship and why he does not require the fins that fish have.

27. Cf. *QDA* 4, *annex*. 18.

Furthermore, there are fewer fumes in the human than in birds. An indication of this is that, just as birds have feathers, so there are hairs on a human, and these hairs are small when compared to birds' feathers. Therefore, compared to birds, hairs do not have the power that feathers have for supporting a body in air. Because feathers are wide, they therefore are held up by air; hairs, however, are long and more earthy, and this is why they cut through the air and seek the ground, which is the [natural] place for heavy things. Thus a capacity for walking is a proper characteristic of the human, just as is the capacity for receiving instruction, and contrary *differentiae* do not suit him well. And the entire reason is that nature acts for an end, and fish seek their nutriment in water, and birds do so in air, but the human does this especially on the ground, and this is why nature better provided him with instruments for walking, like feet, rather than for swimming or flying. This is why, etc.

1. On to the arguments. One must respond to the first that one has more heat in two ways: either quantitatively—and thus a human has more heat than fish and birds—or proportionally in comparison to other qualities, and in this way birds have more heat. Thus their flesh is hot and dry, and this is why, even though these animals are not as hot as the human, nevertheless they have instruments by means of which they can be borne on the air and water. The human lacks these instruments, and this is why, etc.

2. To the second argument one must reply that a live human being is governed by a natural rule [*regimen naturale*], and because there is in his body more earth than other elements (because, according to the Philosopher, mixed bodies are dominated by earth and water), for that reason a live one seeks the bottom.[28] But a container changes the content toward its disposition, and this is why, if the body of a dead human lies in the water for a long time, the water dissolves the earthy parts and converts them to its nature, and this is why such a dead body does not seek the bottom but rather floats.[29]

And, besides this, every contrary naturally flees from its contrary, so far as possible. Now the dead body of a human is most

28. Ar., *DG* 1.1 (314a1f.).

29. Cf. *supra*, *QDA* 4, *annex*. 18.

unclean, and water is a clean element, and this is why water naturally struggles as far as it is able to expel a dead body using the power of contrariety, and this is also why it does not cease to act on that body until the body moves, so that the water may expel the body by this motion.

Some say that this is owing to the fact that a living human being seeks the bottom due to its weight, since, according to the Philosopher, he is dominated like other mixed bodies by earth and water. But in a dead one, bile [*fel*] is separated from the gall bladder and distributed to the individual members, making them light, because it is of the nature of fire. This is why it floats, etc.

Others say that when a human dies and sinks to the bottom, and the water then enters in and fills the porous members—like the stomach, bladder, lungs, and others of this sort—and once the air is expelled, the water enters in, lest there occur a vacuum. As a result the body is lifted up, and it floats, etc.

Still others say that earlier the body had moisture in its own right. But when it is dead, the water surrounding it moves that little moisture toward itself, along with its body. And this is why it floats, etc.

Question 13: Whether in respect to motive parts bloodless animals should surpass those with blood.

Next one inquires whether in respect to motive parts bloodless animals should surpass those with blood.

1. It seems not. Because the more noble a nature is, the more it requires many things. Therefore, animals with blood need more organs than bloodless ones.

2. Moreover, some bloodless animals, like serpents, fish, and many others, lack feet. Therefore, it is not reasonable that blooded animals should have fewer feet.

The Philosopher says the opposite.[30]

One must respond that the more noble a nature is, the more

30. Cf. Ar., *Part. An.* 4.11 (690b11f.); *De incessu animalium* 8 (708a9f.); A., *DA* 14.2.5.54–61 (*SZ* 2: 1069–72).

power it has and the more it needs several instruments. And because blooded animals are better than bloodless ones, this is why they require several parts according to their species. But because those that are bloodless have hard and poor flesh and are exceedingly cold owing to their lack of blood, for this reason it is necessary that they simply be immobile or that they have several motive parts. And this is why the number of feet among animals that move does not exceed the number four, or the number two for wings, whereas animals without blood have to have several feet and several wings, because owing to the weakness of their nature they cannot move with fewer.

1. On to the arguments. To the first argument one must reply that the more noble a nature is, the more parts it demands according to its species, but nevertheless it does not require several parts with respect to number for its operation, because, owing to the perfection of its nature and the strength of its power, it can accomplish this with few with respect to number, whereas the more imperfect can only accomplish this with several.

2. To the second argument one must respond that the serpent and an animal of this type does not need feet owing to the hardness and viscosity of its flesh, and likewise on account of its flexibility. This is why it uses these in place of feet, etc.

BOOK FIFTEEN

Question 1: Whether sex is necessary for the generation of animals.

"We have already determined above," etc.[1] With regard to this fifteenth book, one first inquires into sex, and whether sex is necessary for the generation of animals.

1. And it seems not. For according to the Philosopher generation exists because the individual is corruptible.[2] Therefore, the generative power is implanted in it so that it can be preserved in species, although not in itself. Therefore, generation corresponds to corruption. But corruption can occur without any distinction of the sexes, and therefore so too can generation.

2. In addition, according to the Philosopher in the second book of *On the Soul* and the fourth book of the *Metaphysics,* each one is perfect when it can make one like itself.[3] Therefore, since an animal is more perfect than a plant, and plants can generate one like themselves by themselves, so much more so can an animal.

3. In addition, a male is able to generate in another; a female, however, generates in herself. But these facts suit the elements as well as animated beings, for fire can generate in another and something can be generated in fire. Therefore, just as sex is not required for the generation of the elements, neither will it be required for the generation of animals.

The Philosopher says the opposite.[4]

One must reply that generation is of two kinds. One is the

1. Ar., *GA* 1.1 (715a1f.). Cf. A., *DA* 15.1.1.1 (*SZ* 2: 1084).

2. Cf. Ar., *DG* 1.3 (318a9f.).

3. Ar., *Metaph.* 4.16 (1021b21f.).

4. Ar., *GA* 1.1 (715a18f.). Cf. Avic., *DA* 15.1 (fol. 59rb); Averroes, *GA* 1.1.2; A., *DA* 15.1.1.2–7 (*SZ* 2: 1084–87).

transmutation of one by another, and in this type the one generating disposes the matter of the other and introduces form, and this kind of generation is attributed to the elements. The other type is generation by division or a cutting-off of a part from the whole, or of one from another, and in this type of generation not only is form introduced by the ones generating, but also the matter is provided by them. And because nature always intends what is better, and what is better is nobler, in things that are acted upon, nature separates off the more ignoble. And the agent is more noble than the one being acted upon, just as form is nobler than matter, and this is why, in the generation of animals, nature separates the male as the nobler from the female, because perfect animals generate by the separation of semen, and in this separation the male is just like the agent, and the female is just like the one being acted upon. And this is why sex is necessary for the generation of ones like these.

In addition, sperm is not always suited for generation everywhere but only in the appropriate place, a place, moreover, receptive of sperm, so that it is suitable for generation. This exists only in the female or in a part emitted by the female. And he says this about fish, because the female emits the future offspring or eggs, and the male follows behind and spreads his sperm over them. And this is why sex is necessary for the generation of animals.

1. On to the arguments. To the first, one must reply that generation proceeds by way of nature. But every corruption is in a certain sense contrary to nature, and this is why more things are required for generation than for corruption, because nature's operation proceeds in an ordered way, and an operation that is contrary to nature proceeds without order.

2. To the second argument one must reply that the more perfect something is in the lower realms, the more things are required for its generation, as is said in the second book of *On Heaven and Earth*.[5] And this is why, although an animal is more perfect than a plant, nevertheless it cannot be generated from

5. Ar., *DC* 2.12 (292a22f., b17f.).

only one thing as a plant can, but rather requires several things for its generation owing to its perfection.

3. To the third argument one must reply that although an element generates in another and in itself, nevertheless this is said equivocally with respect to an animal. For the generation of fire, it is not required that there be one fire as agent and that there be another sustaining the action for the one that is generated.

Question 2: Whether nature intends the generation of a female.

One inquires further whether nature intends the generation of a female.

1. And it seems not. For nature intends nothing flawed, because the accidental and the natural are in opposition. But a female is a flawed man [*vir occasionatus*], as is said in the sixteenth book of this work.[6] Therefore, nature does not intend to produce a female.

2. In addition, nature intends that which is better. But a defect and privation are not better than a property [*habitus*]. Therefore, nature does not intend a defect and privation. But a female is generated only from a defect and on account of a privation. Therefore, nature does not intend to generate her.

To the contrary. Nature only fails in a few cases. But a female is generated in many instances. Therefore, she is not generated contrary to the intent of nature.

To this, one must reply that nature is of two types: universal and specific [*natura particularis*]. Universal nature intends to conserve the entire universe and its parts, and because species are parts of the universe and not individuals, this is why universal nature intends principally to conserve the species. But a species of animals cannot be conserved without the generation of individuals, and a female as well as a male are required for this generation. This is why universal nature intends the female as

6. Ar., *GA* 2.3 (737a27f.). Cf. A., *DA* 16.1.14.73.

that without which the species cannot be preserved. Specific nature, however, intends to produce something like itself, and because the power of the male is the agent for the generation of the animal, and not the power of the female, this is why the one acting on a specific thing [*agens particulare*] principally intends to produce a male. If, nevertheless, there is a defect in terms of the material or the heat, which it uses like an instrument, and it cannot generate suitably according to its intent, then it intends what it can produce, and thus a specific nature principally intends a male, but nonetheless it secondarily and in a flawed way [*occasionaliter*] intends a female.

In this way a solution is apparent to the arguments. For the first two [arguments] proceed from a specific nature which does not principally intend a female, just as it does not intend a flaw or defect, but nevertheless it can intend to produce her secondarily.

Question 3: Whether male and female diversify species.

One inquires further whether male and female diversify species.

1. It seems so. Diversity in essence causes diversity in species. But the efficient cause and matter are diversified in essence; and therefore they never coincide, as is held in the second book of the *Physics*.[7] But the male is the efficient cause, and the female is just like matter. Therefore, etc.

2. In addition, diversity in form causes diversity in species. But male and female diversify form, for they are caused by the soul, since they are not found in inanimate things. Therefore, etc.

3. In addition, substance does not come to be from non-substances. But the fetus comes to be from male and female. Therefore, male and female are substances. But male and female are opposites and do not tolerate one another in the same animal

7. Ar., *Phys.* 2.7 (198a22f.).

and cannot be universally received. Therefore, they will necessarily diversify species.

To the contrary. In univocal generation, the one generating generates one like itself. But a human is generated from male and female. Therefore, male and female are alike in the human species.

To this, one must reply that white and black can be compared in various ways—either to the genus of color which they divide, or to the subject which they inform, like a wall, a horse, or a dog. If they are compared in the first way, they diversify species, but they do not if they are compared in the second way, because they follow upon the individual matter. One must speak in the same way about male and female, because if they are compared in terms of sex, they diversify species. For sex is a certain natural potential for generating in a second species of quality, and is divided into masculine and feminine just as color is divided into white and black. If, however, they are compared with respect to the subject in which they exist, then they do not diversify the species, because they are caused only by diversity in the material. For if the heat is tempered, a male can be produced from the same material which produces a female when there is weak heat, and this would not happen if the things which are attributes of masculinity and femininity differed in species.

1. On to the arguments. To the first, one must reply that male and female are not diversified in the essence of the species. And when it is said that they are disposed like agent and matter, one must add that this is not because of the substantial form or matter but because of an accidental disposition, just as hot water acts on cold water through an accidentally acquired disposition. And nevertheless it does not follow from this that they differ in essence, because this action exists only because of accidental dispositions. In like manner the male has the definition of the agent on account of the strength of his power and heat, and the female has the definition of one undergoing the action because of her cold and moistness.

3. To the second argument one must reply that when the

phrase "from non-substances," etc., appears, the preposition "from" indicates the circumstance of the material cause, and thus substance (or a fetus) does not arise from male and female.[8] But if it should indicate the circumstance of the efficient cause, then this has to be understood of the principal efficient [cause] and not of the instrumental efficient [cause]. But male and female are not principal efficients but only instrumental ones, and acting instrumentally during natural generation is an accident.[9] For it occurs in intermediate active and passive qualities just as through instruments.

2. To the third argument one must reply that although male and female are found only in animate things, nevertheless they follow more upon the dispositions of the body than of the soul. For although an organization of parts is found only in animate things, and is nevertheless a disposition of the body, it is owing to the soul. Just so male and female are dispositions of the body, and this is why diverse instruments are attributed to them, yet they proceed from the soul as if from a remote cause.

Questions 4–5: Whether male and female co-exist in one and the same thing.

Why a male whose generative members have been cut off can become effeminate, whereas a female does not become virile.

One inquires further whether male and female co-exist in one and the same thing.

1. It seems not. A contrary acts on its contrary, and this is why contraries do not co-exist in one and the same thing. But male and female are contraries, because they divide sex. Therefore, they cannot co-exist in one and the same thing.

2. In addition, if they were to co-exist in something, this would occur especially in plants. The conclusion is false. For the Philosopher says in the book *On Plants* that male and female

8. Although it comes second in the text, this response clearly refers to the third argument listed at the beginning of this question.

9. "Principal efficients" and "instrumental efficients": Conrad of Austria's shifts of gender throughout this passage make clarity elusive.

are distinct in plants, because plants are hard on account of the male and soft on account of the female.[10]

The Philosopher says the opposite. For he says that sex is confused in plants, but it is distinct in animals.[11]

In conjunction with this, one can ask why a man whose generative members have been cut off can become effeminate, whereas a female does not become virile, when the same reasoning [*ratio*] would seem to be at work in both cases.

To the first, one must reply that sex is confused in plants and as a result male and female co-exist in them, whereas these are distinct in animals. And the reason for this is that a plant does not require a great diversity of parts for the sake of its perfection, but rather a generative power exists in each of its parts; nevertheless, this is for a different reason, because in plants heat is like the father and earth is like the mother. Therefore, the fruit and seed are produced from an earthy and watery moisture by the action of the natural heat. But a sensitive soul is more perfect, and requires greater diversity in the parts, and this is why in all those that are generated by propagation—that is, intercourse—the sexes are distinct.

One can, however, make the distinction that sex is distinct in two ways: either in terms of the thing or in terms of the operation. In terms of the thing it is distinct in animals and confused in plants, but in terms of the operation sex is distinct in plants just as it is in animals, because plants are fertile with respect to the female and sterile with respect to the male. And in like manner they are hot and hard with respect to the male, and cold and soft with respect to the female.

1. On to the arguments. To the first, one must reply that contraries, in their extremes or most intense forms, do not exist in the same thing at the same time; nevertheless, they can co-exist when they are broken up, as is apparent in something dusky, with respect to the white and the black.[12] But yet, for various reasons, male and female can occur in the same one owing to a

10. Ps. Ar., *De plantis* 1.6 (Meyer, 1841).
11. Ar., *GA* 1.1 (715b15f.).
12. "Dusky": *fusco*, may refer to a dark-complexioned individual.

flaw in nature, as is evident among hermaphrodites, who have each member. But this is a monster in nature.

One can respond another way to the argument that the powers of contraries are well able to exist at the same time in the same one, although their acts cannot.

2. To the second argument one must reply that the Philosopher says in the book *On Plants* that sex is distinct in plants in terms of operation. Here, however, he says that sex is confused in them in terms of the thing itself.

One can respond to the question that is asked in conjunction with this in two ways: In one way, one can respond that a female is related to a male just as a privation is related to a property [*habitus*],[13] and a transition can occur from a property to its privation, but not the other way around, as blindness can occur in one who sees, but not the other way around. And this is why a male can be made effeminate, but not the other way around.

One can respond better in another way that principal members can be lost, but once they have been lost, they cannot be restored. But in the first generation a female is deprived of the principal members of generation, namely, the testicles, at least externally, although they exist internally at the base of the womb.[14] Therefore, for this reason the male can lose the principal members, but the female cannot recover those members of which she was deprived during generation, and this is why a male can be rendered effeminate—because once the members in which heat and power flourish have been cut off, he is reduced to the female's complexion, and from that point on he becomes cold and moist like a female. But these members cannot be generated in a female after her generation is completed, and this is why she cannot be rendered virile.

13. "Property:" *habitus*, derived from *habeo*, implies the possession of a property, quality, or characteristic. A privation, *privatio*, is therefore its opposite.

14. On female testicles in relation to the womb, see especially A., *DA* 15.1.6.30 (*SZ* 2: 1098).

Question 6: Whether the male is hotter than the female.

Further one investigates a comparison of the sexes, of male to female. And first, whether the male is hotter than the female.

1. And it seems not. For a hot complexion either generates nothing at all, or generates a female, according to medical authorities. Therefore, a female is produced from a hot complexion, and not a male. Therefore, she will be hotter than the male.

The Philosopher says the opposite.[15]

To this one must reply that one can compare complexion in many ways: either in general, with respect to which it is said that the living is hotter than the non-living. Or it can be in terms of climate or region, and thus we say that those living in a hot region are hotter; or in terms of species, and in this way it is said that among animals the lion is a choleric animal and the human is a sanguine animal. Or it can be with respect to the individual, and in this way one person is said to be choleric while another is phlegmatic, etc. Or comparison can be made with respect to sex, and thus, generally speaking, a male is hotter than a female. And the reason for this is that a female is moister than a male. Excessive moisture blocks heat and extinguishes it. And this is why medical authorities say that the hottest female is colder than the coldest male.[16]

1. On to the arguments. To the first argument one must reply that heat is twofold: either it is tempered or distempered. A hot or tempered complexion or a tempered heat always produces a male, whereas a distempered heat always produces a female. For if the heat is too weak, then the moisture of the menstrual blood can not be digested as much as a male's complexion requires. If, however, it is too excessive, then it causes the natural heat to escape and weakens the power, as is evident in summer. And as a result the menstrual blood leaves behind too much

15. Ar., *GA* 1.19 (726b32–34). Cf. A., *DA* 12.1.3.32, 15.2.6.106, 18.2.5.78 (*SZ* 2: 1132–33; 1318–19).

16. Cf. Avic., *Can. med.* 3.21.1.8; A., *DA* 12.1.7.85 (*SZ* 2: 925–26).

undigested moisture. And this is why if the heat departs from a tempered state—either above or below it—either it generates nothing at all or it makes one effeminate, and this is the understanding of the physicians.

2. To the second argument one must reply that the loose texture of a woman's body is not a result of heat, but results from an abundance of moisture that is not digested, and in the opposite way the density [*compactio*] of a male's body is not the result of cold but results from the strength of the heat, because not only does cold coagulate but heat does so as well. This is apparent in the art of the brick-maker when fire compacts mud, etc.

Question 7: Whether a man has to be larger than a female.

One inquires further whether a man has to be larger than a female.

1. And it seems not. Because where the nutritive power ends, there the augmentative begins. Therefore, where the nutritive power is stronger, there the augmentative power is fortified. But the nutritive power is stronger in a female than in a male, because the female nourishes both herself and the fetus, which the male does not do. And therefore, etc.

2. In addition, simplicity follows form, and quantity, matter. But the female has the role of matter in relation to the male; therefore, she naturally surpasses him in size.

The Philosopher says the opposite.[17]

To this, one must reply that three things are required for a large body: namely, the strength of the power, the intensity of the spreading heat, and a great deal of moisture that can be spread about. The first two are found in a male, but the third is found in the female. And because two that are conjoined can effect more than one alone, in many cases males are larger than females. Nevertheless, the contrary is the case in birds of prey because they have a very strong heat. Therefore, in the males

17. Ar., *GA* 1.19 (727a15–22). Cf. A., *DA* 15.2.6.107 (*SZ* 2: 1133).

of this sort there is an elevated heat that consumes a great deal of moisture whereas in the females the heat is more tempered on account of the moisture that retards the heat, and this is why among these birds the females are larger than the males. And this is the reason why they fall into one article.

1. On to the arguments. To the first, one must reply that the reason the female nourishes both herself and the fetus is owing to the weakness of her digestion, because she cannot digest all the nutriment and convert it into her own substance because very often she is weakened by that superfluity which is later converted into milk and menstrual blood. Therefore, she converts less of the nutriment into her own substance than is converted in the male.

2. To the second argument one must reply that although the female has the role [*ratio*] of matter and is just like matter compared to the male who generates, nevertheless the female's menstrual blood is the matter for the male as well as the female.

Question 8: Whether a male or female has a longer life span.

One inquires further whether a male or female has a longer life span.

1. And it seems that it is the female. This is because moisture *per se* contributes to life, whereas heat does so *per accidens* because in itself heat consumes moisture. Since, then, the female is moister and the male is hotter, the female will have a longer life span.

2. In addition, where illness is slower, life is longer. But the female grows ill more slowly, and therefore, etc.

The Philosopher says the opposite in *On the Reason for Shortness or Length of Life.*[18]

One must reply that the male has a longer life span naturally, but the female *per accidens,* because the length of a natural life span depends on the radical moisture and a tempered heat,

18. Ar., *De long. et brev. vitae* 5 (466b9f.).

whereas shortness of life arises from consumption of the radical moisture by the heat. But a tempered heat and the radical moisture are more abundant in the male than the female, and this is why, etc. Nevertheless, the female has a longer life span *per accidens,* because she does not work as hard and thus does not consume so much, and she is cleansed more by the menstrual flow and is less debilitated by sexual intercourse. And this is why she is conserved more. And these are accidental causes.

1. On to the arguments. To the first, one must reply that it does not belong to heat *per se* to consume, and certainly not tempered heat, and this is the sort that is in the male. And in addition to this, moisture is of two types: namely, airy and watery. Airy moisture contributes to life, as is said in the book *On the Reason for Shortness or Length of Life,*[19] but watery moisture—which is abundant in the female—does not. The first type, however, exists in the male, and this is why, etc.

2. To the second argument one must reply that illness is of two types: namely, natural and unnatural. Unnatural illness occurs less often in females than in males, because their superfluities are purged through menstruation. Nevertheless, natural illness occurs in them more than in men, because in males the power to resist is more powerful, but weaker in females. And this is why, etc.

Question 9: Whether the male's flesh is softer than the female's flesh.

One inquires further whether the male's flesh is softer than the female's flesh.

1. It seems so. Because "those soft in flesh are sharp in mind," according to the Philosopher in the second book of *On the Soul* and in the twelfth book of this work.[20] But the male has a quicker mind, and therefore he has softer flesh.

2. In addition, what is softer is more obedient to motion. Since, then, the male is more suited for motion than the fe-

19. Ar., *De long. et brev. vitae* 5 (466a22f.).

20. Ar., *De anima* 2.9 (421a25–26); *PA* 2.2 (648a3f.).

male, it seems that the parts of the male are softer than those of the female.

The Philosopher says the opposite. He says that the parts of the male are harder, etc.[21]

1. One must reply that softness in an animal can arise in two ways: either from an abundance of moisture and a weakness in power, and this is more common in the female than in the male; or it can arise from the strength of the power and of the heat spreading to the parts, and this type of softness is more common in the male than in the female, and this softness contributes to a good sense of touch and to the mind's discretion, and the first argument proceeds from this.

2. To the second argument one must reply that drier nerves are better suited for motion than moist nerves. Therefore, paralysis very often arises from excessive moisture of the nerves. Now, however, there is superfluous moisture in the females, and this is delegated to the motive members, and this is why it aggravates them. Therefore, females are less suited for motion than males. Therefore, the argument arrives at an opposite conclusion from the one proposed.

Question 10: Whether the male is better nourishment than the female.

One inquires further whether the male is better nourishment than the female.

1. And it seems not. Because the cleaner the nourishment, the better it is. But the female is cleansed more than the male, through the purging of the menstrual blood. Therefore, she is better nourishment.

2. This is evident in many cases from experience [*per experimentum*]. For many animals are better nourishment in their youth, because they are more moist. Since then females are moister than males, they therefore will be better nourishment.

21. Ar., *HA* 4.11 (538b2f.). Cf. A., *DA* 4.2.4.107 (*SZ* 1: 486).

To the contrary. That one is better nourishment that is more digested. But the female has poorer digestion; on account of the weakness of her digestion she does not produce sperm. Therefore, etc.

One must reply to this that some animals are good nourishment by virtue of their species, just as sheep or goats are hot and moist, and this is why they are good nourishment. And in ones like this the male is better nourishment than the female because he is more tempered, and especially one that is castrated.

Other animals are not suitable nourishment, but rather they are incompatible with nourishment owing to their moisture, like the pig, and some are such owing to their dryness, like a bird. Moreover, in those animals which abound with moisture, throughout their species, the males are better nourishment. Therefore, the male pig nourishes better than the female. In others, however, which abound in dryness, the opposite holds. Therefore, female birds are better nourishment than males, as is clear in the case of the hen. Therefore, the male is not simply better nourishment than the female, or vice versa, but it is different in different cases, as has been seen.[22]

1. On to the arguments. To the first, one must reply that although the female is cleansed more by purgation, nevertheless in and of herself she is more unclean; otherwise, she would not be purged. This is why the female is always tainted in some way, etc.

2. To the second argument one must reply that some animals have a moist womb, and others have one that is dry. In animals with a dry womb, the younger the animals are, the better. Therefore, a calf is better nourishment than an ox, and a *capreolus* or a kid is better than a she-goat or a billy goat, and so too for the others.[23] In animals with a moist womb, the younger they are, the worse. Therefore, the flesh of a young pig is the worst as nourishment, because unless it is salted it is too moist. And

22. Cf. A., *DA* 23.1.24.116 (*SZ* 2: 1628).

23. The term *capreolus* here is vexing. A. is clearly pairing young and old versions of the same animal, e.g., calf and ox. Thus *capreolus* seems parallel to *capra* (she-goat) as kid (*haedus*) does to billy goat (*hircus*). Yet *capreolus* is normally a wild goat, and not a term reserved for a juvenile.

for a similar reason the front portions of some animals are better [than the rear], and in other animals the rear portions are better. Therefore, those animals that have front portions that are elevated on account of a lot of movement, are better in the front, and also because superfluities run from an elevated and lofty place downward. Therefore, the front portion of a sheep or cow is better owing to its elevation than the front portion of another—in others with an opposite condition the opposite holds—and in birds who fly the front portion is better, and I say this with respect to the hen.[24] In the former, however, when they have their hind portions elevated or when they work more with the hind portions, the hind portion is better, as in the case of the pig and in birds that are non-flyers, like the goose, because a goose walks more than it flies. And the reason for this is that the portion that is lower and the part that works less is more unclean, because superfluities are sent more to such parts, since flow is more likely toward a part that slopes downward.

Question 11: Whether the male is better suited for proper behavior [mores] *than the female.*

One inquires further whether the male is better suited for proper behavior than the female.

1. And it seems not. For an animal that is more teachable for proper behavior is more suitable. But according to the Philosopher in the beginning of the ninth book, the female is more teachable toward proper behavior than is the male.[25] And this seems to be because of the argument that females resemble children, according to the Philosopher in this chapter.[26] But children are more teachable than old people, as the Philosopher wishes in the second book of the *Ethics*.[27] Therefore, etc.

2. In addition, prudence is an intellectual virtue. Without it,

24. The clause, as printed in the text, literally reads: "the front portion of a flying bird is better in those who fly." It seems clear that the doubling is due to an intrusion of a marginal gloss or a confused manuscript tradition.

25. Ar., *HA* 9.1 (608a9f.). Cf. A., *DA* 8.1.1.1–6 (*SZ* 1: 667–70).

26. Ar., *GA* 1.20 (728a17), 5.4 (784a4f.).

27. Ar., *Eth. Nic.* 2.2 (1104b11f.).

moral virtue cannot be perfected. But females are more prudent than males, as the Philosopher wishes. Therefore, etc.

The opposite is stated in the ninth book of this work, and this is clear. For generally, proverbially, and commonly it is affirmed that women are more mendacious and fragile, more diffident, more shameless, more deceptively eloquent, and, in brief, a woman is nothing but a devil fashioned into a human appearance. Thus I saw one like this at Cologne, who seemed to be a saint and yet, in brief, ensnared everyone with her love.

To this, one must reply that a female is less suited for proper behavior than is a male. For a female's complexion is moister than a male's, but it belongs to a moist complexion to receive [impressions] easily but to retain them poorly. For moisture is easily mobile and this is why women are inconstant and always seeking after new things. Therefore, when she is engaged in the act under one man, at that very moment she would wish, were it possible, to lie under another. Therefore, there is no faithfulness in a woman.

Believe me: if you believe her you will be deceived.
Believe a teacher who has experienced it.[28]

Moreover, an indication of this is that wise men almost never disclose their plans and their doings to their wives. For a woman is a flawed male and, in comparison to the male, has the nature of defect and privation, and this is why naturally she mistrusts herself. And this is why whatever she cannot acquire on her own she strives to acquire through mendacity and diabolical deceptions. Therefore, to speak briefly, one must be as mistrustful of every woman as of a venomous serpent and a horned devil, and if it were allowed to say what I know about women, it would stupefy the entire world.[29]

28. Corner identifies the first line of this couplet as stemming from the *Catonis Disticha,* but the second line is based on a phrase common throughout antiquity and frequently used by Michael Scot. See Corner (1983), 307–8. Cf. Haskins (1928), Otto (1890), 127, and Büchmann (1952), 226.

29. "If it were allowed": A. uses the term *fas est,* which in antiquity meant allowed by divine law. One wonders if he were referring to prohibitions against

1. On to the arguments. To the first, one must reply that instruction is of two types: for one is given concerning things that can be done [*operabilibus*] with respect to affect, and the other is given concerning things subject to cognition [*intelligibilibus*], and this consists in understanding and deliberating. As far as the first is concerned, the female is more teachable than the male, because she is more easily moved to different affects, toward which she is disposed. But as far as the second is concerned, the contrary obtains because the sensible powers are weaker in a woman owing to the coldness of her complexion, since she has a poorer sense of touch and, as a result, a weaker intellect.

2. To the second argument one must reply that a woman is not more prudent than a male, properly speaking, but she is cleverer. Therefore, prudence smacks of good, and cleverness smacks of evil. Therefore, the female is more prudent, that is, cleverer, than the male with respect to evil and perverse deeds, because the more nature departs from the one operation, the more it inclines toward the other. In this way, the woman falls short in intellectual operations, which consist in the apprehension of the good and in knowledge of truth and flight from evil. This is why one who inclines to evil inclines more to sensitive appetite, unless she is ruled by reason, as is apparent in the seventh book of the *Ethics*.[30] Therefore, sense moves the female to every evil, just as intellect moves a man to every good. And this is why, etc.

Question 12: Whether sperm comes forth from the entire body.

One inquires further into the principle of generation. And first, whether sperm comes forth from the entire body.

It seems so. For the generative power is given to every animal subject to corruption so that it may be preserved in one like itself, because it cannot be preserved in itself. But every body

revealing what had been divulged to him in the confessional. But cf. *QDA* 13.18 where A. discloses details from confessions.

30. Cf. Ar., *Eth. Nic.* 7.2 (1145b8f.).

part is corruptible. Therefore, the generative power exists in every part. But generation occurs through the emission of sperm. Therefore, sperm comes from every part.

In addition, according to the Philosopher's argument, pleasure that occurs in one part is great, but is greater in more parts, and is therefore greatest in all parts.[31] But the highest pleasure occurs in the emission of sperm. But this would not be the case unless sperm arose from all the parts. Therefore, etc.

Moreover, evacuations that relieve all the parts arise from all parts. But the evacuations of sperm relieve the entire body. Therefore, sperm proceeds from the entire body.

This is confirmed by the fact that the retention of sperm does not only afflict one part, but the entire body. This would not be the case unless it proceeded from the entire body.

The Philosopher says the opposite.[32] And he proves it with this argument: that whatever proceeds from the entire body does not have a designated receptacle in the body, as is evident for sweat. But sperm has a designated receptacle. Therefore, it does not proceed from the entire body.

To this, one must reply that sperm can proceed from any given part of the body and, at the same time, from a designated part, because sperm is a superfluity of the last food. But the last food for any given member can have some superfluity. Therefore, sperm can be derived from any given member.

Moreover, it can arise from a designated part because nothing prevents one part from existing with perfect power and another one with imperfect power. Therefore, a perfect part can convert the last food totally into its nature leaving nothing superfluous behind. A part with imperfect power, however, can be hindered in this respect. Therefore, nothing prevents the sperm from being derived from one part and not from another.

On to the arguments. To the first, one must reply that it can be conceded. For in some way the arguments on both sides can be conceded, etc.

31. Ar., *GA* 1.17 (721b14f.).

32. Ar., *GA* 1.17–18 (721b7ff.). Cf. Avic., *Can. med.* 1.1.4.1, 1.1.5.1, 3.20.1.1–3; *DA* 15.2 (fol. 60ra–va). Averroes, *GA* 1.18; A., *DA* 15.2.1–3.59–63 (*SZ* 2: 1111–13).

Question 13: Whether the sperm is of the nature of parts or of some superfluity.

One inquires further whether sperm comes forth from every part or from some part by means of its dissolution or of derivation, and this is to inquire whether sperm is of the nature of parts or of some superfluity.

1. And it seems that sperm comes forth by means of dissolution. Because the body is weakened and enfeebled and dissolved by the sperm's emission. But this would only be because the sperm were something that is resolved or released from the parts. This is why, etc.

2. In addition, an accident does not exist without substance. But the power of every part exists in the sperm. Therefore, something of every part is also in the sperm. But this cannot be without the dissolution of the part. Therefore, etc.

3. In addition, the fetus resembles the parents in all its parts, because a cripple commonly generates a cripple and a leper generates a leper. But this would not be so if the sperm were not dissolved from each part. Therefore, etc.

The Philosopher says the opposite.[33]

To this, one must reply that sperm is not something that exits from the parts by dissolution, because nothing is dissolved from the parts of the body without pain. But the ultimate pleasure occurs during the emission of sperm. Therefore, it does not come forth through the dissolution, or separation of a part, because if it did happen with the dissolution of the one containing it, then it would happen with pain.

Moreover, a great dissolution arises from a large body, and a small from a small. But animals with a large body have less sperm than animals with a small one. Therefore, sperm is not some dissolution.

In addition, if the sperm were something separated off from the nature of every part, since each part of flesh is flesh and each part of bone is bone and each part of nerve is nerve, then

33. Ar., *GA* 1.18 (725a21f.). Cf. Avic., *DA* 16.3 (fol. 60va–61ra); Averroes, *GA* 1.18; A., *DA* 15.2.3–5.93–95 (*SZ* 2: 1127–28).

it would follow that sperm should be something composed of flesh, bone, and nerve. But every animal is composed of things like these. Sperm, then, would be a kind of animal. And the argument is the same for the woman's menstrual blood. Therefore, it would follow that any semen or sperm would be two animals. But one thing cannot be made from things like this. Therefore, sperm is not something that is separated off from individual parts.

Moreover, this is apparent by analogy in plants, because the seed [*semen*] is not something separated off from each part in them, because then any part could produce fruit, but fruit is generated from a certain superfluity residing in a certain part. Therefore, the same argument will apply to the animal semen.

1. On to the arguments. To the first, one must say that the emission of sperm is twofold: ordered and disordered. In an ordered emission nothing is emitted except the superfluity of the last food, and this does not weaken the body but rather it cleanses it. And the other emission is disordered, and such a one weakens the body, not because something is dissolved from the parts, but because the last food, which ought to be converted in the parts, is delegated to the seminal vessels, that is, the testicles.

2. To the second argument one must reply that the soul's power is emitted with the sperm. Nevertheless, this power exists in the spirit as in a subject. Therefore, there is a certain separating-off of the spirit with the sperm, but the spirit is not part of the body.

3. To the third argument one must reply that a likeness arises between the one generating and the one generated, not because the sperm is separated off from the parts of the one generating but because the sperm is a superfluity of the last food, which was in close potential to the body of the one generating, so that it may be converted into it. And this is why a flawed one generates a flawed one and a leper generates a leper, because the leper's last food, the superfluity of which is sperm, is corrupted in accordance with the corruption of his whole body. Therefore, the leper's sperm is corrupted and this is why there is a similar corruption in the fetus. The reason for the similarity

of the one generated to the one generating will become evident in the sixteenth book of this work.[34] Nevertheless, so far as it suffices for the premise, reason does not support the opposite view any more than it does the premise, because one who is mutilated often generates one who is not mutilated, and this is when the power of the one generating is capable of producing one that is perfect in species.

Question 14: Whether sperm is derived more from one part than from another.

One inquires further whether sperm is derived more from one part than from another.

1. It seems not, because an animal having all its parts is produced from sperm. Therefore, every part exists in potency in the sperm, and if this is so, then it is not derived more from one part than from another.

2. In addition, sperm is a superfluity of the last food, as I suppose. But the last food can overflow from one part as well as from another. Therefore, sperm is not derived more from one part than from another.

3. To the contrary. Sperm has the role of a mover and of a distinct principle. Therefore, it is particularly derived from parts that are distinct and designated for motion. But parts like this are official parts more than they are uniform parts. Therefore, etc.

4. In addition, sperm is especially hot and moist. Therefore, it especially descends from that part in which the principle of heat and moisture exists. But the heart and the liver are this sort. Therefore, etc.

To this, one must reply that sperm is derived more from one part than from another, because nature inclines more to the preservation of principal parts than non-principal parts, and this is why it sends more of the clean food to the principal parts, and therefore there is more of the superfluity of the last food

34. See *QDA* 16.3; *DA* 16.1.14.72–75 (*SZ* 2: 1194–97).

in these parts. And this is also why more sperm is derived from these parts and especially from the brain, because the brain is white and soft and moist, and in this respect it agrees with the substance of the sperm. Furthermore, the brain shares in the heat from the heart, and the sperm can receive heat in it. Therefore, the sperm comes forth particularly from the brain.

And an indication of this is that the brain is weakened from excessive intercourse and the eyes, which agree with the brain, become sunken and are greatly weakened. Moreover, this can be revealed by many experiments [*experimentis*], because if the testicles of an intoxicated man are placed in water, his drunkenness dissipates, because the testicles particularly agree with the brain; therefore, the fumes withdraw and the drunkenness is allayed, etc.

This is also evident in another way, because hairs begin to grow on the cheeks and groin and testicles from the first generation of sperm, because the sperm descends to these parts and warms them, and this is why hair begins to grow abundantly there. And a certain master, Clement of Bohemia, told me there was a certain hoary old monk who approached a certain beautiful mistress and just like a starving man he demanded her sixty-six times before the striking of matins; the next day he fell down and, on the very same day, was dead. And because he was a noble, his body was opened up and his brain was found to be entirely evacuated, so much so that nothing more of it remained than the size of a pomegranate, and similarly his eyes were destroyed. Nature marvels at this, although it seems to be consonant with reason. This, then, is an indication that intercourse particularly evacuates the brain; this is why, etc.

1, 2. Therefore, one must respond to the argument that although sperm can come forth from all the parts, since some parts are more principal than others and have a greater agreement with sperm, this is why it proceeds more from these parts than from others. And with this a solution is apparent for the first two arguments.

3. To the first argument one must reply to the contrary that the generative power belongs to the vegetative [*vegetativa*] soul and the vegetative [*vegetabilis*] soul operates more or earlier

through uniform more than through non-uniform parts. For a hand or a foot does not grow unless flesh and bone grow first, and it is the same for the others. Therefore, although an animal operation is suited more to official parts than to uniform parts, nevertheless a natural operation occurs earlier through uniform parts than it does through non-uniform parts, and as a result it does occur through non-uniform parts.

Or, in another way, one can say that it arises from uniform parts as far as its first generation is concerned, but arises from official parts as far as its complete digestion.

4. To the second argument one must reply that the first principle of heat is in the heart. Nevertheless, the heat in which the brain participates is more suitable to the sperm, because it is joined to moisture, whereas the heat in the heart is joined to the dry. But sperm is hot and moist, and this is why, etc. Therefore, it flows principally and radically from the heart, but sperm flows more immediately from the brain, etc.

Question 15: Whether sperm is a superfluity of the last food.

One inquires further whether sperm is a superfluity of the last food.

1. And it seems not. For that which is necessarily required for a thing's constitution is not superfluous. But sperm is like this. Therefore, etc.

2. Moreover, every superfluity is either pure or impure. But sperm is not a pure superfluity, because when it is retained in the body it does not cause corruption. Nor is it an impure superfluity, because an animal is not constituted from it.

The Philosopher says the opposite.[35]

One must reply that sperm can be compared in two ways: either for the preservation of an individual in itself, or for the preservation of an individual in another. If in the first way, then

35. Ar., *GA* 1.18 (726a26f.). Cf. A., *DA* 15.2.5.93, 15.2.6.102 (*SZ* 2: 1127; 1131).

it is a superfluity of the last food. If in the second way, then it is not a superfluity but rather it is necessarily something requisite. The reason for the first is: sperm cannot come from the true nature of the members, since then it would not need the final digestion or completion in the generative members, but rather would immediately become an animal.

And in addition, this sort of separating-off does not occur without pain and sadness, even though this happens with the highest pleasure. Therefore, it necessarily follows that sperm be something beyond the nature of the whole and the part. This can only be something that is superfluous with respect to the individual itself.

But one must know that a superfluity is of two kinds: There is one that the individual does not need either for its own sake or for the sake of another. Urine and sweat are superfluities like this. The other is one that the individual does not need for its own sake at a given time, but which it does need for the sake of another in which its likeness is preserved. And such a superfluity must necessarily exist in potency just as the individual itself does, and the sperm is this superfluity, and this is why the Philosopher says, "Sperm is the superfluity of the last food."

1. On to the arguments. To the first, one must reply that although sperm is not superfluous with respect to that which is generated from sperm, it is nevertheless superfluous with respect to that which has already been converted into nutriment for the individual.

2. To the second argument one must reply that although sperm remains in individual members, it is impure with respect to that which has been converted into the members of an individual. But when it is delegated to the generative members, it is completely digested there and made purer, and then it can be called a pure superfluity with respect to that which ought to be generated from it.

Question 16: Whether sperm is generated immediately from blood.

One inquires further whether sperm is generated immediately from blood.[36]

It seems so. For "sperm is a superfluity of the last food." But according to the Philosopher, blood is the last food.[37] Therefore, etc.

Moreover, sperm in men, and the menstrual blood in women, are generated from the same thing. There is no difference except in the digesting power. But the menstrual blood arises immediately from the blood, and therefore so too does sperm.

In addition, after the emission of sperm, blood comes forth from those who engage excessively in sexual intercourse, according to the Philosopher.[38] But this would not be the case were sperm not generated immediately from blood. Therefore, etc.

To the contrary. Good sperm, which is natural, is white and globular. Blood, however, is reddish and moist. Therefore, sperm is not immediately generated from blood.

In addition, blood's generation occurs in the heart and liver, and flows from these through the veins to the individual members. Therefore, in order to be suited for conversion in the members, it requires a further change, and as a result sperm is generated from something which is beyond the nature of blood.

To this, one must reply that sperm is not immediately generated from blood; rather, blood flowing in abundance through the veins is changed at their ends into a thinner moisture, and this moisture crosses through the sponginess of the members and is made even thinner, and the pure is separated from the impure, and what is impure is expelled through sweat or through abscesses or through the sediment [*hypostasis*] in urine. And what is pure is made proportional to the members themselves, and what remains after sufficient conversion of this moisture into the members becomes sperm. Thus blood is the universal prin-

36. Cf. A., *DA* 15.2.6.101–2 (*SZ* 2: 1130–31).
37. Ar., *GA* 1.19 (726a30f.).
38. Ar., *GA* 1.19 (726b7f.).

ciple of the generation of sperm, but its immediate principle of generation is the moisture that is generated from blood. This moisture has a different nature in various parts, because its nature differs in bone and flesh, and this with respect to potency.

By this a solution is evident to the arguments. The first two prove that blood is a universal principle and, as it were, the remote principle for the generation of sperm, and the Philosopher understands this. And this is why after sufficient emission of sperm a lot of blood is also emitted.

The arguments prove the opposite, that the special and immediate principle of the sperm's generation is something beyond blood, like moisture.

Question 17: Whether one who has a defect in some part necessarily generates an animal that is defective in the same part.

One inquires further whether sperm that has been separated off from a certain part may exist in potency to certain other parts. And this is to inquire whether one who has a defect in some part necessarily generates an animal that is defective in the same part.

1. And it seems so. Because nothing is created from nothing. But in one who has been maimed, the matter of the part that has been taken away is as nothing. In such a mutilated one, therefore, there is nothing from which the missing part can be generated.

2. Moreover, everything that is created is created from something proportional to it. But there is nothing proportional to the missing part in the semen separated off from one who has been mutilated. Therefore, this part cannot be created from his sperm.

To the contrary. Frequently one who has been mutilated generates one who is not mutilated, and a cripple generates a noncripple. But this would not be the case were parts of the sperm determined for determinate parts of the animal.

To this, one must reply that the determinate parts of the sperm are not determined for determinate parts of the animal; rather, it sometimes happens that an animal comes from both homogenous and heterogeneous ones. Now because a plant is uniform, almost any of its parts can become a plant, generating it either by grafting or by putting it in the ground, as is apparent in the willow.[39] Nevertheless, plants that are more noble and complete are generated only from seed [*semen*], just like complete animals. But sometimes vile things are generated through putrefaction. Therefore, the generative power in an animal can produce an animal from sperm, which is uniform in its parts, so much so that a whole can produce a whole from either the whole or from the part; otherwise, twins would not be generated. Therefore, although the sperm is not separated off from a part that has been removed, nevertheless the power that is in the generative members can dispose the semen toward a homogenous part to which it wants to move, etc. Therefore, the fact that one that is mutilated is generated from one who is mutilated is owing to imperfection of this power, because if the power were potent it could create one not mutilated from one mutilated, and a non-cripple from a cripple.

1. On to the arguments. To the first, one must reply that although the matter of a part that has been removed is like nothing, because that part is not in the semen, nevertheless the sperm that arises from other parts can be disposed to that part and can be made proportional to a uniform part.

2. And in this way a solution is evident to the second argument.

Question 18: Whether sperm is necessary for the generation of animals.

One inquires further whether sperm is necessary for the generation of animals.

1. And it seems so. Every bodily operation occurs through a bodily organ, as the Philosopher says in that book.[40] But the

39. Cf. A., *De veg.* 1.2.9.183.
40. Ar., *GA* 1.2 (716a23f.). Cf. A., *DA* 15.1.2.8 (*SZ* 2: 1088).

generation and formation of members is a bodily operation. Therefore, it occurs with a bodily organ mediating it. But such a thing is nothing other than the sperm. Therefore, etc.

2. In addition, two things are required for natural transmutation: an agent and matter. But according to the Philosopher sperm serves as the agent, and the menstrual blood serves as matter.[41] Therefore, of necessity sperm is required.

To the contrary. Sperm and milk arise from the same thing and are ordered to the same end, except that milk is for nutrition and sperm is for completion or composition or maintenance. But milk does not exist in animals lacking breasts. Therefore, neither will sperm exist in animals lacking the generative members. But there are many animals like this. Therefore, etc.

In addition, the Philosopher says that sperm exists in some animals and in others it does not.[42] Therefore, etc.

One must reply that some animals are generated from putrefaction and some are generated by propagation. Sperm is not necessary for generation from putrefaction. It is necessary, however, for the generation of perfect animals. The Philosopher proves this clearly through many indications.[43] For wind eggs do not yield chicks because they are not generated from the sperm's power. Similarly, fish eggs do not produce animals until the male has spread his semen over them.[44] Therefore, one can say that just as a heart or something analogous to a heart exists in every animal, so too in every animal that is not generated from putrefaction there exists sperm or something resembling sperm, as in the ringed animals, as the Philosopher says.[45]

1. To the first argument one must reply that there is a difference between sperm and milk. For milk is only ordered for the nourishment of animals that are generated in the womb, and because there are many animals that are generated outside the womb—like those generated by laying eggs—not all animals

41. Ar., *GA* 1.20 (729a24–33). Cf. A., *DA* 15.2.7–8.118–22 (*SZ* 2: 1138–40).
42. Ar., *GA* 1.17 (721a30f.).
43. Ar., *GA* 1.16 (721a2f.).
44. "Animals": perhaps, "live young," as is common in the *DA*.
45. Ar., *GA* 1.16 (721a2f.).

have milk or breasts, which are the vessels for milk. But sperm is ordered for the generation and formation of the members of every animal, whether it is generated in the womb or outside of it, and this is why sperm is more necessary than milk.

Nevertheless, not all animals have testicles, because testicles exist for the sake of well-being and not simply for being, as is said in the text, since nature has arranged the testicles so that sperm can be conserved and retained until a designated time and so that it not be emitted suddenly.[46] Therefore, animals that lack them have coitus quickly and cannot retain their semen. This is just as it is for animals that have short and straight intestines when compared to animals that have intestines that are twisted and not straight.

2. To the second argument one must reply that although the Philosopher may say that sperm does not exist in some animals, nevertheless he concedes that there is something analogous to sperm in them.

Question 19: Whether females produce sperm.

One inquires further whether females produce sperm.[47]

1. And it seems so. For "sperm is a superfluity of the last food."[48] But the superfluity of the last food exists in females just as in males. Therefore, etc.

2. In addition, the white of the egg is compared to sperm, and the yellow or the yolk is compared to the menstrual blood. But the female can produce by herself an egg that is divided into the white and the yellow. Therefore, the female emits sperm.

3. In addition, the greatest pleasure consists in the emission of sperm. But sometimes females take the greatest pleasure in coition, and this could not be the case if they did not produce sperm. Therefore, etc.

46. Ar., *GA* 1.4 (717a12f.).

47. Cf. Ar., *GA* 1.17 (821a32f.); Avic., *DA* 15.2 (fol. 60rb); Averroes, *GA* 1.18; A., *DA* 15.2.1.58, 15.2.6.101, 15.2.11.141–45 (*SZ* 2: 1111–12; 1130–31; 1149–52).

48. Ar., *GA* 1.19 (726a30f.).

To the contrary. Sperm bears the same relationship to the man, as the menstrual blood does to the woman. But the man emits sperm because he does not menstruate. Therefore, the female menstruates because she does not produce sperm.

In addition, matter and agent do not coincide, according to the Philosopher in the second book of the *Physics*.[49] But the female provides the matter, and the sperm is the agent. Therefore, the female does not produce sperm.

One must reply that, properly speaking, the female does not produce sperm. And the reason for this is that sperm is a superfluity of the last, completely digested food. But the completion of digestion results from a strong heat, whereas females have only a weak heat, and this is why there is insufficient power in the female to generate sperm. Instead, just as a strong heat in the male creates sperm, so a weak heat in the female creates the menses, because menses is a crude and undigested blood.

Nevertheless, one must understand that there can be sperm in a female in some manner, and this is when the heat is kindled in them. Sperm like this is found more often in black women, who have sex more than all other women, and more than white women. Because black women [*nigrae*] are hotter and particularly dark [*fuscae*],[50] they are the sweetest for sex, as lechers[51] say, as well as because they have a tempered opening of the vulva so that it embraces the penis very pleasantly.[52] And this is so in thin ones more than in fat ones, since in the fat ones the spermatic superfluity is converted into fat and diminishes the heat, but in thin ones it is the other way around.

And this must be understood concerning digested and globular sperm. If, however, we mean by "sperm" moisture that is white, watery, and undigested, this is how sperm is more abun-

49. Ar., *Phys.* 2.7 (198a21f.).

50. Cf. Song of Songs 1.4–5: *nigra sum sed formonsa . . . nolite me considerare quod fusca sim*—"I am black but beautiful . . . do not consider that I am dark."

51. *Leccatores*: flatterers, lechers. See Petrus Alphonsi, *Disciplina clericalis* 3 (PL 157: 678B): *clerici pro sapientia sua non sunt honorati, unde facti sunt leccatores, et ad magnum venere* [= *venerunt*] *honorem.*

52. "Pleasantly": reading *suaviter* for *suavitur. Suavitur* is in fact a verb, meaning "it kisses," but if it is to be read here we must supply an "and" between the two verbs.

dant in white women than in black ones, as the Philosopher says.[53]

Nevertheless, one must observe that the sperm that is found in women is not suited for generation in accord with the proper sense of "sperm," because in sperm, properly, there is an active power and one capable of forming the members and of introducing soul, and this power exists only in the male's sperm. Therefore, the female serves generation only through her menses. For when it is digested further by means of the power of the male's sperm, it can receive the form of a fetus [*conceptus*] while a second part of it can be converted into its nutriment.

Thus one must respond to the question in brief that females, universally, do not produce sperm as males do and, if they do produce sperm, they do not create sperm that is as digested or as powerful for the sperm's operation, as males do.

1. On to the arguments. To the first, one must reply that the superfluity of the last food can be of two types: either completely digested or undigested. If it is digested, it is sperm; if it is undigested, then it is menses. And this must be understood to concern the superfluity which it needs.

2. To the second argument one must reply that the white of the egg is not compared simply to sperm, because the chick arises from the white and is nourished by the yellow or yolk. Therefore, the female's menses is analogous to or corresponds to both the white and the yellow in the egg, because the fetus is formed materially from one part of the menses, and it is nourished by another part. Thus it is for the egg.

3. To the third argument one must reply that pleasure during coition does not arise only from the emission of sperm. For children and old people, who cannot produce sperm, take pleasure in coition; instead, pleasure arises from the contact of the male's sperm with the womb or from contact of the penis with the vulva, and not because the woman produces sperm.

53. Ar., *GA* 1.20 (728a2f.). Cf. A., *DA* 15.2.6.112 (*SZ* 2: 1135).

Question 20: Whether sperm is a material part of the fetus [conceptus].

One inquires further whether sperm is a material part of the fetus.

1. It seems so. Because whenever something is generated from a mixture of two things, each is a material part for it, just as is apparent in the generation of a mixture made from the elements. But a fetus is generated from a mixture of sperm and menses, and therefore, etc.

2. In addition, if sperm were not a part of the fetus, then I ask, once the fetus is completed, where does the sperm go? If it still remains, then it is superfluous or it will be able to form a new fetus in the course of time without the joining of the male. If it is corrupted, then it has brought about its own corruption, which is false, since nothing intends its own corruption.

3. In addition, some are generated by conversion and others by propagation. One that generates by conversion only transmutes matter and introduces form. But one that generates by propagation prepares the matter, and, once it has been prepared, it introduces form, because one that generates like this generates by separating off something from itself. Therefore, since a male generates by means of propagation, a part of the fetus will be separated from him.

The Philosopher says the opposite.[54] And it seems from the argument that matter and an agent do not coincide. But the sperm is the agent with respect to the fetus. Therefore, it is not its material part.

Controversy exists between the Philosopher and the physicians over this question. For the Philosopher posits that sperm is not a part of the fetus but is only the agent.[55] He says that sperm is related to the fetus as art or an artist is related to what is produced by art. But the artist is not a material part of an arti-

54. Ar., *GA* 1.21 (729a34f.). Cf. A., *DA* 15.2.8.122 (*SZ* 2: 1140).
55. Ar., *GA* 1.22 (730a32f.).

fact. Therefore, neither will the sperm be a part of the fetus, as a carpenter is not a part of a bench.

The physicians posit that the sperm is a part of the fetus.[56] For they posit that a fetus is made from a mixture of each semen, namely, of the male and the female; otherwise, nothing of the father would be present in the son. Very often, however, a son resembles the father in his bodily disposition, which would not be the case were his matter not received from the father.

Nevertheless, it seems that I ought to say that there are two things in sperm, namely, moisture or a superfluity of the last food and a power of the father's soul existing in a certain frothy spirit. Therefore, as far as the moisture is concerned, the sperm can be a part of the fetus just as the menses is, yet as far as the spirit itself is concerned, it cannot, because the moisture is itself first mixed with the menses just as wine is mixed with water or rennet with milk. It is agreed, though, that the rennet is part of the cheese and the water is part of the wine, etc. But the spirit moves to completion and to the complete digestion of the menses through the power of the father's soul that exists in it. Therefore, because the female's menses is sufficient *per se* to provide the matter of the fetus, another part from the sperm is not necessarily required for the fetus's substance, but the power of the sperm or the frothy spirit, which is in the sperm, is necessarily required. This is why the Philosopher says—and properly so—that the sperm is not a material part of the fetus because it is not called sperm except to the extent that it has an active and motive power, and something that is moving and acting to this extent is not part of the one generated. This is why, properly speaking, sperm is not part of the fetus.

Nevertheless, nothing prevents some material part of it from passing over into the matter or body of the fetus, so as a result a mixture is made with the menses, and this is how the physicians understand matters. And perhaps the principal and radical

56. Cf. Hippocrates, *De natura pueri* 4; Galen, *De semine* 1.1 (Kühn 4: 512); Avic., *Can. med.* 1.1.5.1; *DA* 16.1 (fol. 61ra–vb); Averroes, *GA* 1.20; *Colliget* 2.10; A., *DA* 15.2.8–11.122–45 (*SZ* 2: 1140–52).

members are generated from that one, but the flowing members are generated from the menses. This is the understanding the Philosopher had with respect to the power of the sperm.

1. On to the arguments. To the first, one must reply that sperm is not properly mixed with menses in terms of its power, just as an agent is not mixed with matter, or if it is mixed, this is only true for its material moisture and not for its moving and acting spirit.

2. To the second argument one must reply that for every natural agent, the more distant it is from its principle and the more continuous its operation, the more it is weakened and the more it approaches defect, because one acting physically is acted on again while acting and it acts again while being acted on, according to the Philosopher.[57] As to the power that has been separated out from the father and resides in the frothy spirit—the more distant it is from the father, and the more it continues in its operation, the more the one operated on or the object approaches perfection, whereas the spirit itself approaches defect. Therefore, in the end the spirit itself is deficient and vanishes, and the moisture joined to it passes over into some superfluity, which it does not need, or it passes over into some members.

Therefore, one must reply to the form of the argument that it does not act *per se* for its own corruption, but it does act *per se* for the generation of one like the one from which it is generated. This occurs, however, because it acts for its own corruption. Therefore, for this reason it is not said to act, because nature is said only to act for that which it intends, etc.

3. One must reply to the third argument that one generating by propagation embraces both male and female. Therefore, something is separated off from each of them, because the sperm is separated off from the father with a power capable of making an embryo and with a frothy spirit, and the matter is separated off from the mother, namely, the menses, etc.

57. Cf. A., *Phys.* 3.1.7.

BOOK SIXTEEN

Question 1: Whether the soul is in the sperm.

"NOW WE HAVE already declared that the powers," etc.[1] About this sixteenth book he asks first whether the soul is in the sperm.

1. It seems that it is. For everything that has any consequent operations has a principle of these operations. But semen has operations, of which the soul is the principle, because it is nourished and grows and is moved, and these things are found only in animated things. Therefore, etc.

2. In addition, nothing acts except through that which is already in act. But the soul's active power is in the semen, and generation is of like from like. Therefore, the soul is in the semen in an act for which it is the power.

3. In addition, an eye that does not have the form of the eye is an eye only equivocally. By the same reasoning, then, if the semen does not have soul (which is the principle of these operations and the source for these powers [*potentiae*]—the augmentative, nutritive, and generative), then it is semen equivocally because these are found in semen just as sight is in the eye.

4. In addition, "complete" and "incomplete" do not distinguish essence or species. But semen is related to the animal as something incomplete is to one complete. Therefore, these do not distinguish them in essence. But this would not be so if the semen did not have the soul; therefore, etc.

The Philosopher says the opposite, for he says that the soul is not in the semen in act but only in potency.[2]

1. Ar., *DG* 2.1 (731b19f.). Cf. Averroes, *GA* 2.3; A., *DA* 16.1.1–7.1–45 (*SZ* 2: 1133–78).

2. Ar., *GA* 2.1 (735a4f.). Cf. A., *DA* 16.1.8.46–50 (*SZ* 2: 1178–80).

To this, one must say that the soul is not in the semen in act but in potency. Nevertheless, the power of the soul is in the semen in act, because just as the prime mover acts in the motion of projectiles by imparting some power to a second, moving thing, by which the second can move after the first ceases to do so, so too is a certain power separated from the father along with the sperm, and this power moves and operates after the father is at rest. Thus the power in the semen is the power of the soul which is derived from the soul of the father with the sperm, and which produces the soul in the semen itself, and yet is not the soul in act because "soul is the act of a physical, organic body," etc.[3] And with respect to the vegetative [soul], it requires homogeneous parts, and with respect to the sensitive it requires official [parts]. But in the semen, since it is not an organic body, there are neither homogeneous parts (like flesh and bone and things of this sort) nor official ones (like the foot and things of this sort). Thus the soul is not in the semen itself in act. Thus the Philosopher says in the seventh book of the *Metaphysics* that "sperm acts just like those things which come from art."[4] Because just as a house does not exist in the mind of the architect in act, neither is the soul in the semen in act, but only in potency. And in the second book of *On the Soul* the Philosopher says that the soul is the act of a body having life in potency, not in the sense that it casts off the soul as it does the semen, but rather implying that the soul is not in the semen in act.[5]

1. On to the arguments. To the first, one must respond that the operations that are attributed to the sperm or the semen are from the power that is in the sperm. This power, however, is in spirit as if in a subject, yet it acts through the power of a separate agent, namely, the father's soul, from which it is derived.

2. To the second argument one must reply that semen acts through a power that is in semen in act, but it does not act through the soul.

3. To the third argument one must reply that something is

3. Ar., *De anima* 2.1 (412b5f.).
4. Ar., *Metaph.* 7.9 (1034a33f.). Cf. A., *Metaph.* 7.2.11.
5. Ar., *De anima* 2.1 (412a19f., b25f.). Cf. A., *De anima* 2.1.3.

called an eye equivocally if it does not have sight, and is not so called because it does not see in act (for example, the eye of one who is sleeping).[6] So, contrariwise, semen is correctly called semen equivocally if it does not have power, but not if it does not have soul.

4. To the last argument one must reply that "complete" and "incomplete" do not distinguish an essence in the same form, in the way that something is more or less white. Nevertheless, if these are in different forms, they do distinguish the essence, just as they do for substance and accident which are related to one another as complete and incomplete. Thus the premise is correct. The forms of the semen and the animal are different and contraries, as it were; otherwise, the form of the semen would not be corrupted during the animal's generation. But they are said to relate to one another as complete and incomplete, because once the semen's form has been cast off from the matter of the semen, it exists in potency to the form of the animal, and the semen's matter is potentially that which the animal is in act. Nevertheless, this act cannot be introduced while the form of the semen remains, and this is why the argument does not succeed.

Question 2: Whether the power of the soul must be in the semen.

Further, one asks whether the power of the soul must be in the semen.

1. It seems not. For that which reduces another to act is nobler than it, and, according to the Philosopher in the third book of *On the Soul,* a universal agent is nobler than the one undergoing change.[7] But the power that is in the semen reduces the animal to act or introduces the soul. Therefore, this power is nobler than the soul. But the power of the soul is not nobler than the soul any more than potency is nobler than act. Therefore, etc.

6. Cf. Ar., *De anima* 2.1 (412b21).
7. Ar., *De anima* 3.5 (430a18f.).

2. In addition, the further a power extends from its cause, the more it is weakened. But the more the power in the semen operates, the more the fetus's power is strengthened, because the fetus is nearer to act. But if it were the power of the soul, it would be weakened more. Therefore, etc.

3. In addition, every bodily operation occurs through a bodily power. But the formation of the fetus is a bodily operation. Therefore, it occurs through a bodily power and not through the soul's power.

The Philosopher says the opposite.[8]

One must respond that it is necessary for the soul's power to be in the semen, because according to the Philosopher in the eighth book of the *Physics,* it belongs to animated beings to move on their own, whereas inanimate things do not move themselves.[9] But the power that is in the semen moves the semen and inclines it toward organization. Therefore, this power is a power of the soul.

Furthermore, whatever operates on another, acting like a craftsman on a crafted product, is not a power of the body but rather a power of the soul. But the power in the semen operates just like a craftsman, according to the Philosopher.[10] Therefore, it is a power of the soul.

Moreover, the generation of animals is univocal, because a horse arises from a horse and an ass from an ass. But this would not be the case if the soul's power were not in the semen.

1. On to the arguments. To the first, one must reply that an agent is of two types: instrumental and principal. A principal agent is nobler than the one undergoing change, but this is not necessarily so for an instrumental agent. Now, the power of the soul that is in the semen is an instrumental agent whereas one that generates, like a father, is a principal agent. And this is why it is not necessary for the power in the semen to be nobler than the soul.

8. Ar., *GA* 2.1 (735a4–9). Cf. Averroes, *GA* 2.1.3; A., *DA* 16.1.6.35–45 (*SZ* 2: 1172–78).

9. Ar., *Phys.* 8.2 (252b19f.).

10. Ar., *GA* 2.1 (734a33f.; b31f.).

2. To the second argument one must respond that although the fetus's power grows stronger the more the power of the semen operates, nevertheless the semen's power grows weaker the more it operates. Thus the operating power is one thing, and the power that is introduced is another.

3. To the third argument one must respond that two things are required for the generation of an animal, namely, a power of the body and a power of the soul, because an animal is composed from both. From the body comes heat for the sake of life, and spirit for the sake of the motion of generation. To these there correspond two others with respect to the soul, namely, the vegetative power corresponds to heat and the sensitive power corresponds to spirit.

Question 3: Whether this power is from the father's soul.

Next one asks whether this power is from the father's soul or not.

1. And it seems not. Because a power does not exist without its proper subject. But the father's soul is not in the semen. Therefore, neither is its power.

2. In addition, according to the Philosopher's text, the soul in the embryo is similar to one who is asleep, and this soul begins to function when something comes to it from outside.[11] But the soul in the embryo similar to one who is asleep is the soul of the fetus and not of the father. Therefore, it will be its power.

3. In addition, as far as it is able, nature always acts through the fewest intermediaries possible. But the power that is in the fetus after it has been first formed can prepare [*disponere*] the fetus for its further development. Therefore, it can prepare the semen much better for the generation of the fetus.

4. In addition, if the mother's menses prevail over the father's semen, the fetus follows after the mother, and vice versa. But this would not happen unless the mother's power were in the semen. Therefore, this power comes from the mother as much as from the father.

11. Ar., *GA* 2.1 (735a8f.).

On the contrary. The active power is in the sperm. But the sperm is only from the father, and therefore, etc.

One must respond that the power that is in the semen principally, and acting as an agent, is from the father's soul, and not from the mother's soul or the soul of the fetus, because it is the father who provides form and the woman who provides the matter. But form is introduced by the power of an agent, and matter is prepared to receive form by this same one. Therefore, the power that prepares the semen and introduces the form, that is, the soul, flows from the father's soul in the semen and does not come, however, from the mother's soul because she does not govern the preparation of the nature but only adapts the matter for the reception of form. It is clear that this is not a power of the fetus's soul because that which does not yet exist does not have an operating or operative power. But when the semen goes forth, the soul of the fetus does not yet exist. Therefore, its power is not in the semen. This is clear in both natural and artificial creations, because the power that prepares air to receive the form of fire is not the power of the fire that is about to be generated, but is the power of the fire doing the generating. Thus the premise is correct.

1. On to the arguments. To the first, one must say that power can relate to something else in two ways: in one way, effectively or formally and with respect to its origin [*originaliter*], and in another way subjectively and materially, just as the light that is in the air as a medium can be said to be the light of the sun effectively, because it proceeds from it, and can also be said to be the light of the air because it is in the air as in a subject. Thus the power that is in the seed is effectively the power of the father because it derives from him, and nevertheless is a power of the semen as of a subject.

2. To the second argument one must reply that the situations concerning the semen and embryo are quite different. The soul is not present in the semen in the manner of a first act nor in the manner of a second act, but only in potency. But the soul is present in the embryo before it is organized in the manner of a first act, and clearly not in the manner of a second act, because the soul operates through organs. Thus the organs come before

the soul's operations just as instruments come before something that moves them. Thus the soul in the embryo before the organs are completed is like one sleeping, who does not operate in act owing to the weakness of his members, which are at rest at the time, etc.

3. To the third argument one must reply that nature acts through the fewest intermediaries, so far as possible. But the power of the fetus's soul cannot operate before it must be a soul, and this is why the power of its soul is in the semen only in potency, whereas after the generation of the fetus it can operate effectively.

4. To the last argument one must respond that the fetus does not follow the mother, as if the mother's power were in the semen regulating and preparing it. But when the heat is weak in the semen and a great deal of moisture obstructs the heat, then the power which comes from the father's soul cannot achieve its intended end, and then it acts according to the disposition of the matter. This is why this power is a principal agent. The fact that the fetus follows the father or the mother stems from the diversity of its power, (that is, whether it is more or less potent) and its material disposition.

Question 4: Whether this power is corrupted once the fetus is formed.

Next one asks whether this power is corrupted once the fetus is formed.

1. And it seems not. In those having a right order, the power of that which is prior exists in that which is posterior. Therefore, the father's power is in the son that has been formed. But this would not be so if the power flowing from the father's soul were corrupt.

2. In addition, if this power were corrupted once the fetus has been formed, then it would be corrupted either from itself or from the fetus. But it is not from itself because nothing causes its own corruption, nor is it from the fetus, because it introduces form to the fetus, and an effect does not corrupt its cause.

3. The Philosopher says the same thing. For he says that it is inappropriate to say that this power is corrupted.[12]

To the contrary. Every power exists in order to act. Therefore, something that does not exist in order to act is useless. But if this power exists once the fetus is formed, then it would not exist in order to act, and therefore, it would be useless. But nothing like this occurs in nature. Therefore, etc.

One must respond that the power derived from the father is two-fold. One is ordered toward the preparation of matter and the induction of form. Another power is consequent to the introduction of form, and this one is ordered toward operations similar to those of a principal agent. Once the fetus is completed, the first power departs and disappears into itself. This is similar to the motion of projectiles, because as long as a rock is in motion there always remains some power of the one impelling or projecting it, but when it approaches its ultimate end and comes to rest, then the power of the one first projecting it forward disappears. Thus the premise is correct. The power that has been separated from the father's soul along with the sperm resides in a sort of frothy, airy, mossy spirit, and so long as the fetus is being formed, this spirit and, as a result, this power always remain. But when the fetus has been completed, this spirit is resolved into various superfluities, and the power that was in it earlier disappears. The power that follows after the soul, however, remains in the fetus as long as the soul endures, and this power is still said to come from the father's soul because it depends on him as cause and effect. Thus, just as the soul of the one that has been generated is said to be from the father, so the power of this one is said to come from him too.

1. On to the arguments. To the first, one must respond that the father's power is in the fetus. But this is not the same as that one that earlier was ordered toward the soul—that is, which preceded the soul—but is rather the one that follows the soul.

2. To the second argument one must reply that this power is corrupted neither by itself nor by the fetus, but is rather cor-

12. Cf. Ar., *GA* 2.3 (736a22f.). Cf. A., *DA* 16.1.9–10.51–57 (*SZ* 2: 1181–85).

rupted by its container—that is, by the one that is nearest to it—because "the generation of one thing is always the corruption of another," according to the thought of the Philosopher.[13] Thus the embryo's generation brings the semen's corruption.

3. To the third argument one must reply that the Philosopher says this by way of arguing that it is inappropriate for the form to be corrupted, because similarly he also says that it is inappropriate that it [the form] remain. In the same way a solution is apparent for the other.

Question 5: Whether the soul's power operates through a mediating heat.

Next one asks whether it is necessary for heat to be in the semen or whether the soul's power operates through a mediating heat.

1. And it seems not. Because things mixed from earth and water coagulate in the cold, and cold dominates in ones like these. But all things that are generated from earth and water here below, in the sphere of active and passive [things], coagulate; therefore, they are generated by cold.

2. In addition, according to the Philosopher in the fourth book of the *On Meteorology* [*De meteora*], corruption is of two types. One is natural, and putrefaction is like this and occurs principally from heat and moisture, as he says in the same place.[14] The other is unnatural, and this occurs by violence. Since, therefore, the power in the semen works toward life, and since heat is the principle of corruption, as touch is, it does not operate with a mediating heat.

3. In addition, nature aims at what is the better thing among those things that are possible. But although heat conserves a thing, nevertheless it also continuously consumes either the subject or the object, whereas cold preserves it for a longer period, as is evident in things congealed by cold, for example, in a crystal. Therefore, since the soul's power acts toward conservation, it acts more with a mediating cold than with heat.

13. Ar., *GA* 1.3 (319a5f.).

14. Ar., *Meteora* 4.1 (378b28f.).

The Philosopher says the opposite.[15]

One must respond that this power acts in sperm with a heat mediating just like an instrument, because heat and cold are active principles, and the nobler a form is, the nobler the instrument by which it operates. Thus superior or celestial bodies act with a mediating light, which is a nobler thing among active bodies. But among these inferior things, soul is the noblest form. Therefore, it acts with a nobler mediating instrument. But heat is nobler than cold, just as a *habitus*[16] is nobler than a privation. Therefore, etc.

Moreover, according to Aristotle in the book *On the Reason for Shortness or Length of Life,* heat is the principle of life, and its movement occurs from the center to the circumference.[17] Contrariwise the movement of cold occurs from the circumference to the center and consequently it is the cause of mortification. Thus, the soul's power operates both with a mediating heat and cold, yet more often with a mediating heat since, according to Avicenna,[18] the role of heat is common to the entire body and because, according to Hippocrates, heat is nature's friend, whereas cold is like food for heat since cold tempers heat, keeping the heat from fully performing its consuming action.

1. On to the arguments. To the first, one must respond that not all things coagulated from earth and water are coagulated by cold; to the contrary, they can be coagulated by heat. For if something is mixed from earth and water, heat can evaporate parts of the water, and after these are evaporated, that which remains draws together, hardens, and coagulates, as is clear in the case of mud once the water has been evaporated from it by the sun. And this is the case too in the production of the fetus, since the heat existing in it consumes the watery and superfluous moistures and evaporates them, and this is the reason why what remains is more solid and compact and is transformed

15. Ar., *GA* 2.3 (736b33f.). Cf. Averroes, *GA* 2.2; A., *DA* 16.1.13.68–71 (*SZ* 2: 1192–94).

16. A. contrasts here a *habitus*—a quality, characteristic, or relationship that is possessed by an object—with its privation or absence.

17. Ar., *De long. et brev. vitae* 5 (466a18f., b21f.). Cf. *QDA* 12.16.

18. Avic., *Can. med.* 1.1.3.1.

into members. Thus heat causes it to come together by consuming that which is thin and by acting on the power of the soul. The soul operates from an end, and the heat acts in an instrumental fashion under the soul's power as an axe hews under the power of a craftsman, etc.

2. To the second argument one must reply that heat is of two types: natural and accidental. The cause of putrefaction is an extraneous and accidental heat, and this is why "the generation of one thing is the corruption of another." This is why one cannot be generated unless another is corrupted. Thus, since the power in the semen intends to produce a fetus, it must necessarily act with some mediating power that can corrupt the semen, and heat is something very much like this.

3. To the third argument one must reply that although cold things last longer in some things, as in a crystal and things like this, nevertheless in animated beings warmer things last longer, according to the Philosopher in the book *On the Reason for Shortness or Length of Life*.[19]

Question 6: Whether the heat, which is an instrument of the soul, is an elemental heat.

Further one inquires whether the heat, which is an instrument of the soul, proceeds from the elements or is an elemental heat.[20]

1. And it seems not. For it is proper to an elemental heat to consume and destroy. But it is proper to a celestial heat to preserve, and the same is true for a natural heat. Since then the heat, which the soul employs, conserves life and does not corrupt it, this heat is a celestial and not elemental heat.

2. In addition, whatever appears in posterior things flows in from the prior things.[21] But the heat that is in animals is the principle of life. Therefore, it flows in from prior causes of life. But the prior causes of life are either the soul or the celestial

19. Ar., *De long. et brev. vitae* 5 (466a29f.).

20. Cf. Averroes, *GA* 2.3; A., *DA* 16.1.16.86 (*SZ* 2: 1203–4).

21. "Flows in from": perhaps, "is influenced by."

bodies. Therefore, this heat is not an elemental heat but an animal or celestial heat.

To the contrary. Whatever is in a mixed body comes to it from mixables. But this heat is the heat of a mixed body. Therefore, it proceeds from mixables.

To this one must respond that the heat that the soul employs comes to it from mixables, but in some cases it is assigned to a celestial heat. This can be explained from the effects of this heat. For an elemental heat acts on an object and on its own proper subject; thus heat acts on wood, and also consumes other things which it touches by means of its own subject. And the heat that the soul employs acts on an object in the same way, as is clear in the digestion of the nutriment. And it also acts on its subject, since when the nutriment is absent this heat cannot cease from acting, and this is why it then acts on the radical moisture, which is its proper subject. Therefore, in this respect this heat resembles an elemental heat. Yet since this heat is also the principle of life, in this respect it resembles the celestial heat, whose function is to conserve other things. Thus this heat is partly analogous to an elemental heat, and partly analogous to a celestial heat, but it nevertheless differs in number from each of these, because when the elemental heat arrives, this heat departs, and when this heat arrives, the elemental heat departs. And in the same way the celestial heat acts on it and consumes it. And this is why one must say that each is analogous to the other but, nevertheless, also distinct from the other.

1. On to the arguments. To the first, one must respond that although this heat conserves life, it is nevertheless not a celestial heat although it is analogous to it in this respect.

2. To the second argument one must reply that all these bodies below receive the influence of bodies above. Thus this heat receives influences both from the soul and from bodies above, because this heat appears as a result of the commixture of mixables but is governed by the soul's power. And an indication of this is that when the soul departs [the body] at death, the power of this heat changes.

One must respond to the argument to the contrary that this

heat appears from mixables, yet it is nevertheless governed during its operation by the soul's power, and it differs, then, from an elemental heat.

Question 7: Whether spirit is necessary in animals.

One inquires further concerning spirit. And first, whether spirit is necessary in animals.

1. And it seems not. For in every mixed body, heavy things are dominant. Therefore, since spirit appears from a commixture, the heavy dominates in the spirit. But heavy things are dark, and dark and heavy things are really not well suited for the motion and cognitive function of animals. But spirit is not established for any other purpose. Therefore, it is superfluous and unnecessary.

2. In addition, spirit is established for the union of the soul and the body. But just as there is a difference between body and soul, so there is a difference between spirit and soul. Therefore, there must be some other medium to unite spirit and soul, and so on to infinity. But it is not proper to proceed to infinity, as the Philosopher proves elsewhere.[22] Therefore, it is not proper to place spirit as a medium between body and soul as a unifying medium, just as the Philosopher implies.[23]

3. In addition, principles are simpler than things that take their origin from principles. But spirit is not required as a medium between the form of an element and its matter, and therefore neither is the spirit required as a medium between the soul and the body. For if it were, then spirit, which is derived from a principle, would be subtler than its principle because no form of an element can act as a medium.

The Philosopher implies the opposite in the text.

One must respond that spirit is necessary for animals. And the reason for this is that every bodily operation occurs through a corporeal medium. But motion and sense cognition are bodi-

22. Ar., *Phys.* 7.1 (242a28f.).

23. Ar., *GA* 2.3 (736b32f.). Cf. Averroes, *GA* 2.3; A., *DA* 16.1.13.68–69 (*SZ* 2: 1192–93).

ly operations. Therefore, of necessity they occur through corporeal media. But the principles of motion and sensation come from determinate parts. Therefore, it is necessary to posit some medium through which sense and motion flow to the other parts. But spirit is this medium.

Moreover, the sensitive power adds over and above to the vegetative power, just as, as Aristotle indicates in the second book of *On the Soul,* a four-sided figure adds to a three-sided one.[24] Therefore, the instrument through which the sensitive power operates adds to the medium through which the vegetative power operates, and this in the realm of spirit [*in spiritualitate*] since the sensitive power is more spiritual. But the vegetative power operates with a mediating heat. Therefore, it is necessary to posit something more spiritual through which the sensitive power operates. And this is nothing other than spirit. Thus it is necessary to posit spirit, because the motive power and, similarly, the sensitive power, are principally rooted in determinate parts, and yet these are moreover suited for other parts as well. This would not be the case if the influence of the principals did not occur in posterior ones. Yet it is necessary that something convey this influence, and this is the spirit. And this is why, etc.

1. On to the arguments. To the first, one must respond that there are two things to consider in spirit, namely, the manner of the mixture and the impression of the soul. With respect to the manner of the mixture, the heavy dominates in spirit, but with respect to the impression of the soul this spirit receives an aptitude for motion. Thus, although materially it is heavy and dark, because to some extent it comes forth from heavy mixed bodies, nevertheless, formally, it is light and clear from the soul's irradiation.

2. To the second argument one must reply that spirit is not required as a medium uniting soul and body, since they unite themselves to one another. Rather, spirit is required for the execution of the soul's operations, since the execution of motion and sensation occurs with a mediating spirit, and this is why one does not proceed to infinity.

24. Ar., *De anima* 2.3 (414b31f.).

3. To the third, one must respond that spirit is not more subtle than an element but rather is better proportioned for the operation of the soul than is fire or air, and this is why spirit is more necessary than an element.

Question 8: Whether spirit is a body.

Next one inquires whether spirit is a body.[25]

1. It seems not. For it is impossible for two bodies to exist [at the same place] in the same time. But spirit is in the arteries at the same time as the blood. Therefore, spirit is not a body.

2. In addition, according to the Philosopher in the second book of the *Ethics,* the mean is equally distant from the extremes.[26] But spirit is just like a nature midway between the substance of the soul and the body. Therefore, by the same argument by which spirit is a body, soul also will be a body.

3. In addition, spirit is a medium or instrument of the sensitive power, just as heat is of the quickening [*vivificativae*] or vegetative power. But heat is not a body, but is rather an accident. Therefore, by the same argument, neither will spirit be a body.

To the contrary. Everything capable of motion is a body. But spirit is capable of motion. Therefore, it is a body, for it crosses from the brain to other parts of the body.

In response, some say that spirit is not a body but an accident, since just as the celestial bodies operate with a mediating external light, so sensitive bodies operate with a mediating internal light. But that internal light is spirit. Yet this does not seem true, because a thing is neither made active nor moved by the fact that it receives light. But in a body activity and motion are caused by spirits. Therefore, the spirit is not light.

Thus one must respond that one must consider two things in spirit, namely, the nature of clarity and transparence, and spirit

25. Cf. A., *DA* 16.1.13.71 (*SZ* 2: 1193–94); *De spiritu et respiratione* 1.1.3–4; *De somno et vigilia* 1.1.7.

26. Ar., *Eth. Nic.* 2.6 (1107a2f.), 7 (1107b31f.), 8 (1108b23f.).

receives this clarity and transparence from the power and heat of the heart, in which it is generated. One must also consider the subject of this transparence, because an accident does not exist without a subject, and in this way spirit is a body. And this is clear from the determination made by the author of *On the Difference between Spirit and Soul* [*De differentia spiritus et animae*], for he says that "spirit is a subtle body."[27]

1. On to the arguments. To the first, one must respond that spirit is a subtle and very penetrating body. This is evident from thunder and earthquakes. Moreover, the parts of an animal's body are rarefied and porous; otherwise, not every part of the animal would be nourished and grow. Thus when the spirit moves in the body, it passes through the porous passages of the body, just like the spirit that enters or exits from the earth or just like the vapors that ascend in the air. Nevertheless, these do not exist at the same time and place with the air, so that two bodies would exist at the same time, which is impossible. But air cedes its place to the vapors, just as the subtler cedes its place to the coarser. It is the same for the motion of the spirit in the body.

2. To the second argument one must reply that because an animal consists of a soul and a body and its operation is of the whole and not just of the part, for this reason there are some [powers] ordered to its operation with respect to the soul, and some with respect to the body. With respect to the soul there are potencies and powers, and with respect to the body there are heat and spirit. Thus, spirit better conforms to the nature of the body than to that of the soul, and this is why spirit is more a body.

3. To the third argument one must respond that spirit and heat cannot be treated in the same way, since heat signifies a single quality through which an agent operates, and it is not a substance of the quality itself, although some physicians, like Haly, posit that heat is a substance because by heat they understand that in which heat is grounded as well as the heat itself.[28]

27. Costa ben Luca, *De differentia spiritus et animae* 1 (Barach 1878, p. 121).

28. I.e., Haly 'Abbās (d. 1010), one of the most famous of all Muslim scientists and physicians. One of his works, translated into Latin by Constantine the

But spirit denotes that very substance for which there are accidents like motion and other such things.

Question 9: Whether spirit is generated in the heart.

Further one asks whether spirit is generated in the heart.[29]

1. It seems not. For spirit is generated from the subtle parts of blood. But the greatest thinning [*subtiliatio*] of the blood does not occur in the second digestion, which is in the heart, but rather in the third, which is in other parts. Therefore, spirit is generated in individual parts.

2. In addition, in both the first and the second digestion there occurs a weakening of the powers and the spirits. A sign of this is that after taking in food an animal is rendered less fit for motion and sensation. Therefore, the generation of spirits does not occur in either the first or second digestion, but rather their corruption does.

3. In addition, everything that is in a body is nourished. Therefore, spirit can be nourished. But nourishment only occurs after the third digestion, which does not occur in the heart. Therefore, the spirit is nourished in other parts. But the nourishment of a part occurs where its generation occurs. Therefore, etc.

On the contrary. Every power of the body flows out from the heart. Therefore, a medium that conveys a power proceeds from the heart. But spirit is that medium. Therefore, spirit proceeds from the heart.

To this, one must respond that the generation of spirit is of two types. One is its totally new transformation, and such a transformation as this is found principally in the heart because just as the heat by which the vegetative power operates flows from the heart, so too does the spirit by which the sensitive power operates flow from the heart. Another type can be its gen-

---

African, circulated under the title *Pantegni* and was often mistakenly accepted as Constantine's own composition.

29. Cf. Avic., *Can.* 1.1.3.1; A., *DA* 1.2.20.381, 16.1.13.70 (*SZ* 1: 189; 2: 1193); *De spiritu et respiratione* 1.1.9.

eration through a certain restoration after the fashion of a certain nourishment [*fomenti*], strengthening, and reinvigoration, and in this way spirit can be generated in individual members. Now, spirit strengthens every aromatic [quality] even though, because of [their] heat, aromatics frequently impair the brain by filling it with fumes, etc. Thus in the first way the spirit is generated only from the subtle parts of the blood in the heart, but in the second way it can be generated not only in those individual members that serve the last digestion, but also those serving the first and second digestion.

1. On to the arguments. At the first, one must reply that a heat is found in the heart that is sufficient for digesting blood, to the extent that it is also sufficient for generating spirit. Actually heat is especially abundant in the heart, and this is why the generation of the spirits especially occurs there.

2. In the same way one can reply to the second argument. For he proves that spirit can be generated in another way in the individual parts of the third digestion, but this is not properly generation, only a strengthening or restoration of the spirits.

3. To the third argument one must reply that spirit is not nourished owing to its subtlety, but spirit can rather be strengthened or simply corrupted, and then it disappears or simply remains. Thus the generation of spirits is continuous in the heart, which would not be the case if spirit were nourished; instead, spirits disappear successively and continuously, as it were, from the other parts and this is why they are generated anew almost continuously.

Question 10: Whether the vegetative soul comes into being in a natural way.

Next one inquires about the soul's departure. And first, whether the vegetative soul comes into being in a natural way [*per viam naturae*].

1. And it seems not. For nature operates by means of mediating active and passive qualities in which there is opposition. But opposition is repugnant to life. Therefore, since the vegetative

[soul] is the first principle of life, it cannot be produced from active and passive qualities, nor, as a result, in a natural way.

2. In addition, a cause is nobler than its effect. But life is nobler than every natural or material form. Therefore, life cannot be produced through any [natural] production or material power, and it is therefore produced by a supernatural agent.

The Philosopher says the opposite.[30]

To this, one must respond that the vegetative soul is produced in a natural way. For the one that generates makes the one generated like itself. But vegetative souls have three powers [*potentiae*]: the nutritive, the augmentative, and the generative. Since, therefore, the generative is a power of a vegetative soul, and the vegetative is a material form—because it is united to matter and uses a material organ and operates through motion—therefore, whatever is produced by a power of the vegetative soul is produced in a natural way. But the vegetative soul is produced through a power of the vegetative soul, and therefore, etc.

And, in addition, the vegetative soul uses a material body for every operation and for nutrition and growth, because for nutrition and growth it uses an aliment, and for generation it uses a superfluity of the aliment. But this would not be the case if it were not a material form. But a material form is introduced in a natural way, because it is a first principle of nature. For this reason, etc.

1. On to the arguments. To the first, one must reply that a universal agent acts with a particular agent because "a human generates a human, and the sun does as well," as is maintained in the second book of the *Physics*.[31] Thus something that surpasses the nature of the elements proceeds from the commixture of the elements. For the fact that a magnet attracts iron does not occur through the nature of some element that is dominant in it, but rather through the influence of superior or celestial bodies. Thus a soul cannot be produced from the ele-

30. Ar., *GA* 2.3 (736a29f.). Cf. A., *DA* 16.1.10–11.56–64 (*SZ* 2: 1184–89).
31. Ar., *Phys.* 2.2 (194b13).

ments by means of a power that belongs to the elements, but this only can be accomplished well by a power that surpasses the nature of the elements.

One must respond to the form of the argument that although elements that are mutually contrary at their extremes are not receptive of life when set against each other, nevertheless these can be reduced to a middle condition that is not repugnant to life.

2. To the second argument one must reply that life is multiple, because to have intelligence is a certain kind of life. Thus there is a certain life that surpasses every material form, like the life of the separate substances, whereas another life nevertheless follows after the material form, like a vegetative and sensitive soul, and life such as this can be introduced in a natural way, etc.

Question 11: Whether a sensitive soul can be produced through the power of the semen.

Next one inquires whether a sensitive soul can be produced through the power of the semen.[32]

1. And it seems not. Every form that is perfect and not composite is produced *per se* without generation, because generation pertains to composites. But the sensitive soul is perfect; otherwise, it would not move the body. And it is not a composite because it is a form of the body. Therefore, it is produced *per se* without matter.

2. In addition, nothing acts beyond its species. But the generative power [*potentia*], which lies within the sensitive soul, is numbered among the powers of the vegetative soul. Therefore, the generative power does not go outside its species, and therefore, does not extend to the sensitive soul.

3. In addition, each thing is generated from one like itself. If, then, the sensitive soul is produced from the semen, then the sensitive soul or a part of it will be in the semen. But the conclusion is false because neither the sensitive soul nor a part of it

32. Cf. Ar., *GA* 2.3 (736a35f.). Cf. A., *DA* 16.1.10.58–59 (*SZ* 2: 1185–86).

exists without a part of the body, and semen is not a part of the body.

4. In addition, if the sensitive soul were produced by the power of the semen, then the complete sensitive soul or that power would either remain, or not. If it remains, then either it is the same as the sensitive soul, or different. If the same, then the same thing is both generating and generated. If different, then it is useless. If it does not remain, then it has worked for its own corruption, as was said above.[33]

To the contrary. Just as a form that is produced from the elements of the world is related to the power that is in them, so too the form of a perfect animal is related to the power that is in the semen, as the Philosopher implies at the end of this eighteenth chapter of this book.[34] But those that are generated from the elements of the world, like those begotten from putrefaction, proceed from a power that is in the elements themselves. Therefore, even the forms of perfect animals proceed from a power which is in the semen, but this form is the sensitive soul. And so it is correct to argue in this way: Just as the power in the elements of the world relate to imperfect animals, so does the power in the semen relate to perfect animals. But the power in the elements of the world produces a form of the imperfect animals. Therefore, the power which is in the semen produces a form of the perfect animals.

To this, one must reply that the sensitive soul is produced from a power that is in the semen. For if the sensitive soul were a form existing *per se* and having an operation *per se,* it would come to be *per se* because a thing relates to being and operation in the same way that it does to becoming. But this is not the case, because it neither exists nor operates without the body, since it is destroyed when the body is destroyed, and this is why it is necessary that the sensitive soul come to be by means of other corporeal forms which do not come to be *per se* but only *per accidens* in the fashioning of bodies. Thus in the seventh book of

33. *QDA* 16.4.

34. Ar., *GA* 4.10 (777a31f.). Cf. A., *DA* 18.2.9.95, 12.1.4.50, 17.2.5.77–84 (*SZ* 2: 1329; 912; 1276–80).

his *Metaphysics* the Philosopher says that forms neither exist nor not-exist without generation and corruption.[35] This is why it is necessary that the sensitive soul is produced in being through a certain nature and a power that is in the body and in the semen.

But one must understand that the more perfect a form is, the more it extends itself to many things, just as the hotter something is, the more it can heat other things from a distance. Thus some things are generated from themselves and without a medium, and some are generated through a medium and without a medium. The first are the elements and those near to them, as fire only generates fire by means of contact and so without a medium. But living things can generate both through a medium or without one. They act without a medium as when the flesh takes part in nourishment, seeing that flesh generates flesh and bone generates bone. And they act with a medium, as in the case of generation in the proper sense of the word, because in generation they generate through the medium of semen, in which the power of the one generating exists. Nor is it important whether the one generated is said to come into being from the one generating or from the semen, just as it is not important to say that something comes to be from a principal agent or from an instrument, because an instrument acts only by the power of the principal agent.

1. On to the arguments. To the first, one must reply that a sensitive soul is perfect, but it does not subsist *per se,* and this is why it does not come into being *per se.*

2. To the second argument one must reply that one acting on its own power does not go beyond its species, but one acting on another's power can go beyond its own species, but it does not act by that one's power. Thus although the generative power is numbered among the powers of the vegetative soul, if, however, the seed in which the power of the soul exists comes forth from a plant, this power acts by the power of the entire plant, and if it proceeds from an animal, then it acts by the power of the entire animal. And this is why, because it acts by the power of the

35. Ar., *Metaph.* 7.7 (1032a12f.).

whole and not by its own power, it can produce not only the vegetative but also the sensitive soul.

3. To the third argument one must reply that neither the soul nor a part of the soul is in the semen, as was said, but rather all that is in the semen is merely a certain power capable of producing a soul, and this is why it is not necessary that some part of the body be in the semen, since this power is in the semen or the spirit, which is frothy body. And this power is not only the power of the soul itself, but is also a power of a superior agent, like the sun. Thus power is in the semen in two ways: as a power of both a universal and a particular agent, and this is why the Philosopher says that "a human generates a human, and the sun does as well."

4. To the fourth argument one must respond that the active power is in the semen of the father and the matter for the fetus is provided by the female. Moreover, the vegetative soul is in this material immediately as first act, but not as the second act, just as the sensitive soul exists in one who is sleeping, and by means of the power that is in the sperm this material is digested and completed and formed until the introduction of the sensitive soul. But when the sensitive soul is introduced, certain principal parts are already being formed, like the heart and the liver, in which there is the power for disposing and preparing the material for the reception of the forms of the other members. Thus, once the sensitive soul has been introduced, the power that is in the sperm is not needed any longer, because then the embryo can sufficiently develop itself through its own nutritive and augmentative action.

Question 12: Whether the power that is in the semen exerts any power in the production of the intellectual soul.[36]

One inquires further whether the power that is in the semen exerts any power in the production of the intellectual soul.

1. And it seems so. For the agent is one and the same whose action bears on the disposition of the matter and on the intro-

36. "Exerts any power": In this section A. uses the phrase *potest in aliquid*, meaning to produce something through one's own power or ability.

duction of form. For if one agent disposed the matter and another introduced form, then the one that is generated would not really be a unity. Therefore, since the sensitive soul is disposed by means of a power that is in the semen, the intellective soul will be disposed by the same power.

2. In addition, a human generates a human. Therefore, he generates one like himself in species. But the human species consists in (that is, is perfected by) the intellective soul. Therefore, a human produces the intellective soul.

3. In addition, a nobler operation is owed a nobler form. But of all the lower operations the noblest is to generate one like oneself. Therefore, this operation is suitable for the intellective soul. And if this is so, then the intellective soul can be produced by the power of the semen.

The Philosopher says the opposite.[37] He says that the intellect only comes from outside, that is, from an extrinsic soul.

To this, one must reply that the intellective soul cannot be produced through any power that is in the semen. And the reason for this is that a material cause exerts no power on an immaterial effect. But intellect is immaterial, since it does not depend on the body or on matter either for its being or its operation. For, as the Philosopher says in the third book of *On the Soul,* it is separated from these just as the everlasting is from the corruptible.[38] Therefore, it cannot be produced through a material power, and the power in the semen is of this type.

Furthermore, the power works in the semen to produce something like itself, insofar as it is a power of generation. But the power in the semen uses an organ and operates with a mediating body. Therefore, it is derived from a form which uses an organ. But the intellective soul does not use an organ. Therefore, the power of the intellective soul is not in the semen and, as a result, cannot produce the intellective soul. But since each and every thing has its coming into being in the same way as it has its being and its operation, and since the intellect can exist *per se* and, moreover, can operate *per se,* it therefore has its com-

37. Ar., *GA* 2.3 (736b27f.).
38. Ar., *De anima* 3.5 (430a22f.).

ing into being *per se*. And the intellect is itself simple since it is the form of the body. Therefore, it necessarily comes to be without the supposition of matter, and, as a result, it necessarily proceeds into being from the first cause.

1. On to the arguments. To the first, one must respond that in agents that are not properly ordered, if one agent disposes the matter and another introduces form, then the one generated is not a unity. But in properly ordered agents one can dispose the matter and the other can introduce form, just as happens in a ship, where one prepares the lumber and so prepares the material, and another joins the lumber together and thus introduces form. In the same way, all of nature is like an instrument for a superior agent, and for this reason an inferior nature can well dispose the matter and a superior agent can introduce form.

2. To the second argument one must reply that a human is not said to generate a human except to the extent that he can dispose the matter to receive form, but not because there is a power in the semen capable of introducing intellect.

3. To the third argument one must reply that the most perfect operation of these inferior [things] is to generate one like oneself. Nevertheless, this is not so for those separated from matter. Thus the one that is the most perfect in the entire genus of being cannot generate one like itself, because this would imply its imperfection. And in like manner an ability to produce something *per se*, from no one or nothing [*ex nullo vel ex nihilo*], is repugnant to the causative nature, because this belongs only to one whose power is infinite. And this is why, since the intellective power only proceeds *ex nihilo*, its production pertains to the first cause alone.

Question 13: Whether the intellective soul is created and infused into the body simultaneously.

Next one inquires whether the intellective soul is created and infused into the body simultaneously.

1. It seems not. Because the entire universe is perfect, and its principal parts especially are incorruptible things. But if the

intellective soul were created at the same time that it is infused into the body, then it would follow that something can be added to the universe and that it was not perfect previously, but rather something imperfect to which it is possible to make some addition.

2. In addition, the end is that to which the principle corresponds. But the intellect remains after its separation from the body. Therefore, by an equal argument, it existed before it was joined to the body.

3. In addition, a cause precedes the thing caused, and the more universal the cause is, the more it precedes the thing caused. But that soul that is nearest the first cause is the most universal. Since the first cause precedes the thing caused by the longest duration, then that which is nearest the first cause precedes more than one more remote. But every form, no matter how material, is prior to its matter in nature, if not in time. Therefore, soul precedes body more than in nature and, as a result, precedes it in time.

4. In addition, between any two "nows" there falls a middle. But the creation of the soul is in the eternal "now" whereas its conjunction with a body is in the temporal "now." Therefore, between these two "nows" there will be a middle, and with respect to that middle the intellect precedes the body.

The Philosopher says the opposite in the twelfth book of the *Metaphysics.*[39] He says there that although a form can exist after separation from matter, as the intellect does, nevertheless nothing exists before matter.

To this some respond that the intellective soul has the same status [*condicio*] as intelligence, and this is why it existed from eternity just like an intelligence and then it happens that it is united to a body.[40] It was therefore created before being joined to a body.

But this explanation fails in a number of ways. The first is that the intellective soul and intelligence do not have the same status. This is evident from their different operations. For we

39. Ar., *Metaph.* 12.3 (1070a24f.). Cf. A., *Metaph.* 1.1.9.

40. By "intelligence" A. means a separate, incorporeal, rational agent, like the angels.

come to understanding by distinguishing and by means of the receipt of species that have been received from the senses, but intelligence understands without distinguishing and through species that are not acquired from somewhere else. And this is why intelligence is properly said to be an intellectual substance, whereas the human soul is a rational substance.

This position falls short in a second way, since if the soul merely happens to be united to a body, then a human will either be a being [*ens*] *per accidens*, if he is a composite being, or he will only be the soul.

Moreover, if the soul's creation precedes its being joined to the body, then either it is natural for the soul to be united to a body or not. If it is natural, then it remains to ask why it was not united to the body immediately. Now it is not necessary to provide a cause for this, because it is not better for the soul to be separated from the body rather than united to it, because the intellective soul exists for the sake of the completion of the species and it completes the species once it has been united to the body and not when it has been separated from it. Also, it understands in the proper manner when it has been united, not when it has been separated. Therefore, it is better for it to be united than to be separated. If, however, it were contrary to its nature to be united to a body, then this would be a punishment and a burden for the soul, and the opinion of the Platonists would follow. And this is why one must reply, following the Philosopher's meaning,[41] that it is created and infused into the body at one and the same time. For if it existed before its infusion, it would exist for nothing, and nature does nothing in vain, according to the Philosopher in the first book of *On Heaven and Earth*.[42]

1. On to the arguments. To the first, one must reply that it is not individuals but species that form a part of the universe.[43] Thus, although there may occur some addition to or reduction in individuals, the universe is not more or less perfected as a result of this, and this is why, although the intellective soul is cre-

41. Cf. Ar., *GA* 2.3 (736b12f.).
42. Ar., *DC* 1.4 (271a33).
43. Or: "of the universal."

ated *de novo,* it does not follow as a result that the universe was imperfect before this.

2. To the second argument, one must respond that the end and the beginning correspond to the extent that something is resolved into the same thing from which it was constituted. For just as a mixed body is composed from four [elements], so it is resolved into the four, and just as each and every one comes to be from matter, so then each is resolved into matter. Thus if intellect were composed out of nothing, then it would return to nothing, just as it comes to be *ex nihilo.* But the end is not proportioned to the beginning in every thing, and this is especially so among those beings that depend upon an agent as far as will is concerned, because the thing can begin or cease to be according to his will.

3. To the third argument one must reply that although the first cause precedes the thing it caused in duration, nevertheless it is not necessary that every posterior cause precede the thing it caused in duration, and this especially holds true for a formal cause.

4. To the last argument one must respond that a middle point in time [*tempus medium*] falls between two "nows" of the same measure, although two "nows" of a different measure can exist simultaneously. For the same eternal "now" corresponds to the entire succession of time, since eternity exists all at once and without succession. Now just as the soul, although it is indivisible, still exists wholly in any part of a divisible and extended body, so too, although eternity is a simultaneous whole, nevertheless it subsumes time (which is successive) beneath itself.

Question 14: Whether all members are generated simultaneously.

Next one asks about the formation of the members. And first, whether all members are generated simultaneously.

1. And it seems that they are. For according to the Philosopher in the second book of *On the Soul* "the acts of active agents are present in the one undergoing action or being disposed for

it."[44] Thus if two materials are equally disposed when an agent is present, they will receive act at the same time. But the power in the sperm acts on the whole menses, in which all the members exist in potency. Therefore, it will form one member for the same reason as another, since it acts equally on every part of the menses.

2. In addition, the fact that an agent is acting on something now and did not do so earlier is because it was waiting for something that was lacking. If, then, a power forms one member now and not another, it is necessarily waiting for something that is needed for the formation of the other member. But this is not the case, for the semen of male and female are sufficient to produce an animal.

The Philosopher says the opposite.[45]

To this, one must say that not all members are generated simultaneously. This is manifestly clear to the senses, because the interior members are formed before the exterior ones, and the heart is formed before others. It is also seen by reason. For everything acting by means of natural transmutation needs time for its action, and everything acting by intention produces earlier what it most intends to do. The exception is those that are ordered for the sake of another, because in such as these that which is first in intention is last in execution. Now then, the power in the semen acts both by transmuting and through intention. It is necessary, then, that nature first stabilize something upon which it will establish the thing it has acted upon [*suum operatum*] (that is, its object). And this is why it first produces the member that is more necessary and lastly the one that is less necessary.

Or in another way one can say that generation intends two things: either the last goal [*terminum*] of an entire process of change, or the entire change preceding the last goal. If generation occurs in the first way, then not all members are generated simultaneously because the nature of each member is not com-

44. Ar., *De anima* 2.2 (414a11f.).

45. Ar., *GA* 2.3 (737a18f.). Cf. Avic., *DA* 15.3–16.1 (fols. 60va–61vb); Averroes, *GA* 2.4; A., *DA* 16.1.14.72–80 (*SZ* 2: 1194–1200).

pleted simultaneously. If generation occurs in the second way, it can be said that all members are generated simultaneously because all members are on the path to completion simultaneously.

1. <On to the arguments>. To the first, one must respond that although all members exist in potency in the semen simultaneously, nevertheless the agent first intends that which is more necessary, and this is why it acts more toward its production and produces it sooner.

2. To the second argument one must reply that it does not produce all the members simultaneously because it needs something, since the matter is not best prepared [*disposita*] for all members. And this is why, before it introduces the form for all the members, it requires that the matter be prepared for all the members, and the agent cannot do this at the same time; rather, first it prepares [the matter] for one and then for another. In this way a solution to this question is apparent.

Question 15: Whether the heart is generated first.

Further one inquires whether the heart is generated first.

1. It seems not. For a natural power operates by means of similar parts, but the natural power is prior to all other powers, for the fetus is nourished first, etc. Therefore, a homogenous part is generated first, and therefore, not the heart, which is an official member.

2. In addition, the Philosopher argues that the heart is the first member because, once the other members have died, the last operation remains in the heart.[46] But one can make the same argument for the liver, because, once all the other powers have ceased, a power operates last in the liver, as is evident among paralytics.

3. In addition, those that exist for an end precede the end in generation. But the brain exists for the sake of tempering the heart, and the liver for its nutrition (that is, to restore what has

46. Ar., *GA* 2.5 (741b15–19). Cf. A., *DA* 16.2.2.105 (*SZ* 2: 1213–14).

been lost). Therefore, the liver and the brain precede the generation of the heart.

4. In addition, a moisture is badly bounded by its own boundary, but well by that of another. In order for moisture to sustain its action, it is necessary that it be bounded elsewhere and be contained in something else. But the semen is contained in the placenta [*secundina*].[47] Therefore, the generation of the placenta precedes the generation of every other member.

The Philosopher says the opposite.

To this, one must say that a part is of two types. One is necessary to a thing, resulting in the thing's creation [*constitutio*], and another is a part that does not result in the thing's creation but is said to be a part because it is necessary for generation and is formed first from the matter of generation. If we are speaking in the first way, the heart is generated first. If in the second way, then it is the placenta, which is a sort of membrane containing the semen until it is digested and until the form of the fetus is introduced and completed. The placenta arises from a certain portion of the semen itself, but the placenta exits last with the fetus and does not result in its creation. Thus it is the first thing generated from the semen, but only the heart is the first part of the animal generated from the semen. And the reason for this is that the first power is the vital power, for nothing is nourished, grows, or has sensation unless it is alive. Thus all the powers have their roots in the vital power. For the liver only operates with a mediating heat, and sensation cannot occur without heat and motion, as is said in books two and three of *On the Soul*.[48] But the root of heat is in the heart, and this is why the heart is the first part that is generated in an animal, and after the heart comes the liver. Thus Avicenna says that two bubbles appear first in the semen: out of one arises the heart, and out of the other the liver.[49] Nevertheless, the soul is received first into the heart, and this is why the heart is said to be the seat of

47. The origin and function of the *secundina* are fully discussed by A. in *DA*. See for example *DA* 7.2.2.115 (*SZ* 1: 638–39).

48. Ar., *De anima* 3.1 (425a6), 2.5 (416b32f.).

49. Avic., *Can. med.* 3.21.1.2; *DA* 9.5 (fol. 43rb A), 16.1 (fol. 61ra–rb).

the soul. The Philosopher demonstrates this many times in the text.[50]

1. On to the arguments. To the first, one must reply that a part is called "official" because it is deputed to some task [*officium*], and thus both the heart and liver are official parts. A part is called "official" in another way, because it is composed of diverse parts, and thus these dissimilar parts are called "official." Therefore, the heart can be called a homogenous part in one way, because it has the same nature as the whole and the part, and it can be called official in another way. When it is said in the first argument that a natural power operates by means of homogenous parts, one can say that in one way the heart is a homogenous part. Or one can respond in another way to this argument that a natural power is not first, but the vital power is prior.

2. To the second argument one must respond that every power has its own corresponding operation. Thus, although motion and sensation, which are operations of the animal power, are lacking in paralytics, they are not absent by virtue of a lack in the natural power, which is in the liver. Still this one is not absolutely first itself, because when the natural power is absent the vital power can remain in the heart.

3. To the third argument one must reply that some things are ordered to other things as if to an end in two ways. They can be ordered either for generation, and ones such as these must precede, or for conservation, and ones such as these can follow. But the liver and the brain exist for the sake of the heart, not for its generation but for its conservation, and this is why it is unnecessary that they be generated first.

4. To the last argument one must respond that although the placenta is generated from the semen, nevertheless it is not part of the animal, etc.

50. Ar., *Part. an.* 2.10 (656a27f.), 3.3 (665a10f.); *GA* 2.4 (740a3f.), 6 (743b25f.).

Question 16: Whether the informative power for the other members is in the heart after it has been formed.

One asks further whether the informative power [*virtus informativa*] for the other members is in the heart after it has been formed.

1. It seems that it is. For among those observing a proper order, the prior is the cause of the posterior. But there is an order among the members. Therefore, since the heart is the first, it seems to be the cause of the others.

2. In addition, the same situation exists for the act of influencing as seems to exist for forming [*informatio*]. But all the members receive the influence of the heart's power, and therefore also its power of informing.

3. In addition, the formative power operates through motion, heat, and spirit. But the principle for motion, heat, and spirit is in the heart. Therefore, the formative power is principally in the heart.

On the contrary. For one effect there is one cause. But an animal is one, and therefore, there is one informative power for the animal. Because it forms the other members as a power, it also forms the heart. If, then, the power that forms the other members were in the heart, then the heart would form itself.

One must reply that the formative power does not thrive in the heart. And the reason for this is that the power required for forming something else and for operating before it has been constituted does not thrive in that thing, because being would then exist in non-being. But the formative power operates before the heart's formation, and this is why, etc. Nevertheless, once the heart is formed, it disposes itself to the formation of the other members because, once the heart is already formed, the members are not in the same power [*potentia*] as they were before. This would not be the case if the formative power [*virtus formativa*] of the heart were not disposed to the formation of the other members. Thus one must say that the informative power [*virtus informativa*] is derived from the father's heart and follows after him, but precedes the heart of the actual fetus. Now the power in the male's sperm does not cease to sustain and arrange the matter from the female until form or the sensitive soul is introduced. It is not so much the case that that power, or the formative power, becomes the sensitive,[51] because then the one generating and the one generated would be the same, which would be more nutrition than generation. But since the sensitive soul is in the embryo as its principal part, it then arranges the body in which it exists by means of the augmentative and nutritive power. Thus the informative power is in the sperm until the formation of the embryo, but after the embryo's formation its informative power is in the embryo itself. This question, however, is inquiring into the informative power before the embryo's completion.

1. On to the arguments. To the first, one must reply that in those observing a proper order, the prior is not always the cause of the generation of the posterior, for cause is spoken of in many ways, since that which is prior can be either the final or the formal cause and it is not necessary for a cause such as this to be acting on or informing [what is posterior].

2. To the second argument one must respond that influencing is one thing and informing is another, since influence proceeds from something perfected whereas formation does not proceed from a perfected part. To the contrary, formation ends with a perfected part.

3. To the third argument one must reply that the motion, heat, and spirit, which the informative power uses, are derived radically from the father's heart and follow the father's heart, but precede the heart of the actual fetus.

Question 17: Whether sterility is a natural affliction [passio].

Further one inquires into sterility. And first, whether sterility is a natural affliction [*passio*].[52]

51. "Sensitive": the grammar here allows for two possibilities: that A. is referring to either the sensitive soul or the sensitive power. We have chosen not to remove the ambiguity.

52. Ar., *GA* 2.8 (746b12–747b24); Avic., *Can. med.* 3.21.1.8; Averroes, *GA* 2.6; A., *DA* 16.2.1.92–94 (*SZ* 2: 1207–8).

1. It seems that it is. What follows natural causes is natural. But sterility comes from natural causes; therefore, it is a natural affliction.

2. In addition, something that is contrary to nature is repugnant to the being of the actual object and does not contribute to its preservation. But sterility contributes to the preservation of life, seeing that its opposite, like pregnancy or fertility (that is, coition), shortens a lifespan. Therefore, sterility is not unnatural.

3. In addition, whatever is contrary to nature is either a sickness or a cause of sickness or an accidental trait of sickness. But sterility is none of these, since these are ordered to corruption and sterility is not. Therefore, etc.

To the contrary. The generative power is a certain natural power. Whatever impedes the generative power is contrary to nature. But sterility is of this sort, and therefore, etc.

Moreover, nature always strives toward the universal and operates with respect to the individual. Therefore, whatever impedes the operation of the individual impedes the operation of nature and, as a result, frustrates nature's intent. But sterility is of this sort, etc.

One must say that nature is spoken of in many ways. In one way nature refers to another type of accidental trait, according to which something is said to be prior to another thing by nature, as a cause is to the thing caused, and in this way nature does not refer to the object [*res*] but to the mode of the object. In another way, a nature is said to be any object outside the soul.[53] And thus it is said that a chimera does not exist in any nature. In a third way nature is said to be the principle of motion and rest, and this is nature as treated by the Philosopher in natural science.[54] But this nature can further be understood in two ways: either universally or particularly. Universally, as we say, because nature acts for the one and not the many. If particularly, this too can be in many ways: In one way, as it falls under the

53. "Soul": a term that includes the intellective power or what we would more commonly call "mind."

54. Ar., *Phys.* 2.1 (192b11f.).

rule of the vegetative soul, whose works are nutrition, growth, and generation. Thus the nutritive power is therefore called a natural power. In another way, nature falls under the rule of the vegetative and sensitive souls, and this nature is said to be the animal power.

Thus, when one asks whether sterility is a natural affliction, one must respond that to the extent that nature is ascribed to a vegetative potency, sterility is not a natural affliction, because it impedes its natural operation, which is the generation of one like itself. If, however, nature is understood universally, then it can be said that in one sense it is a natural affliction and in another sense not, since one form of sterility is curable, and this one comes about from fat or eating too much, but the other is incurable, like that in a woman who is mannish [*virago*], even from her first creation. Nevertheless, when speaking in a literal sense, I say that sterility is not a natural affliction, since each one is perfected when it can generate another like itself.[55] But sterility impedes this perfection, and a proper perfection is natural to each and every one, and this is why, speaking universally, sterility is not natural. As a result, the first arguments have to be conceded to the objection.

1. To the first argument one must reply that whatever follows natural causes ordered to the course of nature is itself natural; the other is not.

2. To the second argument one must reply that coition is of two types: ordered and inordinate. Inordinate coition shortens the lifespan, but the ordered does not.[56] Now sterility is opposed not only to ordered coition but also to inordinate coition. And this is why sometimes it causes a shortening of the lifespan, and sometimes it preserves life.

3. To the third argument one must respond that sterility can be both a cause of disease and an accidental trait of disease, and sometimes it leads to corruption and sometimes it merely comes from corruption. This is why, etc.

55. See *QDA* 2.1, 5.2.

56. See *QDA* 15.14 for a dramatic example.

Question 18: Whether sterility has its root in the soul or the body.

Next one inquires whether sterility has its root in the soul or the body.[57]

1. And it seems have its root in the soul. For the generative power is a power of the soul. But sterility results from a defect in the generative power. Therefore, it comes from a defect that has its root in the soul.

2. In addition, error and falsehood can come from all things except the first [thing] as is maintained in the ninth book of the *Metaphysics*[58] and the eleventh of this book. Therefore, the defect, which is sterility, has its root in the soul.

3. In addition, sterility stems from a lack of heat, spirit, and things of this sort which contribute to generation. But the soul is the principle of all of these. Therefore, it has its root in the soul.

On the contrary. Privation has its root in matter, as is said in the sixteenth book of this work.[59] Sterility is a certain privation. Therefore, it has its root in the body.

One must respond that sterility is of two types. One is moral, and the other is natural. The moral one is found among those having discretion, as when a person wills abstinence. Thus sterility of this type has its root in the soul. But natural sterility does not have its root in the soul *per se,* since the powers of the soul are not weakened *per se* because, according to the Philosopher in the first book of *On the Soul,* "if an old person were to receive a youth's eye, he would see just as a youth does."[60] Thus the visual power of the old man is not lacking in itself, but only owing to its union with an organ. And in the same way a sense is not corrupted *per se* by an excessive sensible, but rather owing to loss of proportion in the organ.

Thus, with respect to these arguments, one must understand

57. A., *DA* 16.2.8.130–33 (*SZ* 2: 1226–28).
58. Ar., *Metaph.* 9.10 (1051b25f.); A., *Metaph.* 9.4.6; Ar., *Part. An.* 1.1 (639b23f.), 1.5 (644b22f.). We read here "IX" for the text's first "XI."
59. Ar., *GA* 2.1 (732a3f.), 2.8 (748a31f.).
60. Ar., *De anima* 1.4 (408b21f.).

that three things must be considered in terms of the soul: namely, the soul's own power; its connection with the body; and its operation through organs. Now a defect does not occur in the power itself nor, as a result, does sterility, but there can nevertheless be a defect in the soul with respect to the body and the organs.

1–3. As a result, responses are clear to the arguments, proving that sterility has its root in the soul. For the responses proceed from the body and the organs, and this *per accidens,* and this is why, etc.

An argument proceeds against the objection, because [sterility] is not from the soul in itself or through itself.

Question 19: Whether generation of a mule from a mare and an ass is natural.

Further one inquires whether generation of a mule from a mare and an ass is natural.[61]

1. And it seems not. Because natural generation is of like from like. But this type of generation is not of this sort. Therefore, etc.

2. In addition, natural generation comes to an end at something perfected. But this generation comes to an end at something imperfect; therefore, etc. Here is the proof of the assumption: that something is imperfect that departs from the customary operation of nature. But a mule departs from the customary operation of the generative power, and therefore, etc.

3. In addition, if this generation were natural, species would be capable of mixing together; but the conclusion is false, because species are, according to Porphyry, unmixed.[62]

The opposite is apparent to the senses. And one can prove by reason that unnatural generation is monstrous. But this generation is not monstrous; therefore, etc.

One must say that the mule's generation can be compared to another one generated *per se,* and thus it is unnatural and

61. A., *DA* 16.2.8–11.130–51 (*SZ* 2: 1226–36).

62. Porphyry, *Isagoge* 5.18, trans. Boethius (Brandt, 1906), 334, 11–12.

flawed because the horse's semen is in potency to a perfected animal, and the mule is from the genus of imperfect animals. The same is true for the ass's sperm. Nevertheless, if the mule's generation is considered as a mixture of the ass's sperm with the horse's semen, this is natural, because, just as a middle color is naturally produced from a mixture of white and black, so too from the semen or menses of the mare, which are exceptionally hot, and from the sperm of the ass, which is exceptionally cold, a middle one can naturally come to be departing from each of the extremes. Thus, when the ass's sperm acts on the mare's semen, it does not entirely prevail over the matter, nor it is entirely overcome by it. And this is why, although it cannot produce one entirely like itself, it does, so far as it can, produce one similar to itself. With respect to the fetus's shape, it resembles its mother more, because size is consequent upon matter. But with respect to its complexion, it resembles the father.

1, 3. With this a response to the arguments is clear. For they prove that this generation is unnatural, and this is true when considering it in relation to each of the ones generating *per se*.

2. To the second argument one must reply that although the mule belongs to the genus of imperfect animals, its generation is nevertheless natural in a certain respect because there are many animals that are imperfect when compared to others whose generation is natural because they are perfect in their own genus.

Question 20: Whether the mule and she-mule are sterile by nature.

Further one inquires whether the mule and she-mule are sterile by nature.

1. It seems not. Because an offspring is generated from a dog and a wolf that is capable of generating, and so too from a hen and a partridge [*cubeg*], and so too this one will be capable of generating for the same reason.[63]

2. In addition, one temperate [in complexion] is generated

63. Cf. *DA* 16.79.130 (*SZ* 2: 1226).

from one that is temperate. But the ass's matter is tempered by the mare's semen, and contrariwise. Therefore, etc.

The Philosopher says the opposite.[64]

One must say that the mule and she-mule are naturally sterile. And the cause for this is threefold. One is the proportion of the semen, which the Philosopher touches upon, because the ass's sperm is very cold and even a little of the mare's semen is hot.[65] Thus, when the ass's sperm is joined to the mare's menses, each withdraws from its own nature since the sperm becomes warm and the menses become cold beyond their own nature, and this is why an animal is generated that departs from nature. The second cause is a defect in moisture. For the mule and she-mule are so deficient in spermatic moisture that, if the she-mule were to conceive and become pregnant, she could not nourish the fetus [*conceptus*] until the time of its birth. And likewise, if the mule should produce sperm, it would be dry because the moisture of the she-mule's menses would consume a great deal of it. Another cause is the indisposition of the parts serving generation, because the mule is an animal with a harsh complexion. Thus the instrument of both the male and the female hardens past any use for generation.

1. On to the arguments. To the first, one must respond that although an animal is generated from a dog and a wolf, or a hen and a partridge, the one does not differ in species from the other, and this is owing to the suitability of the semen. But this is not so in the case proposed owing to the disagreement of the semen between the ass and the mare.

2. The other argument proves that a mule can be generated from an ass and a horse, but it does not prove that a mule generates.

64. Ar., *GA* 2.8 (747a24f.). Cf. Averroes, *GA* 2.6; A., *DA* 16.2.8–11.130–51 (*SZ* 2: 1226–36).

65. Ar., *GA* 2.8 (748a22f.).

BOOK SEVENTEEN

Questions 1–3: On birds' eggs.

"We have already discussed the cause of sterility," etc.[1] In this seventeenth book one inquires first whether there ought to be a distinction among the parts in birds' eggs.

1. And it seems not, because there is a greater heat in birds' eggs than in fish eggs; but there is greater heat in a woman than a bird. Therefore, since there is no distinction among parts in a woman's seed [*semen*], a distinction will exist all the less in the seed of birds.

2. In addition, the middle savors the nature of the extremes. But in the first generation of the egg the whole egg is yellow, and in the last generation the whole becomes liquid. Therefore, it will be uniform in the middle just as it is in the extremes.

The opposite is evident to the senses.

Further one inquires whether the parts of the egg are rightly ordered.

1. It seems not. For things below imitate things above, as far as possible. But in this universe the earth is in the middle. Therefore, it is true in others as well that that which is more earthy will seek the middle more. But the shell is earthier than the yolk. Therefore, the shell is in the middle and the other parts will surround it, just like the airy or watery parts.

2. In addition, whiteness is a sign of coldness, as is apparent in snow, and redness a sign of hotness, as is apparent in bile and honey. Since, then, the hot should seek the circumference and the cold should seek the center, it seems that the white of the

1. Cf. Ar., *GA* 3.1 (749a10f.); Avic., *DA* 17 (fol. 62rb–62vb); Averroes, *GA* 1.1; A., *DA* 17.1.1–2.1–15 (*SZ* 2: 1237–44).

egg will be inside toward the center, and the yoke will be toward the circumference, since it is red and hot.

The opposite is apparent to the senses.

Third, one inquires whether the larger part of the egg ought to come out first.

1. It seems not. For that which is more pointed penetrates another more easily. Therefore, the pointed part is better suited for coming out than is the larger part.

2. In addition, in the ninth book of this work the Philosopher says that animals come out on their head.[2] But the chick's head is toward the pointed part of the egg. Therefore, the pointed part will precede the other when coming out.

The Philosopher says the opposite.[3]

To the first, one must reply that there are distinct parts in eggs. And the reason for this is that the nature that produces the one generated provides for it whatever things are appropriate for its generation. Now, however, an animal does not exist without nutrition, and this is why, when an animal is generated from eggs outside a uterus, nature provides not only that there is something in the egg from which the animal can be generated, but also provides something by which it can be nourished while it is enclosed within the eggshell. First, there is the egg white; second, there is the yolk. Therefore, the egg white provides something in place of the sperm as matter for the fetus, and the yolk provides its nourishment.

Nevertheless, the reason why these things are distinct in birds' eggs but not in fish eggs can be threefold: one is the incomplete nature of the eggs in fish; another is the heat's weakness, because it cannot separate from each other those things which are distinct; and the third cause is that the fish egg does not result merely in nutriment, or that it is not nutriment simultaneously with the semen, as it is in birds' eggs.

2. Ar., *HA* 7.8 (586b1f.).

3. Ar., *GA* 3.2 (752a15f.); Averroes, *GA* 3.2; A., *DA* 17.1.3.16–23 (*SZ* 2: 1244–48).

1. On to the arguments. To the first, one must reply that a bird's egg and a woman's menses are quite different, because a human is formed inside the uterus, and this is why it is unnecessary that there be two parts to the menses that is received into the woman's womb, one of which is equivalent to the matter for the fetus and the other for its nutriment. This is because the menstrual blood on which the fetus is nourished is continually replenished as the woman is nourished. But a bird is formed outside a uterus, and this is why it is necessary that each exists in the egg: both something in place of the menses and something in place of the nutriment.

2. To the second argument one must reply that since the heat is weak in the beginning, the parts are not distinct and the whole appears to be yellow. But when the heat is strengthened, it distinguishes the parts, whereas at the end, when nature intends to delineate and shape the matter by the power of both the internal and external heat, the matter liquefies, so that it may receive the intended form more easily. And this is why, etc.

To the second question one must reply that nature acts for the sake of an end.[4] Therefore, it acts from intention, because it is not possible to be absolved from the intention toward an end. Since, then, the chick is formed from an egg outside a uterus, nature arranged a hard shell in order to protect what is inside from corruption. And in addition, heat seeks the circumference, and cold the center: this is why the white, which is hotter, is outside the yolk. And, besides this, the yolk is earthier. An indication of this is that if it is cooked, it appears dry and earthy after it has been cooked and the moisture has evaporated. And this is why the shell is naturally situated on the outside as a protection from things contrary to it, and the white is nearer to it, because it is hotter and more airy, and the yolk is in the middle owing to its earthiness.

1. To the first argument one must reply that in some cases nature operates among things below by imitating things above. In other cases it does not operate by imitating things above,

4. Reading *secundum* for the text's *secundo*.

because it operates from an intention to a determined end at which it may not arrive by imitating things above. But if the exterior part of the egg were soft, it would easily be corrupted when the egg is deposited on the ground. And this is why it is hard outside.

2. To the second argument one must reply that whiteness arises from two things. Sometimes it arises from cold and from the heat's weakness, as is apparent in snow and frost and many other things that are undigested. And sometimes it arises from a complete digestion afforded by heat,[5] as in the case of charcoal ashes and in milk. Since, then, the egg white is more completely digested than the yolk, and is therefore whiter than the yolk, it is not from coldness, as the foolish physicians say, but from the completion of digestion, etc.

To the third question one must reply that the larger part of the egg comes out first because it is heavier and the heavier descends more quickly, because the more pointed part is joined to the womb and for this reason it is separated from it more slowly, and also because the more pointed part is hotter. And an indication of this is that females are generated from round eggs and males are generated from oblong ones, because heat always tends to form a pyramidal figure.

1. On to the arguments. To the first, one must reply that the pointed part divides more easily, but the larger part is more inclined to descend.

2. To the second argument one must reply that there is a difference between the exit of animals generated in the belly and the exit of birds, since animals generated in the belly exit onto their heads, but birds exit upon their feet.

With respect to what is said in the ninth book, one must have an understanding of those that generate animals in the belly.

5. Cf. *DA* 1.2.23.439 (*SZ* 1: 212–13).

Questions 4–6: On chicks.

Further one inquires whether the chick is nourished by the yolk.

1. And it seems not. For an animal with blood is nourished by menstrual blood. Therefore, in order for the yolk to nourish the chick, a member is required to convert the yolk into menstrual blood. But this member is the liver, which is absent when the chick's embryo is first nourished.

2. In addition, menstrual blood is to the human as the yolk is to the bird. But menstrual blood does not nourish a human unless it was digested earlier in the liver. Therefore, neither does the yolk. I inquire, then, by what the heart is nourished before the generation of the liver and the generation of the other members.

The Philosopher says the opposite.[6]

Second, one inquires whether the chick is generated with a mediating, external heat.

1. It seems so. Because the egg does not produce a chick without incubation. But incubation is only to warm the egg. Therefore, etc.

2. In addition, the womb bears the same relation to the embryo as that which contains the egg does to the egg or the egg's contents. But the embryo is created only with the womb's mediating heat. Therefore, the chick is created only with a mediating external heat.

To the contrary. An agent and matter are sufficient for generation. But these exist in the egg that is surrounded by an external heat. This is why, etc.

Third one inquires whether birds of prey have few eggs.

1. It seems not. Because a large number of eggs is a sign of strong power. But there is stronger power in the birds of prey than in others. Therefore, etc.

6. Ar., *GA* 3.2 (752b26f.). Cf. A., *DA* 17.1.3.19 (*SZ* 2: 1245–46).

The Philosopher says the opposite.[7]

To the first, one must reply that the fetus is nourished by the yolk. But the fetus has a double status: one is before the completion of the members, and the other is after. Before the members' completion the fetus does not receive nutriment through the stomach or the liver, but through pathways that resemble veins, and at that time it is not necessary for its nutriment to be menstrual blood, since it is possible for the moisture to resemble it. Nevertheless, after the completion of the members it is nourished by menstrual blood, and this is why at that time it is nourished only by that which passes through the stomach and liver. And the first arguments prove this. In fact, at the same time that the heart is generated the menstrual blood is generated in the heart's ventricles, and the heart is nourished on it until the completion of the other members. But before the heart's generation there is no generation of menstrual blood, and the same is true in the egg.

1–2. With this a solution is evident to the arguments.

To the second question one must reply that there is a natural heat in the egg; but the heat is hampered [*ligatus*] as the senses are in one who is asleep, and it can therefore not perform its proper operation unless it is moved by an external heat. Now, just as a seed cannot generate by itself unless it fall to the earth (because when it falls to the earth its power is aroused by the earth's heat), so it is for the egg, since its internal heat is aroused by an external heat, and this is why it does not produce a chick unless it has been incubated or warmed by another. Therefore, one must reply to this question that the principal instrument is the internal heat, which, led from potency to act, makes all things. The external heat, however, is a secondary instrument. Therefore, each heat is necessary, and one alone does not suffice in itself.

With this a solution is apparent to the arguments.

To the third question one must reply that a large number of eggs arises from an abundance of moisture at the same time

7. Ar., *GA* 3.2 (753a30f.). Cf. A., *DA* 17.1.3.22 (*SZ* 2: 1247).

that there is a strong heat. But moisture does not abound in birds of prey. This is because they are good flyers and flight consumes moisture, and also because they have curved talons and large beaks and many feathers, and the superfluity of the menses is converted into these, and, as a result, they have few eggs. And it is because they drink very little, and are therefore dry. Birds with small bodies, however, because they need only a modest amount of nourishment, abound with a great deal of moisture, and this is why they lay several eggs, because they have a lot of material for the eggs. Birds that are good flyers, however, and have, in addition, a heavy body and a lot of moisture produce few eggs at one and the same time (because they are good flyers), but they produce many eggs in succession because they have a great deal of moisture. Pigeons are like this, and they lay eggs many times but lay only a few at the same time, because they only produce two at one and the same time.

To the argument one must reply that although birds of prey have a strong power they lack moisture. Therefore, etc.

Question 7: Whether wind eggs are generated from a superfluity of sperm.

Further one inquires whether wind eggs are generated from a superfluity of sperm.

1. And it seems so. In every work of nature there is something superfluous. But an animal's generation is a work of nature. Therefore, something superfluous remains behind. But nature is not lazy and it therefore can create something from this superfluity. It can generate some kind of egg from what is left behind after the generation of true eggs, but this is only a wind egg. Therefore, etc.

2. In addition, that superfluity (that is, the residue from generation) is more digested than is the residue remaining from nourishment. But an egg can be formed from that. Therefore, one can be formed even more readily from the residue remaining from generation. But the one generated from this residue is not a true egg. Therefore, a wind egg is generated from the residual sperm after the generation of a true egg.

The Philosopher says the opposite.[8]

One must reply that wind eggs are not generated from residual sperm after generation, for wind eggs are found in many birds that never produce sperm. This would not be so if wind eggs were generated from the residue of the sperm left after generation. But when female birds desire coition and take pleasure in it, they emit the menses at the opening of the womb, and although the male's sperm does not make contact with this semen, it is still received into the womb, and the female's nature works on it as far as she can. Therefore, she can form an egg from the menses. But because the agent power consists in the male's sperm and the male's sperm does not make contact with the egg or its matter, this egg is not suited for generation and does not have the power for generating something itself.

1. On to the arguments. To the first, one must reply that something can be left behind after the formation of a fetus or a true egg, but this residue is not the material for a wind egg but instead is suited for the generation of a true egg.

2. And one must reply in the same manner to the second argument.

Question 8: Whether in a wind egg the white has to be distinguished from the yellow.

Further one inquires whether in a wind egg the white has to be distinguished from the yellow.

1. It seems not. The white is to the yellow just as the sperm is to the menses. But females do not always menstruate [*mundificant*], as is said in the fifteenth book of this work.[9] Therefore, since the female's power alone generates a wind egg, there will be no white in a wind egg.

2. In addition, the more perfect a power is, the greater differentiation there is. But a woman's power is stronger than a female bird's. Therefore, since the woman's power cannot so dis-

8. Ar., *GA* 3.1 (751a9f.).

9. Cf. Ar., *GA* 1.20 (727b31f.); A., *DA* 15.2.6.112 (*SZ* 2: 1135).

tinguish her own menses so that it may resemble the parts of a true fetus, therefore neither will a bird's power distinguish the parts of a wind egg so that it resembles a true egg.

The Philosopher says the opposite, and it is evident to the senses.[10]

One must reply that there is a distinction between the white and the yellow in a wind egg. This is that, just as in a woman, a part of the menses at least is double, one of which results in the material part of the fetus and the other in its nourishment. The first part is more digested than the second because the second will be digested further before it is suited for conversion. The same holds true for a wind egg: there is something that results in the material part of the fetus if there is an approximately analogous agent, and there is something that results in nourishment. The first one is the egg white and the second is the yellow. Therefore, since there is a potent power operating with sufficient heat in a [female] bird's semen, it can distinguish these two in a wind egg without union with a male.

1. On to the arguments. To the first, one must reply that the white is not to the yellow just as the sperm is to the menses, but it is rather more the relationship which the more digested part of the menses has to the less digested.

2. To the second argument one must reply that it occurs analogously in birds and in women, because, just as the female bird's power can distinguish parts of the egg, so the woman's power can, in some manner, distinguish the parts of the menses. But just as an animal cannot be formed from a wind egg, so neither can a human being be formed from a woman's menses by her power alone.

Question 9: Whether a wind egg, if the male's sperm comes into contact with it, is suited for generation.

Further one inquires whether a wind egg, if the male's sperm comes into contact with it, is suited for generation.

10. Ar., *GA* 3.3 (754b21f.); A., *DA* 17.1.3.23 (*SZ* 2: 1245–46).

It seems not. Because it is necessary for the male's sperm and the female's menses to mix so that they are suited for generation. But when a wind egg is formed, the male's sperm cannot mix with it, and therefore, etc.

In addition, should a wind egg make contact with some power outside the womb, it is not suited for generation of a fetus; therefore, for the same reason, should it make contact with some power inside, it will not be made fit for generation.

The Philosopher says the opposite.[11]

One must reply that the wind egg has three states: one is before the distinction of the white from the yellow; another is when the white is distinguished from the yellow but yet it is still not enclosed within a hard shell; the third is when the shell has hardened and it is separated, as it were, from the womb. When the wind egg is in the first or second state, however, it can be made fit for generation if it makes contact with the male's sperm, because then it can still be mixed with the sperm in the womb just as if they came together simultaneously in the beginning. If, however, it makes contact with the male's sperm in the third state, then it cannot be made fit for generation, just as it cannot be made fit once it is outside the womb.

With this a solution is apparent to the arguments. For they prove that when the egg's shell is hardened, then contact with the male's sperm cannot make it fit for generation.

Question 10: Whether female birds can be impregnated by the male's kiss or odor alone.

Further one inquires whether female birds can be impregnated by the male's kiss or odor alone.

1. It seems not. An animal is not generated without a generating and moving cause. But the male's sperm is the moving cause. Since, then, neither a kiss nor an odor is the male's sperm, a female cannot be impregnated by them.

2. In addition, everything that enters the mouth passes into

11. Ar., *GA* 3.7 (757b1f.). Cf. A., *DA* 17.1.6.40–45 (*SZ* 2: 1257–59).

the stomach, where it is changed into another nature.[12] What is pure results in nutriment; what is impure is excreted through the bowels. Therefore, if the sperm were emitted into the female's mouth by the male's kiss, and if, along with an odor that came up to the female, the sperm were multiplied, it would still not be fit for generation because the sperm cannot do this before it reaches the womb. But there is no direct pathway either from the stomach or from the sense of smell to the womb. Therefore, etc.

The Philosopher says the opposite about partridges and many other animals, which are impregnated by a kiss or odor alone.[13]

To this, one must reply that birds can be impregnated by a kiss or by an odor alone, but this manner of conception is not suited to generation. The reason for the first [claim] is that the female experiences pleasure from the male's odor or touch or kiss, and her desire for coition is aroused, on account of which she emits menses to the womb's opening. From this menses a wind egg or something like it can be generated in the womb, and so birds can be impregnated by either a kiss or an odor. The reason for the second [claim] is that the male's sperm is to the female's semen just as a potter is to the material for a clay vessel. But the material for a clay vessel does not move itself to the form of a fabricated object, like a pot, without the potter's assistance. Therefore, neither will the female's semen lead itself to the perfected act of [creating an] animal without the male's semen.

1–2. With this a solution to the arguments is apparent. For the first arguments prove that birds cannot be impregnated by a kiss or an odor in such a way that what it has received will achieve an animal's form.

Reason, however, shows with respect to the objection that wind eggs can be generated in this way, and that birds can be impregnated by them.

12. See Mt 15.17.

13. Ar., *GA* 3.1 (751a13f.). Cf. Avic., *DA* 17 (fol. 62ra–vb); Averroes, *GA* 3.1; A., *DA* 17.1.2.10 (*SZ* 2: 1241–42).

Question 11: Whether there is a mixture of semen and nutriment in fish eggs.

Further one inquires whether there is a mixture of semen and nutriment in fish eggs just as there is in birds' eggs.

It seems so. For the generative power produces whatever is necessary for an effect. But nutriment (that is, the fetus's nourishment) is required for the completion of generation. Therefore, in fish the informative power for the egg mixes the nutriment with the matter for generation.

In addition, everything that grows is nourished. Since fish eggs grow once they are cast outside the uterus, they must therefore be nourished. But they are not nourished by anything outside them when they are outside the uterus. Therefore, they provide nutriment from themselves.

To the contrary. The nutritive power is distinct from the generative power. Therefore, where they are distinct by nature, then the principles of generation and nutrition do not exist simultaneously. But there is no distinction of parts in fish eggs. Therefore, etc.

To this, one can reply in two ways. The first is that in a fish egg there exist both the material principle of the fetus [*conceptus*] and a nutriment appropriate for it. Nevertheless, these are not distinct, owing to the weakness of the power. But this indistinctness or confusion does not prove that each principle is not there, because in the beginning birds' eggs are also indistinct and yet each principle exists in them, which becomes clear when they are digested further. And in this way a solution is evident to the argument in the objection.

One can respond in another way that there is no distinct nutriment in fish eggs because fish eggs have only a modest size, since the egg's material stretches just so far as barely to suffice for the generation of the fish. Therefore, just as one can say of a seed that there is material for the one germinated in the seed, and when the seed is cast to the ground it receives nourishment from outside itself and does not have its nutriment joined to it inside itself, so one can reply concerning a fish egg. Thus one can say of the fish egg that when it is emitted on the ground in still water and makes contact with the male's sperm, it receives nutriment from what contains it (that is, from the ground or the water), just as plants or seeds do. And for this reason fish eggs are incomplete, as the Philosopher often says.[14]

With respect to the arguments it is clear that fish eggs do not have nutriment mixed into or joined with them, but receive it instead from outside.

Question 12: Whether all the elements concur for the generation of an animal generated by putrefaction.

Further one inquires about the generation of animals by means of putrefaction.[15] And first, whether all the elements come together for the generation of an animal generated by putrefaction.

1. It seems not. Nature operates through the fewest things it can. But celestial heat and fiery heat do the same thing. Therefore, fiery heat is not required for the generation of an animal like this.

2. In addition, if fire were to enter into the constitution of an animal like this, either it would descend from its sphere, or not. If it descended, then, when it has descended it will be unnatural, and this animal's generation will be unnatural. If it did not descend, but were generated here below, then the mixables would not be separate before their mixture, which is contrary to what the Philosopher says.[16]

3. In addition, that whose nature is to consume does not cause the generation of anything.[17] But it is fire's very nature to consume, and therefore, etc.

To the contrary. The power of the mixables is preserved in the mixed body. But the power of fire is in every mixed body. Therefore, fire is numbered among the mixables, and as a result it enters into the generation of every mixed body.

14. Ar., *GA* 3.5 (755b1f.; 756a21f.).
15. Cf. A., *DA* 6.3.3.123 (*SZ* 1: 584).; *Meteora* 4.1.11.
16. Cf. Ar., *DG* 1.6 (324b6f.).
17. Reading the *ipsius* of *T* over the printed but vague *eius*.

One must reply that all the elements come together for the generation of the animal, [present] either in themselves or in their effects. In themselves, as when the four divided elements approach one another in such proportion that no one totally dominates another but in such a way that they are expelled by the mutual action and passion of the form of the individuals, and in such a way that one form having the power of each of the four is introduced into the matter, which earlier existed under four distinct forms, even if it is introduced in some confused manner.

Moreover, the elements enter the animal's constitution in their effects, just as blood is generated from one aliment and semen from the blood. Since, then, an animal like this is a mixed body, the four elements come together for its generation [present] either in themselves or in another. This can be briefly stated. For acts of generation arise from contraries, and earth and water are required for the generation of all mixed bodies; therefore, so too are their contraries. But the proper contrary for earth is air, and for water it is fire. Therefore, etc.

1. On to the arguments. To the first, one must reply that although a celestial heat is required for generation such as this, nevertheless, because a universal agent only acts particularly, the fire's heat is therefore required in addition to this heat.

2. To the second argument one must reply that fire does not enter a mixture from its own sphere, because then its advent would be unnatural; rather, through the power of a superior body, like the sun, another can be created here below from any element. Therefore, fire can be generated here below, and this fire is adequately suited for generation, and that which is capable of being mixed can be separated. The argument supposes the contrary to this. Or one can say that it is not necessary for the generation of any mixed body that the elements, previously divided, enter into the generation of any mixed body, because something mixed can be made from something mixed.

3. To the third argument one must reply that it is the nature of a disproportional fire to consume, but a fire that is proportional and regulated has [the power] to preserve. Therefore, it

does not exist in a mixed body in its ultimate state and excess, but in a somewhat restrained state.

Question 13: Whether putrefaction is a path to generation.

Further one inquires whether putrefaction is a path to generation.

1. It seems not. For putrefaction is a final corruption. But corruption is not a path to generation, since these are contraries. Therefore, etc.

2. In addition, art imitates nature to the extent that it is able. But art casts off or rejects the superfluous and creates a work of art [*artificiatum*] from the residue. Therefore, so too does nature, in the same way. But something made through putrefaction is superfluous and is expelled by nature. Therefore, putrefaction does not become generation.

3. In addition, the corruption that proceeds from heat is of two types, namely, combustion and putrefaction. But an animal is not generated from combustion, and therefore neither is it generated by putrefaction.

The Philosopher says the opposite.[18]

To the first, one must reply that an animal can be generated from putrefaction. For just as the semen in the womb is digested and completed during an act of digestion by an internal power, and whatever is impure is expelled and whatever is pure that remains is converted into the matter for the fetus, so the putrefied matter in the earth's belly can be digested correctly by the heat of a celestial body, of the earth's body, or of the containing body, and what is impure can be expelled, and what is pure and remains can be in potency to the animal's form, because, just as an internal heat disposes the animal's matter for the generation of something animate, so the external heat can dispose the matter for the generation of an imperfect animal.

18. Ar., *Meteora* 4.1 (379a2–18). Cf. A., *DA* 6.3.3.123 (*SZ* 1: 584); *Meteora* 4.1.11.

1. On to the arguments. To the first, one must reply that, according to the Philosopher in the first book of *On Generation*, "the generation of one thing is the corruption of another."[19] Therefore, although putrefaction represents the final corruption of one thing, nevertheless it can be a pathway into the generation of another.

2. To the second argument one must reply that just as art does not form a work of art out of a separated superfluity, but out of what remains, so too does nature operate in putrefaction.

3. To the third argument one must reply that life depends on the hot and the moist, and not the hot and the dry. But in putrefaction the hot operates with the moist, whereas in combustion the hot operates with the dry. This is why combustion does not cause generation as putrefaction does.

Question 14: Whether animals like this are generated by a superior power.

Further one inquires whether animals like this are generated by a superior power.

1. It seems that they are not generated by a superior power. For a cause and effect are so proportioned that the universal is proportioned to the universal and the particular to the particular. But a superior power has the nature of a universal agent, but the one generated is a particular, and therefore, etc.

2. In addition, motion impedes the generation of animals like this. An indication of this is that they are not generated in running water, whereas they are generated in standing water. Since, then, celestial bodies are in continuous motion, it seems that animals like this are not generated by the power of superior bodies.

The Philosopher says the opposite.[20]

One must reply that two powers are required for the generation of animals like these, namely, a superior power and an inferior power. The inferior power disposes the matter for putrefac-

19. Ar., *DG* 1.3 (318a23f.).

20. Ar., *Phys.* 2.2 (194b13).

tion, into which, once it has been disposed, the celestial power is introduced, operating on the matter just as sperm operates on the menses. And this is why, just as the power of the sperm disposes the menses to the form of a perfect animal, so the celestial power operates through an elemental power on matter that is disposed to the form of an imperfect animal. Therefore, just as "the human generates a human as does the sun," so it can be said that the sun and the surrounding material generate this sort of imperfect animal. The fact that an animal like this can be generated without a seed arises from its imperfection, since the more perfect an inferior thing is, the more things are required to produce it. And this is why, although an animal like this can be produced without seed by a celestial body's power and by the power of the body containing it, nevertheless a horse and an ass and a human cannot be produced solely by a universal agent without the seed of a particular animal agreeing with it in species.

1. On to the arguments. To the first, one must reply that a universal is of two kinds: one for causing and the other for disposing. Aristotle's remark must be understood to apply to the second universal and not to the first, and a celestial body is a universal agent for causing and not for disposing, and this is why it can produce a particular effect well. And this is the reason why a universal agent or celestial body acts only with a particular agent mediating it instrumentally, since all inferior things are instruments of superior ones.

2. To the second argument one must reply that the motion of the surrounding material prohibits putrefaction, and this is why water's motion thins and purifies more than it putrefies, because its own motion continually incorporates the superior power into the impure water. But this is not so for the motion of superior bodies, because their motion causes heat here below. And likewise motion continues here below because, if the first mover should happen to cease to move, then all motion here below would cease. Therefore, the motion of a superior body does not impede putrefaction, but rather arouses the heat needed for generating, although the water's motion impedes the putrefaction of the surrounded body.

BOOK EIGHTEEN

Question 1: Whether the power's strength causes the generation of a male.

"We have already spoken above about the generation of animals," etc.[1] One first inquires in this eighteenth book into the differentiation of the sexes. And first one asks whether the strength of the power causes the generation of a male.

1. And it seems not. Power proceeds in a radical way from the heart. For the power is stronger on the side to which the heart inclines more. But the heart's location inclines to the left side; therefore, the power will be stronger there. If, then, the power's strength were the cause of a male, a male would be generated more often on the left side than on the right side. The Philosopher says the opposite.[2]

2. In addition, the male's power prevails over the female's power, because the male's sperm is like a form and an agent [*agens*] whereas the female's semen is like the matter and that which is acted upon [*patiens*]. Therefore, if the power's strength were the cause of a male, since the male's power is always stronger, then males would always be generated, since like generates like. This, however, is not true.

The Philosopher says the opposite.[3]

To this some say that the generations of male and female differ with respect to the parts of the womb, since for the most part a male is generated on the right side and a female on the left. But this is not true since the opposite is sometimes found

1. Ar., *GA* 4.1 (763b20f.). 2. Ar., *GA* 4.1 (763b32f.).

3. Ar., *GA* 4.1 (765a34f.). Cf. Avic., *DA* 18 (fols. 62vb–63ra); Averroes, *GA* 4.2; A., *DA* 18.1.1–2.1–21 (*SZ* 2: 1281–91).

to occur, namely, that a male is on the left side and a female on the right. And sometimes twins are found in one and the same part of the womb, one of which is male and the other female. And besides, if an embryo were generated in the middle of the womb, it would follow, according to this position, that one and the same fetus would be male on the right side and female on the left side, which is unsuitable.

Others say that heat is the principle generating a male and cold is the principle generating a female. But this still cannot be true since, according to this explanation, if an embryo close in disposition to the male were introduced into a cold womb, it would change into the nature of a female, and if a female embryo were introduced into a warm womb, the contrary would occur.

This is why one must respond in accordance with the Philosopher's notion that the strength of the power is the principal cause of a male and its weakness is the cause of a female. For when the power is strong, then the semen is completely digested, and when the power is weak, the menses is rendered almost indigestible. But a male is generated from well digested semen, and this is why the power's strength is the cause of the generation of a male. Thus the positions of all of the others can be reduced to this cause, because very often a male is generated on the right side of the womb, since the heart's influence is greater on the right side and, as a result, its power is stronger there.

Furthermore, strong heat accompanies the strength of the power, and it is the heat's [role] to digest, and this is why some say that heat is the cause of the generation of a male, because heat is an instrument of the power that is a principal cause. This is also why youths (that is, young boys) and old men very often generate females, seeing that youths have not yet achieved strength in their power and old men are deficient in strength.

Moreover, this is also the reason why the north wind causes the generation of a male, and the south wind causes the generation of a female, because the north wind is pure and cleanses and purifies the air and vapors and sharpens the natural powers.[4] But a south wind is humid and rain-filled and is a cause of

4. See *QDA* 6.22. Cf. Avic., *Can. med.* 3.21.1.12.

better digestion. Thus many factors come together as secondary causes [*accessoria*] for the generation of a male; nevertheless, the strength of the power is the principal cause.

1. On to the arguments. To the first, one must respond that although the heart inclines toward the left side so far as concerns its location, nevertheless it tends toward the right side so far as its influence is concerned.

2. To the second argument one must reply that although the man's sperm is related to the female's semen as an agent relates to matter, nevertheless sometimes the female's semen is disobedient and resists the agent too much, and then the male's power cannot entirely prevail, and this is why it cannot perfectly assimilate the fetus to itself. Instead, it then acts according to the exigency of the matter, and a female is generated.

Question 2: Whether the generation of one that shares each sex is natural.

Next one inquires whether the generation of one that shares each sex—like the hermaphrodite—is natural.

1. It seems not. For just as there is a mixture of elements, so too is there a mixture of the sexes. But the mixture of the elements is uniform, and therefore so too is the mixture of the sexes.

2. In addition, the male's power either prevails or is overcome. If it prevails, a male is generated; if it is overcome, a female. It is not possible, then, for something to be generated in the middle.

3. In addition, if nature could produce such a fetus, then such a thing could generate on its own, since it would have the members appropriate to each sex.

The Philosopher says the opposite.[5]

To this, one must reply that nature acts with intention because it acts for an end. But intention is of two types: first and

5. Ar., *GA* 4.4 (772b26f.). Cf. Avic., *Can. med.* 3.20.1.43; Averroes, *GA* 4.4; A., *DA* 18.2.3.65–69 (*SZ* 2: 1312–14).

second. It belongs to nature's first intention to produce the best that it can. But it belongs to the second intention that, if it departs from the best, it produces one that it is nearest to it. This is why, when the natural power is strong, it produces a male. When, however, the production of a male is impeded by the recalcitrance of the matter—if the dispositions of the matter surpass or simply overpower—then it produces one like the one from which the matter is derived, so that it produces a female. If, however, the power partially prevails and is partially overcome, to the extent that the power prevails it produces members suited to a male, and to the extent that it is overcome it produces members suited to a female. Nevertheless, this does not occur without a superfluity of matter, for otherwise it would not produce a penis and a womb in the same fetus. Thus, if we consider nature's first intention, the production of a fetus such as this is unnatural; however, if we consider nature's second intention, this production is natural because it proceeds from a natural cause.

1. On to the arguments. To the first, one must respond that the mixture of elements is uniform because elements that have been brought close to one another act upon, and are acted upon, one to another. Thus if each part of the mixture were not mixed, but one part were simple, and then this simple part acted on another, then it would be either simple or mixed.[6] But such opposition does not exist between the parts of the male and the female.

2. To the second argument one must reply that it is not the case that the male's sperm always prevails or is overcome in an absolute sense; it can partially prevail, or be partially overcome.

3. To the third, one must respond that such a monster cannot reproduce on its own, because, although it would have the members of each sex, nevertheless it is unable to cast sperm through the male member into its own womb. Thus, while it might be able to cast its sperm and generate in another, nevertheless it cannot become pregnant, because, as was touched on

6. The translation proffered entails a slight repunctuation of the ungrammatical Latin text as printed.

in the fifth book, all mannish women [*viragines*] are sterile because they do not have wombs suited to conception, and this is why, etc.[7]

Question 3: Whether some power in the semen is the cause of resemblance.

Next one inquires into the cause of the resemblance of children, and whether some power in the semen is the cause of resemblance.

1. It seems not. Because only the power of the father or the mother is present in the semen. But the fetus sometimes resembles neither the father nor the mother, and therefore, etc.

2. In addition, imagination contributes to resemblance. For it is said that when a woman imagines the form of a certain demon depicted on a curtain above the bed she always conceives children like the picture, and the same thing is read about Jacob.[8] But it is clear that the power of such a picture was not in the semen.

The Philosopher says the opposite.[9] For he says that variation among the progeny occurs as a result of differences in the powers in the semen.

To this, one must reply that a natural agent generates one like itself, to the extent that it is able. Thus there always remains some power of the one generating in the one generated. If, then, the semen is derived from the one generating, the power of the proximate generator will be in the semen, and, since the power of its generator is in the one generated, the power of both the proximate and remote generators will be in the semen. Now then, it frequently is the case that the male's sperm dominates the matter of the female's menses absolutely, and then the

7. No reference is found in the fifth book, but see Ar., *GA* 2.7 (746b20f.); A., *DA* 18.2.9.93 (*SZ* 2: 1327–28).

8. Gn 30.37–38. For the persistent influence of the notion that imagination can affect the appearance of the fetus, see Paul-Gabriel Boucé (1988).

9. Ar., *GA* 1.18 (723a21–b3). Cf. Avic., *Can. med.* 1.2.2.14, 3.21.1.2, 14; A., *DA* 15.2.3.76, 18.2.9.94–97 (*SZ* 2: 1118–19; 1328–30).

one generated will perfectly resemble the male. For insofar as the animal power is in the semen, it produces an animal; and insofar as a human power is there, it produces a human; and insofar as a power dominates in it with respect to maleness, it will produce a male; and insofar as a power dominates with respect to the accidental and individual dispositions, it produces a child like itself in its characteristics.[10] Thus the complete victory of the male semen or sperm over the female's matter is the cause why the fetus resembles the father. If, however, the sperm's power dominates with respect to the species' characteristics, but does not dominate or prevail with respect to the male's characteristics because its heat is weak and the cold and moisture of the matter of the female are abundant, then, if this matter that is resisting the agent dominates absolutely and with respect to maternal characteristics, then a female resembling the mother is produced. If the power of the father, however, dominates in characteristics common to a male but not in its own characteristics (that is, insofar as he is Socrates or Plato), then this power will produce a male but not one similar to the father. But it does produce a fetus like the father to the extent that it is able, since the power of the grandfather and great-grandfather are also in the semen, although more remotely than the father's.

Thus when it cannot produce one resembling the father with its own power, it produces one like the grandfather or great-grandfather with an even more remote power. If, however, this power departs from the characteristics of the male, yet prevails with respect to its own individual characteristics and dispositions, then it produces a female, yet one that resembles the father. And the contrary obtains if it prevails with respect to the male's characteristics but is overcome with respect to its own characteristics, because then it produces a male resembling the mother. And if it falls short of a resemblance to the mother, it produces one, insofar as it can, that resembles the grandmother or great-grandmother. If, however, it prevails with respect to the characteristics of the male but prevails neither in its own in-

10. "Characteristics" = *conditiones,* more commonly qualities or peculiar features.

dividual and proper characteristics, nor in those of distant forebears, then it will produce a male human being that sometimes will resemble neither of the parents [in any respect]. Thus, with respect to the various modes of prevailing or dominating, the fetus will bear a resemblance in a variety of ways.

1. On to the arguments. To the first, one must respond that the power of the father and the mother is in the semen. But the power of the father is the power of the grandfather, since the father came forth from the grandfather, and this is why the fetus can resemble the grandfather. And the explanation is similar for the more remote forebears.

2. To the second argument one must reply that the imaginative power leaves its impression on the natural power because it is superior to and dominates it. Thus the entire body is altered by various imaginings. And this is how it happens that the eye of an enchanter alters the bodies of children (and this is how it is said that the eye of an enchanter once threw a serf into a ditch), because children are amazed at the shapes or aspects of some things more than others, and, on account of this, fear and trembling are produced in children. And it is this way for others as well. Thus various types of imagining cause a change in every part of the body and this is why that royal woman, whom Avicenna described,[11] imagining the shape of a demon or a dwarf (or, according to others, an Ethiopian) bore children resembling them, because her own power succumbed at the moment of conception[12] owing to the violent imagination, and the natural power was altered according to the kind of thing she was imagining. And it was the same for animals looking upon things of various colors, because a different impression on the soul results in a change in the entire body. And it is similar for the semen. In any case, the power of a superior agent contributes a great deal to these particular causes.

11. See *QDA* 7.3.
12. Or: "in the fetus."

Question 4: Why wise men and philosophers very often generate foolish children.

Next one inquires why wise men and philosophers quite commonly generate foolish and ignorant children.

1. And it seems the opposite is the case. For nature is more solicitous of nobler things. But it is true for brute beasts that nobler parents generate better and nobler young, as is evident among horses. Therefore, this will be even more true for the human, who is nobler than these others.

2. In addition, not only the power of the body but even the power of the soul is in the sperm; otherwise, it would not produce something animated. Therefore, since children generally resemble their parents in the bodily dispositions, they will, for the same reason, resemble them in the soul's dispositions, in terms of wisdom, knowledge, etc.

The opposite is clear to the senses.

One must respond that wise men very often do generate foolish children since, according to Avicenna, the operation of one very intense power prevents another from operating.[13] Now, then, wise men are very prone to imagination and reason, and they engage in very deep thoughts and meditations. Thus the natural power in them is usually weakened, and this is why their sperm is frequently bad and undigested, and this is why they produce poor children, since one who is good at study is bad at business and at generation, that is, at making love. This is why, etc. Other animals are more prone to the natural power and they are not impaired by the depth of their thoughts, and this is why good progeny come from good [animals]. This is why the offspring of one horse is more desirable than that of another.

1. And with this a response to the first argument is clear, since it is not the same for brute animals and the human and because humans are entirely straightforward with respect to the natural power, and this is why they digest sperm well. Thus an intelligent offspring is generated.

13. Avic., *De anima* part 1, 4–5; part 5, 2.

2. To the second argument one must respond that wise men think very deeply about noble matters, and for this reason a measure of their power is transmitted with the sperm. And this is why, etc.

Question 5: Whether anything like a monster exists in nature.

Next one inquires about monsters. And first, whether anything like a monster exists in nature.

1. It seems not. For a monster does not come forth from that whose operation always tends toward the good. But nature's operation is always for the good, because an end possesses the definition of the good and nature operates for the sake of an end. Therefore, etc.

2. In addition, there can be no monster where nothing is superfluous and nothing is diminished. But nature is such as this, because "nature does nothing in vain, nor does it lack anything necessary." Therefore, there is nothing superfluous in nature, and nothing is diminished. Therefore, etc.

The opposite is evident.

One must say that it is possible for a monster to exist in nature. And the reason for this is as follows: there are certain agents, like the superior agents, that do not accept any hindrances to their operations, and there are some agents, like all those which act on matter supplied to them, that can accept hindrances. This is because an agent can be frustrated with respect to its appropriate end by the recalcitrance of matter. But natural agents are of this sort, and they can thus be hindered from accomplishing the end principally intended. But that which happens beyond the appropriate end is an evil in nature and possesses the definition for a monster. Therefore, a monster is possible in nature.

1. On to the arguments. To the first, one must reply that nature's operation is ordered toward the good unless it is hindered, but it can be hindered, and this is why, etc. Or one can say that although a monster is an evil when compared to the ef-

fect intended *per se,* nevertheless the monster itself, insofar as it possesses the definition of being [*ens*], is good.

2. To the second argument one must respond that nature does nothing in vain intentionally, nor does it lack things that are necessary. But if something such as this comes to be, then this occurs outside of its intent and owing to some hindrance it encounters.

Question 6: Whether the cause of a monstrosity stems from the matter or from the efficient cause.

Further one asks whether the cause of a monstrosity stems from the matter or from the efficient cause.[14]

And it seems to stem from the efficient cause. For when completion proceeds from something, then incompletion comes from the same thing. For the Philosopher says that just as a completed effect comes from the agent's strength, so too does an incomplete effect come from its weakness.[15] Therefore, the agent is the cause of monstrosity.

Moreover, nothing leads itself into act. Therefore, matter (whether it be superfluous or diminished) will never produce a monster unless it is led by another into act. Therefore, the agent will be the cause of monstrosity.

The Philosopher says the opposite.

One must say that the primary source of monstrosity is rooted in matter, and the secondary source of monstrosity comes from the efficient cause, because a natural agent acts on the matter supplied to it and acts for the sake of an end. If, then, it finds matter proportioned to the end, it produces the appropriate effect; if, however, it finds the matter either superfluous or diminished, it will produce an effect that is either superfluous or diminished.

Just as it is with a large number of children, so is it for a large number of members. For just as an agent, when it finds matter proportionate to a large number of children, produces many children, and when it finds that the matter is insufficient even

14. Cf. A., *Phys.* 2.3.3.

15. Ar., *GA* 4.4 (770a6f.). Cf. A., *DA* 18.1.6.46–52 (*SZ* 2: 1303–6).

for one it will cease to act, so it is for the generation of members, for example, the hands. When the formative power has completed the five fingers for the hand, if there is still sufficient matter for one more finger and nature is not idle, it will produce a sixth finger. Thus nature first intends only five fingers, but after it finds abundant matter, it intends a sixth. And it is the same when there is diminution, as when matter is lacking after four fingers have been produced.

And if one inquires: Why does nature not divide the matter of the sixth finger into the other five, so that each finger would receive a part of the matter—for this would be better than to produce six—one must reply that nature proceeds in an ordered way and does not produce just anything from just any matter, but produces a proportional thing from proportional matter. Thus, once nature has formed the principal parts, it then produces accompanying proportional parts. Thus it forms the fingers and increases their size according to the proportion of the remaining part of the hand and the arm, and this is why it does not send more matter to the fingers than the fingers demand. So if it were to divide the matter of the sixth finger into the other five, then it would make the five too big and exceed all proportion. And besides, nature works in stages. Thus five fingers are formed before the material for the sixth is distributed, and this is why that matter cannot be divided partially among the five fingers.

It is the same for other monsters. For in those who generate several children in the womb, each child is separated from another by an intervening membrane [*pellicula*]. Thus if it happens that this membrane is ruptured above and not below, then the material for the two children comes together above but is kept apart below, and then there will be generated a human being (or a horse) having one body and one head but having the feet of two individuals. If, however, contrariwise, the membrane is ruptured below and remains intact above, then the fetus will be one below and it will be divided above, having two heads. And it happens the same way for other monsters. Thus a monster's first flaw stems from matter, but because matter does not lead itself into act, but is led into act by an agent, and for this reason the agent is therefore the secondary cause of the monstrosity. Nevertheless, if the matter and the agent are proportional, or diminished or superfluous, then the effect will be partially monstrous, because it will not be complete. But still monstrosity stems from matter, because although it is proportionate *in se,* it nevertheless is disproportionate compared to the agent, because matter and agent are not enough, but proportion is required as is the obedience of one to the other.

With this a response to the arguments is clear. For they prove that the agent is in some measure the cause of the monstrosity, and this we concede, and for this reason, etc.

Question 7: Whether an animal having parts of animals different in species can be generated from two animals that are different in species.

Next one inquires whether an animal having parts of animals different in species can be generated from two animals that are different in species.

1. It seems not. For according to Porphyry, species are unmixed.[16] This would not be the case, however, if a generation of this type were possible.

2. In addition, the womb of one species is not proportioned to the sperm of another. Therefore, if they should have intercourse, they would not produce a fetus.

The Philosopher implies the opposite.[17]

One must respond that the production of such animals having the parts of animals different in species does not always result from animals that are different in species, since the mating seasons of different animals are different, as the Philosopher suggests.[18] For one animal is impregnated at one time when another is not.[19] Thus an animal having different parts, especial-

16. Porphyry, *Isagoge*, trans. Boethius, (Brandt, 1906), 334, 11.
17. Cf. Ar., *GA* 4.3 (769b13f.).
18. Ar., *GA* 4.3 (769b22f.).
19. A. uses the term *tempus impregnationis* and is apparently speaking of mat-

ly parts of these animals, is not generated from the joining of these two, but rather an animal like this is generated from a defect in the agent. For if the proper power is not present in the male's semen, and the power of the species does not dominate, but only the power of the genus dominates, then an animal is produced, and it is not like the male in species. Likewise, to the extent that the power either variously prevails or is overcome, to that extent the fetus will vary. If, however, two animals have suitable times for impregnation, then it is possible that the womb of one will share properties common to the sperm of the other, even though they may differ in their own individual properties. And because they are proportionate in common properties, an animal is therefore produced, and yet it will resemble neither the father nor mother, as is evident in the generation of a mule. If, however, the power in the sperm of the male prevails over some parts of the matter and is overcome by others, then, to the extent that it prevails, it causes the fetus to resemble the father, and to the extent that it is overcome, the mother. And in this way it is possible for such an animal, diversified in its parts, to be generated from different ones.

1. On to the arguments. To the first, one must respond that species are unmixed in such a way that one is not in another, as some propose, and yet, a third species can come to be from two different species, as something pale or dark can be produced from black and white. Thus the premise is correct. The definition of a species arises from its form, and two forms are not conjoined in such an animal, and therefore neither are species. But the matter derived from each can be mixed, and the form is introduced according to the disposition of the matter and the nature of the agent power. Nevertheless, the production of such an animal departs greatly from the order of nature, and so it happens rarely.

2. To the second argument one must reply that if the womb and the sperm are not proportionate in their own properties,

---

ing seasons, although his language sometimes seems to allow the reader to understand "gestation period" [*spatium impregnationis* in *DA,* e.g., 18.2.8.88 (*SZ* 2:1324)].

they can nevertheless be proportionate in their common properties. And this is why, etc.

Question 8: Whether milk is necessary for the nourishment of the fetus.

Next one inquires whether milk is necessary for the nourishment of the fetus after birth.

1. It seems not. After birth the fetus is complete, and therefore it needs a complete nutriment. But milk is not such a thing, whereas something born from the ground is, and therefore, etc.

2. In addition, the offspring (that is, the fetuses) of oviparous animals are not more complete than the offspring of those that produce milk. But the offspring of oviparous animals do not require milk for their nourishment after birth. Therefore, neither do the fetuses of walkers.

3. In addition, nourishment preserves life. Therefore, it especially should be hot and moist. But according to physicians, milk is cold, which is apparent in things produced from milk, like cheese and whey.

The Philosopher says the opposite.[20]

One must reply that nature proceeds in an ordered manner when it operates. Now, however, the fetus is nourished in the uterus by menstrual blood, and when it is born the fetus is more complete outside the uterus than it was earlier. Therefore, it requires a more complete nourishment. But there are parts of the fetus that are not so solid or complete as to be able to digest or convert earthborn materials perfectly, and this is why, immediately after birth, nature gives a nourishment that is more digestible than menses and more refined than earthborn materials. But milk is like this. An indication of its digestion is its whiteness, because at first, during digestion, heat proceeds by whitening, and then later by darkening. And this is why milk appears

20. Ar., *GA* 4.8 (776a15f.). Cf. Averroes, *GA* 4.8; A., *DA* 18.2.7.84–87 (*SZ* 2:1322–24).

in parts of the body that are hot, like the breasts, which are located on the chest in the human and on the belly in other animals, where heat is more abundant, and this is why hairs on the belly are whiter.

1. On to the arguments. To the first, one must reply that immediately after birth the fetus is not so complete or constructed that it can digest earthborn things adequately, but because it is nourished by the menstrual blood before birth, it is therefore nourished afterward by milk, which is somewhat removed from menstrual blood, because it is more digestible.

2. To the second argument one must respond that those born from eggs are nourished by the yolk since oviparous animals do not give birth to offspring inside the uterus but outside of it, and this is why they are not nourished by milk.

3. To the third argument one must reply that milk is naturally cold, because the earthy and watery abound in milk. But it nevertheless participates in heat with regard to having a more complete digestion and with regard to its container. It is unnecessary for the nutriment to be hot and moist at the beginning [of digestion][21] since at the beginning, the nutriment has to be dissimilar to that which it nourishes, but has to be similar to it at the end, as is said in the second book of *On the Soul.*[22]

21. Perhaps, "in principle."

22. The logic seems to be that if our food were similar to ourselves when we ate it, then, logically, our bodies would routinely digest themselves. Rather, the act of digestion must convert something unlike us into something like us. Cf. Ar., *De anima* 2.4 (416a29f.).

BOOK NINETEEN

Question 1: Whether the embryo is nourished before its formation.

"CONSIDERATION MUST be given to the accidents by which [the members] are diversified," etc. [1] In this nineteenth book, first one inquires whether the embryo is nourished before its formation.

1. It seems not. For nothing is nourished unless it is alive. What is not alive in act, but solely in potency, is not nourished in act. But the embryo is like this before its formation.

2. In addition, the nutritive power has organs designated for it, but the embryo does not have organs before its formation. Therefore, etc.

3. In addition, the last nutriment or food of the members is blood, according to the Philosopher.[2] But there is no generation of blood in the embryo before its formation. Therefore, etc.

To the contrary. Nutrition precedes growth. But the embryo grows before its organization. Therefore, it is nourished, since growth does not occur without nourishment.

To this, one must reply that the semen is digested after its reception in the womb, and what is suited to the fetus is separated from what is not, and is converted into a certain fleshy mass, in which there is a regulative power [*virtus regitiva*] and natural heat. Therefore, just as a seed that falls to the ground receives, by its own power, nutriment from the ground, so too does the fetus's matter receive nutriment by its own power from the womb or from what is contained in the womb.

Nevertheless, it is nourished in different ways before and af-

1. Ar., *GA* 5.1 (778a16–17). Cf. Avic., *Can. med.* 3.21.1.2; A., *DA* 18.2.8.88 (*SZ* 2: 1324).

2. Ar., *GA* 1.19 (726b1f.).

ter its organization, because before its organization it is nourished only by a power derived from its father, but after its organization it is nourished by a power of its own soul.

1. On to the arguments. To the first, one must reply that nothing is nourished unless it is alive in act or has the power of living. Now, however, although the embryo does not have life in act before its formation, it still has a power of life derived from the father.

2. To the second argument one must reply that after its formation the embryo is nourished by means of organs, but before its formation it is nourished due to its spongy nature [*spongiositas*].[3] It is analogous to whatever is committed to the ground, since it grows in different ways before and after its formation, because before its formation it exists as a seed [*frumentum*]. In the seed there are compacted earthy parts, and the intrinsic heat extends these earthy parts and spreads them out. This is the right way to understand the embryo. This is why, etc.

3. To the third argument one must reply that menstrual blood, which nourishes the embryo, is not generated in the embryo but in the womb; etc.

Question 2: Whether the embryo emits superfluities in the womb.

Further one inquires whether the embryo emits superfluities in the womb.[4]

And it seems that this is so after its formation. For the formed embryo is nourished by the menstrual blood. But this blood is not entirely uniform. Therefore, it has some impurity.

In addition, everything that can take in food beyond its own capacity has something superfluous in it.[5] But the embryo can do this. Therefore, etc.

3. "Spongy nature": The plural *spongiositates* is quite odd, "spongiosities." Based on context, what is implied is that the matter of the fetus is less dense at this stage, and, lacking formal organs for the reception of nourishment, it relies on a sort of wicking action to draw nourishment into itself.

4. Cf. Avic., *Can. med.* 3.21.1.2; A., *DA* 9.1.7.70 (*SZ* 1: 800).

5. "Capacity": the printed text of *ultra compactionem* is difficult since the word

To the contrary. A superfluity is generated from what enters through the mouth into the stomach. But this does not occur in the embryo, and therefore, etc.

One must reply that there are three types of superfluity (that is, what it does not need), namely, excrement, urine, and sweat. The first superfluity precedes the generation of blood; the second superfluity is generated with blood, because it is strained from blood; and the third is generated in the individual members, after blood. The first one does not exist in the embryo, since it is not nourished on crude or earthy things or through a mouth, but is nourished by the mother's menstrual blood. Therefore, that superfluity, which is the excrement, exists in the mother but not in the embryo.

Similarly, if urine is properly understood to be a moisture that is separated from the blood in the liver, before the blood is relegated to the members then, taken this way, urine does not exist in an embryo, because the embryo is nourished through the umbilical cord, at which the veins converge, and before the nutriment arrives at the umbilical cord it is purified in the mother's liver by separating out the urine. But if urine is understood to be a moisture which flows with the blood and is separated out by the members and emitted internally, then, taken this way, urine is in the embryo since this very same moisture is called "sweat" if it is expelled through the skin's pores. Therefore, the third superfluity, properly speaking, does exist in the embryo but not the first or the second digestion, because the first and second digestions occur in the mother, but the third occurs in the fetus.

With this, a solution to the arguments is clear. For they prove that some superfluity exists in the embryo, and this we concede.

---

means "construction," or "framework." One MSS variant reads *complexionem*, "complexion." Our translation of "capacity" is a compromise between meaning and a possible textual corruption whereby *compactionem* derived from an original *capacitatem*, perhaps through abbreviation.

Question 3: Whether the embryo breathes in the womb.

Further one inquires whether the embryo breathes in the womb.

1. And it seems it does. A proper operation arises at the same time as the power to which it belongs. But respiration is the operation of the lungs and heart, since it is bestowed upon an animal to temper the heart's heat. Therefore, since the heart and the lungs appear in the embryo, then respiration will be present in the embryo.

2. In addition, the heart's heat is assisted in two ways: through nutriment and through respiration. Since, then, it is nourished in the womb, it will, by the same argument, breathe in the womb.

To the contrary. Respiration occurs through the nose and the mouth. But these parts are not open on the fetus in the womb. Therefore, it does not breathe.

One must reply to this that the embryo does not breathe in the womb, since respiration is the drawing-in of air; but this cannot happen in the womb. Nor does it need this drawing action, since in the womb the embryo abounds with a great deal of moisture and can adequately temper the heat of the heart, and this is why it does not have to be tempered from another source. But when the fetus is outside the uterus, its moisture does not cease to be diminished, nor do its exterior parts cease to be dried out by contact with the outside.[6] Therefore, at that time the heart's heat begins to burn, and this is why it then needs respiration.

1. On to the arguments. To the first, one must reply that respiration is not the heart's proper operation, since many animals that have a heart do not breathe, but the heart's motion, or pulse, is more appropriate to the heart, and thus pulse can exist in the embryo by means of the natural spirits that are generated

6. "By contact with the outside": *per continens. Continens* has been used in this work to indicate anything which surrounds and contains something else, e.g., an eggshell. It may therefore indicate the external environment here.

in the arteries and perpetuated by being continuously received by the womb.

2. To the second argument one must reply that the reasons for the embryo's nutriment and for its respiration in the womb are not the same, because, while the fetus is in the womb, the nutriment there suffices for two. With this a solution is apparent for this argument.

Questions 4–5: Whether sleep occurs in the embryo. Whether waking precedes sleep.

Further one inquires whether sleep occurs in the embryo in the womb.

1. It seems not. For sleep is a hindrance to the external senses. But in the womb the embryo does not operate with external senses. Therefore, properly speaking, it neither dreams nor wakes nor sleeps.

2. In addition, "the embryo lives a life like a plant in the womb."[7] But a plant neither sleeps nor wakes. Therefore, etc.

The Philosopher says the opposite.[8]

Further one inquires whether waking precedes sleep, or contrariwise.

3. And it seems that waking precedes sleep. Because the first act precedes the second and a path precedes an end. But waking is to sleep just as a first act is to a second, and just as a path is to an end, since it is related to sleep just as motion is to rest. Therefore, waking precedes sleep.

4. In addition, the deprivation of some characteristic is a privation. Therefore, since sleep is a kind of privation, it necessarily presupposes a wakefulness held earlier.

5. To the contrary. What crosses from one extreme to the other approaches the midpoint before it reaches the extreme. But sleep is just like the midpoint between living and non-liv-

7. Ar., *GA* 5.1 (779a1–2).

8. Ar., *GA* 5.1 (778b20f.). Cf. Avic., *DA* 19 (fol. 63va–vb); Averroes, *GA* 5.1; A., *DA* 19.1.2.5–7 (*SZ* 2: 1334–36). See *QDA* 9.24–28.

ing. Therefore, crossing from non-living to living, one must first pass through sleep just as through a midpoint.

To the first question one must reply that "sleep is a hindrance to the first sensitivity [*primum sensitivum*]," according to the Philosopher in *On Sleep and Waking*.[9] But the first sensitivity exists in the embryo, which is apparent because, if it is pricked or approached by something harmful, it moves. Therefore, this sensitivity can labor in response to excessive [stimulus]. This [stimulus], however, brings on tiredness born of labor. Tiredness in an animal's sensitive parts induces sleep, and this is why sleep occurs in the embryo in the womb.[10]

1. On to the arguments. To the first, one must reply that act is of two types, first and second, and the second is also of two types, namely, internal and external. In the embryo there is the first act of sensation, and there is a second act there also with respect to the internal but not to the external. Because although a person's eyes close voluntarily, he nevertheless is not always sleeping when his eyes are closed, and, likewise, if he actively sees nothing at all at night, he nevertheless is not always sleeping. Rather, he can be asleep at one time and awake at another. And thus the embryo exists in the womb just as the eye exists in darkness or when veiled by the eyelids. Therefore, the fact that it does not operate with respect to what is external is owing to a defect of its eyes.

2. To the second argument one must reply that the embryo lives the life of a plant before it lives the life of an animal, and it is said to live the life of a plant because it does not employ its sensitive power with respect to what is external, just as a plant does not. This is why, etc.

To the other question one must reply that sleep is properly caused by the rising of the vapors to the brain, since some vapors clog and block the spirits' pathways to the individual senses. Therefore, one can say that sleep precedes waking in the embryo, because the senses of the embryo are blocked and hin-

9. Ar., *De somno et vigilia* 3 (458a28f.). Cf. A., *De somno et vigilia* 1.2.10.
10. The Latin of this portion is exceptionally vague.

dered for a long time by the multitude of fumes. And this is why sleep precedes waking.

Or one can say to the contrary that waking precedes sleep, because an act is of two types, namely, first and second, and a proper waking state corresponds to each, and sleep corresponds to each waking state as its state of rest. Therefore, if a uniform comparison is made, a waking state precedes its own proper sleep state. Now it is clear that as soon as the fetus is born it screams and cries, which are acts of a waking state. And it is the same if one were to compare the waking state of a first act to its sleep state.

3, 4. With this a solution is apparent to the first arguments.

5. To [this] argument one must reply to the contrary that, although it might attain sleep before it lives perfectly or completely, nevertheless it does not attain sleep before it attains the waking state, because the large size of its head and the multitude of vapors confer sleep upon it, and these exist in children after their first birthday. And this is why, when children are in their first stage of growth, they sleep a great deal, but when they attain an older age, they sleep less because the vapors are consumed.

Or one can say in another way that sleep is not the midpoint between living and non-living, since these are contradictories. But one who is dreaming or sleeping cannot be said to be non-living, nor can he be said to be living in terms of a second act. And because it is not a true midpoint, it is not necessary that he attain sleep first.[11] This is just as matter can be said to be the middle between being and non-being, and it is nevertheless not necessary for something crossing from non-being to being to pass through matter first. And this is why, etc. Through this a solution to this one is apparent.

11. I.e., he does not have to pass through sleep before he can become awake.

Question 6: On the color of the eyes.

Further one inquires into the color of the eyes, namely, whether blackness arises from an abundance of moisture and grayness arises from a scarcity of it.[12]

1. And it seems not. For in phlegmatic individuals sometimes the eyes appear to be gray, and in melancholic ones they appear black, although in phlegmatic ones nevertheless moisture abounds and in melancholic ones it is lacking.

2. In addition, the visual power consists in the glacial humor.[13] If, then, blackness were caused by an abundance of moisture and grayness by a scarcity of it, then black eyes would see better at night than during the day.

The Philosopher says the opposite.[14]

One must reply that diverse eye colors arise from a diversity of humors. When there is an abundance of moisture in the eye, a black color appears; and when there is a little, a gray color tending to white appears. And when the moisture varies in different parts [of the eye], a varied color appears. This is also evident in things. For where water is deep, it appears black; and where it lacks depth, there the water appears white. And it is the same for the eye, that the eye colors vary according to the diversity of the moisture.

1. On to the arguments. To the first, one must reply that although moisture abounds in phlegmatic people, it nevertheless does not do so in every part. For then it would also be abundant in bone. Therefore, although a phlegmatic complexion is moist and a melancholic complexion is dry, nevertheless nothing prevents a melancholic person's eye from having more glacial humor than does a phlegmatic's eye. And this is why, etc.

2. To the second argument one must reply that a small mo-

12. It must be remembered that the Latin for "black," *niger,* also carries a sense of "dark." Likewise, "white," *albus,* can often merely imply "light."

13. On the humors in the eye and their function in vision, cf. *DA* 19.1.3–4.9–20 (*SZ* 2: 1336–42).

14. Ar., *GA* 5.1 (779b12f.). Cf. Avic., *DA* 19 (fol. 63vb); Averroes, *GA* 5.1; A., *DA* 19.1.3.9–12 (*SZ* 2: 1336–38).

tion is not perceived and ceases in the presence of a larger one. This is evident in many cases, because stars do not appear in the day the same way they do at night because their modest light is concealed and softened in the presence of the greater light of the sun. Now, however, there is less moisture in gray eyes than in black eyes and more visual light, and this is why that light is hidden in the presence of a greater external light. Therefore, gray eyes see better at night and worse during the day, like a cat's. But black eyes, owing to the abundance of moisture, receive the [visible] species only when accompanied by great motion, and this is why they see only in the presence of a great external light. Therefore, they see more poorly at night and better during the day.

Nevertheless, one must understand that good or sharp vision consists in two things, namely, in the ability to distinguish between two visible objects and in the perception of something visible at a distance. The first is caused by the humor's purity and by the skin's softness surrounding the eye. For just as on a clean cloth or piece of parchment a small-sized spot is apparent, but is not apparent on a dirty cloth, so too a faint image and a distinguishing feature of that image are perceived immediately in a clean eye. And the lightness and softness of the skin contribute to this, because a wrinkle in the eye's tunic causes a sort of shadow on its inside. The second is caused by the eyes' location and by the disposition of the skin that closes the eyes. For those [animals] that have eyelids see from a distance better than those who do not, because those who lack eyelids have hard eyes, and the ones that have eyelids have soft eyes, and this is why they have eyelids, so that the eyes will not be injured by things that are outside it. Likewise, deep-set eyes see better from a distance than prominent eyes, because when an eye stands out it is surrounded on all sides by an external light and its ray (that is, its sight) is better dispersed. But sight is more concentrated in deep-set eyes, and a unified power is stronger than one that is dispersed. And according to this argument, if someone were in a deep well during the day, so that he did not see the brightness of the sun, he would see the stars in the heaven, which he would not see while on the earth's surface because his sight would not be diffused by an external light but would rather be better concentrated.

Questions 7–9: On hair.

One inquires about hair. And first, one inquires whether hair is generated from skin.

1. It seems not. For where the same cause is present, so too is the same effect. If, then, skin were the cause for the generation of hair, then all animals that have skin would have hair. But the conclusion is false, as is evident in birds and in fish, and therefore, etc.

2. In addition, the hard cannot be connected to the soft. But the first part of hair is hard and earthy, but what is inside the skin is soft. Therefore, they cannot be connected.

The Philosopher says the opposite.[15]

Second, one inquires whether hair is nourished.

3. It seems so. Growth presupposes nutrition. But hairs grow. Therefore, they are also nourished.

4. In addition, that which is maintained with the arrival of nutriment, and fails when it is lacking, is nourished. But hair is like this. Therefore, etc.

To the contrary. Everything that is nourished or that has grown through nourishment is nourished throughout and grows in every direction. But hair is not like this. Therefore, etc.

Third, one inquires into the cause of the diversity in hair.

5. And it seems that hair ought not to be diverse. For whatever has the same matter and the same effect is not diversified. But hair is like this. For hair is generated from the vapor and humor in the skin and by the heat pushing it out.

6. In addition, hair is generated from the work of the third digestive power, which is evident from the fact that hair can be generated in every part. But the third digestive power proceeds

15. Ar., *GA* 5.3 (782a30f.). Cf. Avic., *DA* 19 (fol. 64ra); Averroes, *GA* 5.3.6; A., *DA* 19.1.6.25–33 (*SZ* 2: 1344–49).

by whitening. Therefore, all hair will be uniform in color and, by the same argument, uniform in other respects.

The opposite is apparent to the senses.

To the first, one must reply that hair is generated in the skin. For when the nutriment is dispatched to individual members, it is dispatched to the skin. But there is often a weak digestive power in the skin, owing to its distance from the principle of digestion. Therefore, there are often fumes and earthy humors in the skin, and when these are dried out and expelled through the skin's pores, hair is generated. And this is why in all animals except the human, hair bears witness to the skin, that if it is white, then the hair is white, and if it is black, then the hair is black. But skin is finer in the human than in other animals. Therefore, human hair cannot be made hard, and the hair's matter cannot be made continuous on a person's skin;[16] rather, the matter is concentrated under the skin, and this is why hair is not diversified in a human according to the nature of the skin, but rather according to the diversity of the nature of the humors under the skin.

1. On to the arguments. To the first, one must reply that only animals that generate have hair, because ones like this are warmer than the others. Therefore, they have more matter suited to hair and, likewise, to producing it. This matter is thicker in birds and fish, and this is why it is converted into scales and feathers, but it is thinner in other animals owing to the dominance of heat and blood, and this is why it suits the generation of hair better.

2. To the second argument one must reply that when a hair exits from the skin its matter is dried out by heat and then it is constricted by cold. And this matter is viscous and sticky and can be continued into a long hair. Therefore, insofar as something like this is concerned, the hard is not connected with the soft until the soft is hardened.

To the second question one must reply that something is nourished or grows in two ways. It can grow by a power that

16. That is, in humans, hair does not convert into anything like quills, shells, or scales.

is present inside it, and this is how flesh and bone are nourished and grow—by the nutritive or augmentative power. Or something can be nourished or grow in another way, by adding something to it, and this is how fire is said to be maintained and to grow—by adding wood to it. Hair is not nourished in the first way, since it does not have a nutritive power within itself, nor does it grow in this way, since it then would grow in every direction. It is nourished and it grows in the second way, however, but to be nourished or to grow in this way is nothing other than to be generated, because, when the hair grows, the first part, which was present first, does not grow, but does acquire something new.

3, 4. With this a solution to the arguments is clear.

To the third question one must reply that differences in the matter and in the skin are the causes of diversity in the hair. Nevertheless, differences in the matter and the skin proceed from an agent. When the skin is thick and has wide pores and there is a lot of thick matter, then thick hair is generated. When, however, the skin is fine and the pores are narrow, then fine hair is generated. When the matter for the hair is abundant and very oily, then long hair is generated. And this is why hair on a person's head is longer than the other hair, because the skin on the head is thicker than the skin on other parts, and a viscous humor abounds in the head owing to its nearness to the brain, and this is why the hair is lengthened, and the hair is short for the contrary reason. If the hair's matter has been dried out a great deal, then the hair is curly and twisted. For, owing to the matter's dryness, the hair is hard, and, owing to the heat, it turns back almost to its place of origin. According to physicians the bending and twisted nature of the pores causes this. And an indication of this is that hair is curly in hot regions, as among Ethiopians and Indians, but in cold and moist regions the hair is smooth. If, however, the hair's matter is very dried out and the expelling heat is absent, then hair loss occurs. And this is why those who enjoy sexual intercourse too frequently go bald quickly, since coition dries out and cools the body, and, owing to this, old men go bald more than young men, because old men are cold and dry. Therefore, the ages [of man] resemble the times of the year. Because, just as in a hot and moist season, spring, for example, leaves grow on the trees, and feathers grow on birds—which is why they particularly sing then, owing to the very great expansion of their lungs. Now these lungs contract in other seasons, and this is why they do not sing then but are sad—and during a cold and dry season, like winter or the period close to it, like autumn, these things fall off (namely, tree leaves), and the feathers on birds decrease somewhat. In much the same way do these things occur in humans according to their different ages, with the difference that the seasons turn in a cyclical manner—this is why the leaves, after they fall off, can grow and return in a cyclical fashion and after their regrowth they fall off again—but the ages of man do not return, and this is why animals like these cannot naturally restore their hair.

Again, hair color varies according to diversity of this humor. For when the humor is undigested and is rather watery, or when it is putrefied by means of an alien heat, the hair grows white and gray. This is why a person goes gray on the temples before he does so on other parts of the head, because there is so much humor in the front portion of the head that it cannot be easily corrupted. On the rear portion of the head it is so modest that it does not increase the heat but rather is quickly digested, and on the side of the head it is in-between these two. If, however, this humor begins to burn, then reddish hair is generated, and if is completely burned, then black hair is generated. Therefore, it is clear from this that differences occur in the hair according to differences in the skin and the humor.

5. On to the arguments. To the first, one must reply that the matter and the efficient cause are not related uniformly, because the matter is sometimes thick and sometimes thin; it can be fragile, viscous, putrid, and burned, and so too for the others. Likewise, the efficient cause can be a strong heat and can be overcome by weakness, etc.

6. To the second argument one must reply that although the third digestion proceeds by whitening, nevertheless at the end it proceeds by reddening, by burning, and by blackening. Or one can say that hair does not receive its color from the digestive

power, but, rather, generally receives it from an external power such as an external heat. And this is why, etc.

Questions 10–11: On voice.

One inquires further into the voice. And first, whether the power of breathing in is the formative power of the voice.

It seems that it is not. For whatever is impeded by forming the voice is not a formative power of voice. But the power of breathing in is of this sort. Therefore, etc.

In addition, a person can desist from forming vocalizations by his free will. But he cannot cease to inhale or exhale. Therefore, etc.

To the contrary. Only animals that have respiration give voice. And the organs of respiration are designated as the organs of voice. Therefore, etc.

Second, one inquires whether different voices follow upon the disposition of the tracheal artery or upon the strength or weakness of the power.

And it seems that they follow upon the disposition of the tracheal artery. Because this is the way it is in musical instruments: a thick string makes a deeper sound, and a thin string makes a higher sound. Therefore, it will be the same with respect to the voice from the tracheal artery.

In addition, a high-pitched voice is a function of subtlety and thinness. But these are found on the part of the tracheal artery. Therefore, etc.

The Philosopher implies the opposite.[17]

To the first, one must reply that the power of breathing in is not the vocative power, but it is necessarily required for it. First it is clear that when air is being inhaled or exhaled, the voice cannot be formed, but instead it is necessary that the air be at rest and retained for the formation of voice so that it may be struck by the tongue. Therefore, when the tongue strikes the air against the teeth or the palate, a voice is formed and it varies in its shape according to the difference in the striking. Therefore,

17. Ar., *GA* 5.7 (786b7–25). Cf. Avic., *DA* 19 (fol. 64rb); Averroes, *GA* 5.7; A., *DA* 19.1.9.43–47 (*SZ* 2: 1353–56).

there is one power that forms the voice and another that draws in air. This is clear among mutes who draw in air and yet do not form a voice.

Therefore, one must reply to the argument to the contrary that although only animals that have respiration give voice, and they accomplish the formation of the voice by the organs of respiration, nevertheless it does not follow that this is a natural power. And besides this, the respiratory organs accomplish the formation of a voice only on account of the air, and the formative power of the voice is in the tongue, whereas the power of breathing in is in the lungs.

To the second question one must reply that differences in the voice follow upon the disposition of the tracheal artery and the disposition of the agent power, because, just as the Philosopher says,[18] a high voice is one that quickly moves what is heard, and a deep voice does so slowly. Now, however, a quick movement is caused by the victory of the thing moving over the movable, whereas a slow movement is caused by the mover's weakness and the victory of the movable over it. When, then, the vocative power forms a voice by moving a lot of air, the movement will be slow and as a result the voice will be deep. When, however, a modest amount of air is moved, then the voice will be quick, and, as a result, it will be high. And this is why children and women have a high voice, because they have weak powers that move little air. But this power is strengthened in men and old people, and they move a lot of air, and this is why their sound is deeper. Moreover, this motive power lies in the nerves, which are stronger in men than they are in women, and stronger in old people than in children. Now some differences follow upon the disposition of the tracheal artery, since a voice is rendered harsh from its dryness (since dry things are very often hard and unbalanced), and a harsh voice is like this. And when the tracheal artery is infused with a viscous humor, the voice is rendered hoarse, because then the air that is struck against the artery is analogous to a piece of lead struck against lead, in comparison to other metals. The artery's breadth and the power's strength contribute to a loud voice, and its narrowness and

18. Ar., *GA* 5.7 (786b25f.).

the brevity of power contribute to a small voice. Therefore, the voice varies according to the diversity of the agent and of the organ, like the tracheal artery, the tongue, and the palate.

Therefore, the arguments arrive at a true conclusion in part, on each side. Nevertheless, the Philosopher proposes that a voice varies because of the agent, because diversity in the organ stems more from matter, and diversity in the agent stems more from form, as is self-evident. This is why, etc.

Questions 12–13: On teeth.

Further one inquires into the teeth: whether they arise from heat or from cold.

1. And it seems it is from cold. Because whatever heat dissolves, cold coagulates, according to the Philosopher in the fourth book of *On Meteorology*.[19] But teeth are like this, and therefore, etc.

2. In addition, teeth arise from the spermatic matter, which is moist. But this sort of matter is coagulated by cold. Therefore, etc.

The Philosopher implies the opposite. For he rejects the view of those who propose that nursing causes teeth, saying that nursing is not the cause of the teeth, but rather the milk's heat is. He intends, then, that they are generated from heat.

Second, one inquires whether the front teeth fall out more than the molars.

3. And it seems not. For the larger something is, the more nutriment it needs. Therefore, nutriment for larger things naturally runs out more quickly than for smaller things, and this happens in the same way. Since, then, the molars are larger than the front teeth, the molars run out of nutriment more quickly than the front teeth.

4. In addition, nature preserves more what it needs more. But it needs [teeth for] cutting more than [teeth for] grinding, as is said in the text. Therefore, it preserves these more, etc.

19. Ar., *Meteora* 4.6 (383a25–32).

The Philosopher says the opposite.[20]

To the first, one must reply that since teeth and bone have the same nature, teeth are generated from heat. Now bones are generated from heat, which is apparent from their hardness and dryness, and so too, therefore, are teeth. This is also evident from the fact that in the hot age [of man], if some teeth fall out, they grow back. But if they fall out during the cold age, they do not grow back.

1. On to the arguments. To the first, one must reply that dissolution is of two types: one takes the form of liquefaction, as occurs in lead, tin, metals, and silver, etc., and the other takes the form of grinding, incineration, or cremation. Those which are dissolved by heat in the first manner are coagulated by cold, but those that are dissolved in the second manner are coagulated by heat, and it is in this way that teeth, bones, and plants of this sort are coagulated, etc.[21]

2. To the second argument one must reply that although the spermatic matter is moist, it can nevertheless be evaporated by a moist heat, and, once it has been evaporated, then coagulation occurs. Nevertheless, if this spermatic matter should be coagulated before the exhalation of the moisture, then this will occur due to cold.

To the second question one must reply that the front teeth are more disposed to fall out than the molars because the front teeth are formed in a thin bone, and, because they are fragile and sharp, they are worn down a great deal as they cut up food. Therefore, because they formed in a weak base, they can collapse and fall out more easily than others, and, because they are worn down a lot, it is expedient that new ones grow back. But molars are formed in a broad, strong bone, and this is why they can dry out more slowly and are worn down more slowly, owing to their breadth. Therefore, the front teeth arise from a more fluid matter, and the molars arise from a more solid matter; and this is why the front teeth grow more quickly than the molars, because the molars do not grow until the child's parts begin to consolidate.

20. Ar., *GA* 5.8 (789a4f.). Cf. A., *DA* 19.1.10.49 (*SZ* 2: 1357).

21. Reading *secundum* for the text's *secundo.*

3. On to the arguments. To the first, one must reply that although what is larger needs more nutriment, what is larger can nevertheless resist dangerous or harmful causes better, and this is why the things that bind the molars to the gums are worn down or broken with more difficulty than those of the front teeth. And in addition to this, more nutriment can be preserved in what is large and ample than in what is narrow and thin, and the molars are formed in a large and wide bone, and the front teeth are formed in a narrow and fragile bone, and this is why the molars are nourished longer, etc.

4. To the second argument one must reply that during youth the path of generation needs the front teeth earlier than it does the molars, but it is the opposite in old age. And the reason for this is that children are first nourished with milk. Therefore, they do not need teeth then. But afterward they begin to be nourished by some thicker, more moist foods, which require dividing up, but not a lot of mastication—things like bits of meat, eggs, drinks, and things of this sort, and this is why they need the front teeth, which are made for dividing, before they need the molars, which are suited to grinding and pressing. But men, and adults, are nourished on coarse and earthy foods, which require grinding, and this grinding is the end of division, as it were, and the end is nobler than those that exist for the sake of an end. Therefore, food is first divided by the front teeth, so that it is more suited for grinding by the molars, and this is why nature insists more on preserving the molars than the front teeth. Moreover, falling out is a defect, and on account of this, when a defect occurs earlier or to a greater degree, then the falling out occurs more. Now, defects occur more in front teeth because they are sharper and more fragile, and this is why they are injured more easily, and the molars are harder and more solid. And this is as nature ordained it, as if it were an art of the first cause, ordering each thing according to what is best. And this is the blessed God, to whom be honor and glory through the infinite and unending ages of ages.[22] Therefore, wise and provident nature, taught to obey its Creator, accomplished all these things in his power. Therefore, the molars are just like the

22. Cf. Averroes, *GA* 5.8.

millstones that grind grain. The front teeth are like those that prepare the grain for the millhouse, and the tongue is like the miller's hand, which puts the grain back to be ground when it falls away; so too does the tongue turn and tumble the food until it is ground or completely chewed, etc.

Here end the questions concerning the *On Animals* that Brother Albert disputed while going over the book of animals for the brothers in Cologne. A certain brother, named Conrad of Austria, who heard the questions from him, collected and reported on the aforementioned book. This was done in the year of the Lord 1258.

INDEX

INDEX

The index lists proper names and individual works cited within Albert's text, as well as general topics or subjects.